CROP PRODUCTION SCIENCE IN HORTICULTURE SERIES

Series Editors: Jeff Atherton, Professor of Tropical Horticulture, University of the West Indies, Barbados, and Alun Rees (retired) former Head of Crop Science, Glasshouse Crops Research Institute, Horticultural Consultant and Editor of the *Journal of Horticultural Science and Biotechnology*.

This series examines economically important horticultural crops selected from the major production systems in temperate, subtropical and tropical climatic areas. Systems represented range from open field and plantation sites to protected plastic and glasshouses, growing rooms and laboratories. Emphasis is placed on the scientific principles underlying crop production practices rather than on providing empirical recipes for uncritical acceptance. Scientific understanding provides the key to both reasoned choice of practice and the solution of future problems.

Students and staff at universities and colleges throughout the world involved in courses in horticulture, as well as in agriculture, plant science, food science and applied biology at degree, diploma or certificate level, will welcome this series as a succinct and readable source of information. The books will also be invaluable to progressive growers, advisers and end-product users requiring an authoritative, but brief, scientific introduction to particular crops or systems. Keen gardeners wishing to understand the scientific basis of recommended practices will also find the series very useful.

The authors are all internationally renowned experts with extensive experience of their subjects. Each volume follows a common format covering all aspects of production, from background physiology and breeding, to propagation and planting, through husbandry and crop protection, to harvesting, handling and storage. Selective references are included to direct the reader to further information on specific topics.

Titles Available:
1. **Ornamental Bulbs, Corms and Tubers** A.R. Rees
2. **Citrus** F.S. Davies and L.G. Albrigo
3. **Onions and Other Vegetable Alliums** J.L. Brewster
4. **Ornamental Bedding Plants** A.M. Armitage
5. **Bananas and Plantains** J.C. Robinson
6. **Cucurbits** R.W. Robinson and D.S. Decker-Walters
7. **Tropical Fruits** H.Y. Nakasone and R.E. Paull
8. **Coffee, Cocoa and Tea** K.C. Willson
9. **Lettuce, Endive and Chicory** E.J. Ryder
10. **Carrots and Related Vegetable Umbelliferae** V.E. Rubatzky, C.F. Quiros and P.W. Simon
11. **Strawberries** J.F. Hancock
12. **Peppers: Vegetable and Spice Capsicums** P.W. Bosland and E.J. Votava
13. **Tomatoes** E. Heuvelink
14. **Vegetable Brassicas and Related Crucifers** G. Dixon
15. **Onions and Other Vegetable Alliums, 2nd Edition** J.L. Brewster
16. **Grapes** G.L. Creasy and L.L. Creasy
17. **Tropical Root and Tuber Crops: Cassava, Sweet Potato, Yams and Aroids** V. Lebot
18. **Olives** I. Therios

OLIVES

Ioannis Therios
School of Agriculture
Aristotle University
Thessaloniki
Greece

CABI is a trading name of CAB International

CABI Head Office
Nosworthy Way
Wallingford
Oxfordshire OX10 8DE
UK

CABI
745 Atlantic Avenue
8th Floor
Boston, MA 02111
USA

Tel: +44 (0)1491 832111
Fax: +44 (0)1491 833508
Email: cabi@cabi.org
Web site: www.cabi.org

Tel: +1 617 682-9015
Email: cabi-nao@cabi.org

©Ioannis Therios 2009. All rights reserved. No part of this publication may be reproduced in any form or by any means, electronically, mechanically, by photocopying, recording or otherwise, without the prior permission of the copyright owners.

A catalogue record for this book is available from the British Library, London, UK

Library of Congress Cataloging-in-Publication Data
Therios, Ioannis Nikolaos.
  Olives / I. Therios.
    p.cm. — (Crop production science in horticulture ; 18)
  ISBN 978-1-84593-458-3 (alk. paper)
  1. Olive. I. Title. II. Series.

  SB367.T44 2009
  634'.63 — dc22

2008015354

ISBN-13: 978-1-84593-458-3

First published 2009
Transferred to print on demand 2015

Printed and bound in the UK by Marston Book Services Ltd, Didcot, Oxon.

Contents

Preface — ix

Acronyms and Abbreviations — xi

Chemical Symbols — xv

1
History of Olive Growing — 1

2
The Olive: Origin and Classification — 9

3
Morphology and Taxonomy of the Olive — 13

4
Structure and Composition of the Olive Fruit — 25

5
Hectareage, Number of Trees and Production of Olive Oil and Table Olives — 33

6
Rootstocks — 43

7
Major Trends in Olive Farming Systems — 49

8
Climatic and Soil Conditions — 51

9
FLOWER BUD INDUCTION AND DIFFERENTIATION 81

10
FLOWERING, POLLINATION, FERTILIZATION AND FRUITING 93

11
ALTERNATE BEARING 105

12
FRUIT THINNING 109

13
SYSTEMS OF PLANTING AND CANOPY TRAINING 113

14
PROPAGATION OF OLIVE TREES 125

15
IRRIGATION OF THE OLIVE 151

16
WATER USE EFFICIENCY BY THE OLIVE 167

17
STRESS-INDUCED ACCUMULATION OF PROLINE AND MANNITOL 175

18
MINERAL NUTRITION OF THE OLIVE 179

19
GROWTH AND SALT TOLERANCE OF THE OLIVE 211

20
PRUNING 221

21
OLIVE RIPENING 229

22
OLIVE FRUIT HARVESTING 245

23
OLIVE VARIETIES 255

24
TABLE OLIVES 271

25
OLIVE OIL 279

26
OLIVE MILL PRODUCTS AND ENVIRONMENTAL IMPACT OF OLIVE OIL PRODUCTION 295

27
OLEUROPEIN, OLIVE LEAF EXTRACT, OLIVE OIL AND THE BENEFITS OF THE MEDITERRANEAN DIET TO HUMAN HEALTH 303

28
BIOLOGICAL AND INTEGRATED OLIVE CULTURE 319

29
CHEMICAL AND INTEGRATED WEED MANAGEMENT IN OLIVE ORCHARDS 325

30
PESTS AND DISEASES 335

31
BIOTECHNOLOGICAL ASPECTS OF OLIVE CULTURE 353

REFERENCES 357

INDEX 399

PREFACE

Olive culture is rapidly growing in terms of both production volume and geographical spread. The crop is now grown in countries not traditionally associated with its cultivation, although in such countries olive culture is not yet economically significant. Furthermore, climatic changes have resulted in the expansion of olive culture to areas more northerly than the Mediterranean countries. The olive is not only a significant food source but it also contributes to human health, with olive products representing an important part of the Mediterranean diet, which an increasing number of people all over the world are embracing for reasons of health.

This book of 'olives' is the result of many years' endeavours in collecting valuable information from the existing literature concerning the olive tree and its culture; a proportion of this information, and experience, has originated from scientific projects of the author and his scientific team. Topics include all aspects of olive culture, from its history, through traditional practices to modern techniques and horticultural procedures. Furthermore, this book covers the basic physiological and horticultural principles of olive culture in both theory and practice.

My objective is to provide knowledge appropriate for students, scientists, both experienced and inexperienced horticulturists and, in general, for anyone wishing to obtain knowledge and experience of olive culture in order to increase productivity and improve product quality. The content is written in a style accessible to anyone and is also appropriate as a textbook at various levels of education. To enhance understanding the text includes illustrations, figures and tables featuring experimental data, and it is a recent addition to the literature on 'olives'.

Ioannis Therios
Professor of Pomology
Aristotle University
Thessaloniki
Greece
E-mail: therios@agro.auth.gr

Acronyms and Abbreviations

2, 4, 5-T	2, 4, 5-trichlorophenoxyacetic acid
2, 4-D	2, 4-Dichlorophenoxyacetic acid
2iP	6-(γ, γ-dimethylallylamino)-purine
3-CPA	3-chlorophenoxy-α-propionamide
ABA	Abscisic acid
ACC	1-aminocyclopropane-1 carboxylate
ACE	Acetylcholinesterase
ADP	Adenosine diphosphate
AFLP	Amplified fragment length polymorphism
AMP	Adenosine monophosphate
AOS	Active oxygen species
ATP	Adenosine triphosphate
BA	6-benzylamino purine or benzyladenine
BAP	Benzylamino purine
BMI	Body mass index
BN	Nutrient medium
BNOA, NOA	β-naphthoxyacetic acid
CAP	Common agricultural policy
CBF1	Regulator of cold acclimation
CCC	(2-chloroethyl)trimethylammonium chloride or Chlormequat
CEC	Cation exchange capacity
CHA	Chlorogenic acid
CK	Cytokinin
CLRV	Cherry leaf roll virus
CMS	Cytoplasmic male sterility
COR	Cold Regulated protein
CTP	Cytidine triphosphate
CuMV	Cucumber mosaic virus
DNA	Deoxyribonucleic acid
DNOC	4, 6-dinitro-ortho-cresol

DNP	2,4-Dinitrophenol
DTA	Differential thermal analysis
ECe	Electrical conductivity
ECw	Electrical conductivity of irrigation water
EDTA	Ethylenediamine tetraacetic acid
ERD	Early responsive to dehydration
$ET_{c'}$	Actual water requirements
Ethephon	2-chloroethyl phosphonic acid
Ethrel	see Ethephon
ET_o	Potential evapotranspiration
FASN	Fatty acid synthase (human gene)
FDF	Fruit detachment force
Fenoprop	see 2,4,5-TP
FMN	Flavin mononucleotide
FTIR spectroscopy	Fourier transform infrared spectroscopy
Fts	Filamentous temperature sensitive proteins
F_v	Variable fluorescence
GA_3	Gibberellic acid
GA_4	Gibberellin
GAs	Gibberellins
GC	Gas chromatography
GDH	Glutamic dehydrogenase
GGA	2-chloroethyl-tris-2-methoxy-ethoxy silane or Alsol
gMS	Genic male sterility
GOGAT	Glutamate-2-oxoglutarate amido transferase
GS	Glutamic synthetase
g_s	Leaf conductance
HER2	Cancer genes
HPLC	High pressure liquid chromatography
Hsp	Heat stress proteins
IAA	Indole-3-acetic acid
IBA	Indole-3-butyric acid
iNOs	Inductible nitric oxide synthase
ISSR	Inter-simple sequence repeats
K_c	Crop coefficient
KGA_3	Potassium gibberellate
KIN	6-furfurylamino purine, <u>syn</u> N-furfuryladenine or Kinetin
KIN	Cold induce
$KMnO_4$	Potassium permanganate
$K_{p'}$	Coefficient related to tree transpiration
K_{S1}	Coefficient related to soil evaporation
LAI	Leaf area index
LAR	Leaf area ratio

LC	Liquid chromatography
LDL	Low density lipoprotein
LEA	Late embryogenesis abundant proteins (prevent protein aggregation)
LFDI	Low frequency deficit irrigation
LLC	Liquid–liquid chromatography
L-Name	N(G)nitro-L-arginine methyl ester
LOX	Lipoxygenase
LPS	Lipopolysaccharide
LSC	Liquid–solid chromatography
LT1	Low temperature inducer
LVI	Low volume irrigation
M	Manganese
MI	Maturity index
MH	Maleic hydrazide
mOM	modified olive medium
mRNA	messenger RNA
MS medium	Murashige and Skoog medium
MSMA	Veliuron
MYB	Gene of cold tolerance, drought and salt stress
N	Newtons
NAA	1-naphthaleneacetic acid
NAAm	naphthaleneacetamide or NAAmide
NADP	Nicotinamide adenine dinucleotide phosphate
NADPH	Reduced NADP
NAR	Net assimilation rate
NIF	Near infrared
NMR	Nuclear magnetic resonance
NPA	N-1-naphthylphthalamic acid
NR	Nitrate reductase
OD	Optical density
OLE	Olive leaf extract
Oleaster	Wild olive
OM	Olive medium
OMWW	Olive mill waste water
OP	Osmotic pressure
PAR	Photosynthetic active radiation
PCD	Programmed cell death
PDO	Protected designation of origin
PEA 3	A transcriptional repressor of HER 2 gene amplification
PEP	Phosphoenol pyruvate
PGI	Protected geographical indication
Pn	Photosynthetic rate
PPF	Photosynthetic photon flux

PRD	Partial rootzone drying
PWP	Permanent wilting percentage
Q_{10}	A temperature coefficient
RAPD	Random amplified polymorphic DNA
RBC_s	Red blood cells
RD	Responsive to dehydration
RDI	Regulated deficit irrigation
RGR	Relative growth rate
RI	Ripening index
RNA	Ribonucleic acid
ROS	Reactive oxygen species
rRNA	Ribosomal RNA
RuDP	Ribulose 1, 5-diphosphate
SA	Salicylic acid
SADH	Succinic acid; 2, 2-dimethylhydrazide
SAM	S-adenosylmethionine
SAR	Sodium adsorption ratio
SCAR	Sequence characterized amplified regions
SDI	Sustained deficit irrigation
SLRV	Strawberry Latent Rinspot Virus
SNP	Single nucleotide polymorphism
SOD	Superoxide dismutase
SOS	Salt overly sensitive genes, SOS1, SOS2, SOS3
sRNA	Soluble RNA
SSR	Simple sequence repeats
TDZ	Thidiazuron
TIBA	2, 3, 5-triiodobenzoic acid
TLC	Thin layer chromatography
UV	Ultraviolet absorption
V_{max}	Maximum rate of absorption
VPD	Vapour pressure deficit
WHO	World Health Organization
WPM	Woody plant medium
WUE	Water use efficiency
Zeatin	6-(4-hydroxy-3-methyl-2-butenylamino purine)
$\Delta^{13}C$	Carbon isotope discrimination
Ψ	Total water potential
Ψ_{leaf}	Predawn value
Ψ_p	Turgor or pressure potential
Ψ_S or Ψ_π	Osmotic potential
Ψ_{soil}	Soil water potential

Chemical Symbols

B	Boron
C	Carbon
Ca	Calcium
$CaCl_2$	Calcium chloride
$CaCO_3$	Calcium carbonate
Cl	Chloride
Cu	Copper
Fe	Iron
H	Hydrogen
$H_2PO_4^-$	Dihydrogen phosphate anion
H_2SO_4	Sulfuric acid
HPO_4^{2-}	Monohydrogen phosphate
K	Potassium
KCl	Potassium chloride
Mg	Magnesium
Mn	Manganese
Mo	Molybdenum
N	Nitrogen
NH_4^+	Ammonium cation
O	Oxygen
P	Phosphorus
S	Sulfur
SO_4^{2-}	Sulfate anion
Zn	Zinc

1

HISTORY OF OLIVE GROWING

The Olive Tree

I am the daughter of the sun

I am the olive tree

The blessed one.

K. Palamas

O olive tree, blessed be the earth that nourishes you and blessed be the water you drink from the clouds and thrice blessed He who sent you for the poor man's lamp and the saint's candle-light.

Folk song from the Greek island of Crete

The olive tree is surely the richest gift of heaven.

Thomas Jefferson

The olive tree marks the route leading towards a higher stage of civilization. As a native plant the olive tree was well known in the Mediterranean basin many thousands of years ago. Hence, in the city of Kymi on the Greek island of Evia, leaf fossils of *Olea noti* were found in carboniferous deposits dating back to the Oligocene period.

Archaeological excavations have revealed the existence of olive leaf fossils dating back to the Palaeolithic and Neolithic eras (37,000 BC) on Santorini Island, Greece (Therios, 2005b). Furthermore, some findings are reported for the cretaceous strata of Provence, France and for North Africa. Olive pollen was found in Greece, indicating its culture in this area. Hence, olive pollen grains appear around the year 6000 BC (Epirus, western Greece) and around 3200 BC in Thessaly and eastern-central Greece.

Olive culture is very old. Fossilized remains of the olive tree's ancestor were found near Livorno in Italy, dating back 20 million years ago, although actual cultivation probably did not occur in that area until the 5th century BC. Olive trees are the oldest and one of the most important fruit trees (Standish, 1960). Olive culture has been closely linked to the glory and decline of advanced civilizations. The olive tree is the greatest gift of God, since it symbolizes peace with its leaves and joy with its golden oil (see web site: History, Olives and Mythology).

The olive is regarded as a symbol of euphoria, purity, victory and honour, and olive shoots have been used to honour victory, peace and wisdom (Greek olive oil). An olive shoot was returned by a dove to Noah on the ark, proving the end of the great cataclysm. During the Olympic Games in Ancient Greece olive wreaths were given to winners. Since olive trees offered food supplies, agricultural communities were stable, evidenced by population growth, and such societies existed for many years. Based on their productive olive orchards, the Greek and Roman Empires developed into great economic forces; destruction of olive orchards resulted in the decline of these once-great establishments. Olive trees and olive oil have engaged the intellect, the senses and the passions of the Greeks for as long as 4000 years; olive oil maintained a sacred place in the Greek religion. Hence, in the Greek Orthodox religion olive oil was used for both baptism and illumination of churches and houses; the sacred lamp that was used in ancient Greece to light houses at night was fuelled by olive oil. Furthermore, aged olive oil was used in sacred rituals of the church at weddings. According to Herodotus (500 BC), the growing and export of olive products were so sacred that olive culture was allowed to be performed only by eunuchs and virgins (see web site: World Mythology Encyclopedia).

According to Greek mythology (Psilakis, 1996) Athena, the goddess of wisdom, struck her spear into the earth and an olive tree appeared (see web site: Stories of Athena). Athena's gift of the olive, which was useful for food, medicine, perfume and fuel, was considered to be a more peaceful gift than Poseidon's horse. Athena planted the original olive tree in the Acropolis, Athens (see Fig. 1.1).

Thus, this location, where the first olive tree grew, was named Athena (Athens) to honour the goddess Athena. According to the mythology this original olive tree still exists at the ancient sacred site and all Greek olive orchards originated from leafy cuttings from the original tree. As stated by Homer, the ancient olive tree in Athens was already 10,000 years old, and it was forbidden to cut down any olive tree; the penalty for this was execution. Athletes who competed in the ancient Olympic stadium (775 BC) were crowned with a wreath made of olive shoots. The first plantings and hand harvesting of olive trees date back to 3500 BC, during the Minoan civilization (Evans, 1903; Cadogan 1980) on the island of Crete (see Fig. 1.2).

Extensive plantings in Greece started in the year 700 BC. The significant Cretan civilization was active until 1450 BC, when it was devastated by an earthquake after the explosion of the volcano of Santorini. In the ruins of Knossos and Festos palaces large olive oil jars were found, and other discoveries provide evidence of the commodities traded between cities and between neighbouring countries (see Fig. 1.3).

The Minoan civilization (Hood, 1971) and its dominance in the islands and the continental area of Greece could have been partly due to the existence

Fig. 1.1. The goddess of wisdom, Athena, with the olive tree she planted in the Acropolis, Athens, according to Greek mythology (Psilakis and Kastanas, 1999).

Fig. 1.2. Hand-harvesting of olive trees on Crete during the period of the Minoan civilization (Psilakis and Kastanas, 1999).

Fig. 1.3. Large earthen jars for olive oil storage in the Minoan palace at Knossos, Crete (Psilakis and Kastanas, 1999).

of olive orchards. Archaeological studies supply circumstantial evidence for the existence of olive trees in the main civilizations of the Aegean, i.e. Minoan, Mycenaean (Nilsson, 1972) and Cycladic. Later, when the Minoan civilization was in decline, the Phoenician civilization originated in the coastal area of Lebanon. This civilization expanded beyond the Mediterranean and later introduced olive trees to Carthage (North Africa) and to Spain (Terral, 2000).

Also, the Phoenicians spread the olive tree to the Mediterranean shores of Africa and Southern Europe. Therefore, olive cultivation expanded to southern Italy, Sicily, Corsica and Sardinia. Olives have been found in Egyptian tombs dating from 2000 BC. Later, the Roman emperors protected olive cultivation by law, and the countries of Morocco, Spain, Tunisia and Italy were the most important sources of olive oil for the Roman Empire. The same protection for olive trees was provided by the emperors of Byzantium, as well as by the Arabs.

Olives were first cultivated in the Eastern Mediterranean region and expanded westwards during the next millennium. Olive cultivation spread from Crete to Syria, Palestine and Israel, to Cyprus, southern Turkey and Egypt. With the expansion of the Greek colonies, olive culture reached southern Italy and North Africa during the 8th century BC and subsequently spread into southern France. In Israel, King Solomon and King David expanded the practice of olive cultivation. Mohammed, the prophet of Islam, suggested 1400 years ago to his believers that they apply olive oil to their bodies. The use of olive oil is common in many religions and cultures. During baptism in the Christian church, holy oil may be used as anointment. Like the grape, the Christian missionaries brought the olive tree with them to California, North America, not only for food but also for ceremonial use. Olive oil was used to anoint the early kings of the Greeks and the Jews, and has also been used to anoint the dead in many civilizations.

Evidence indicates that the olive was also cultivated in Syria and Palestine from the 4th millennium BC. In Egypt the olive tree was also cultivated, as indicated by a papyrus written around 1550 BC. Furthermore, mummies from the 20th to the 25th dynasties discovered in Egypt were found to be wearing olive wreaths.

More recent archaeological investigation has provided considerable evidence concerning the existence of olives in ancient Israel. The olive tree was already being cultivated by the Neolithic period, as evidenced by the recovery of olive stones from excavations in Jericho. At most Palestinian sites, olive wood is not found in large quantities until the Early Bronze Age, when extensive olive cultivation coincided with the onset of urbanization (Goor, 1962).

Archaeological remains of olive wood provide more specific information. The earliest wood remains yet found in ancient Israel belong to *Olea europaea*, dating to 42,950 BC. The olive was one of the basic agricultural products in the economy of Israel. In Egypt, excavation at the site of Karanis in the Fayoum region has revealed that olive trees thrived in the area and that olive oil production was carried out there on a large scale, at least in the 3rd century BC.

During the colonial period olives were dispersed by explorers and colonists. In 1560 olive cuttings were carried to Peru by Spanish explorers. In the early 1700s, Jesuits established missions in Mexico and Baja California. Franciscans founded their first mission in California at San Diego in 1769. The olive tree was introduced to California by the Franciscans as they marched north establishing missions. A visitor to mission San Fernando in 1842 saw an olive orchard in healthy productivity. Subsequently, over the intervening years, olive trees have been planted in several waves and many of these older groves (80–150 years old) still exist in California, mostly in the northern part of the state. However, in southern California population pressure has resulted in the land being very expensive for olive growing.

Early olive plantings were established at a number of sites in New South Wales, South Australia, Western Australia, Victoria and, to a lesser extent, in other Australian states (Smyth, 2002). In the second half of the 20th century, interest in olive growing coincided with the post-war migration of people from European countries who were familiar with olive cultivation. Production of both olive oil and table olives in Australia is increasing, especially of quality products.

Today, the Australian olive industry is a significant player in the domestic market but not yet at an international level. Many changes during recent decades have contributed to this agricultural success story. First, cultural changes have occurred as a result of migration from Europe, due to the dietary changes introduced by immigrants; this has resulted in a rise in olive oil consumption. Other factors include the suitability of southern Australia for industrial-scale olive cultivation, the rise in the Australian economy and the newly gained horticultural experience (Kailis and Considine, 2002). It is expected that Australia will soon produce about 1% of the world's olive products.

The olive tree has been, for a very long time, a cultural element and a common reference point and trademark for people leaving the Mediterranean area (Simantirakis, 2003). We find representation of the olive tree in ancient coins (Greek and Roman), in the modern Greek drachma (20 cents coin) and in the French franc (now superseded by the euro). Furthermore, decorations depicting the olive are found on ancient Greek vessels, silver and golden items and Egyptian sarcophagi. Famous painters like Renoir, Picasso, Dali and Van Gogh included olive trees in their paintings. Also, major writers like Homer, Pindarus, Virgilius (Virgil), Palamas, Seferis, Elytis, Lorca and others glorified the olive tree in their work.

In our everyday routine the olive tree decorates towels, tablecloths and gardens with Mediterranean landscapes. Hence, the olive tree is used to determine the border of the Mediterranean zone: a zone that ends where olive trees cannot grow and produce. Olive oil has been used both as currency and a means of paying labourers or professionals; and olive trees were once used as a dowry given to girls for their marriage.

Olive oil has many health benefits when used in the culinary arena. It has beneficial effects on the digestive and cardiovascular systems due to its low cholesterol content. Other actions include the prevention of wrinkles, dry skin and acne, lowering of blood pressure, reduced muscular pain and strengthening of the nails. The belief that olive oil conferred strength and youth has long been widespread. In ancient Egypt, Greece and Rome it was infused with flowers and herbs to produce both medicines and cosmetics. Excavation in Mycenae, Greece revealed a list of aromatics (mint, rose, juniper, sesame, etc.) added to olive oil in the preparation of ointments.

The earlier ways of crushing olives included the use of a mortar and pestle, and rolling a stone roller over them in a crushing basin. The second step in the production of olive oil is pressing of the mass resulting from crushing. The first significant advance in the production of olive oil was the use of a lever in a lever-and-weights press, the use of which became common during the first Iron Age. Olive oil production became a mass production industry during Iron Age II.

WEB SITES

Greek olive oil, mythology and cultural inheritance: http://www.helleas.com/lire
History. Olives and mythology: http://www.itlv.com/encyclopedia/history/myths
Stories of Athena. 1. Greek mythology: http://www.theoi.com/Olympios/AthenaMyths
World Mythology Encyclopedia. Greek mythology: http://www.worldmythology.ws/greek-mythology/birthof-hermes

2

THE OLIVE: ORIGIN AND CLASSIFICATION

ORIGIN

The olive is native to the Mediterranean region, tropical and central Asia and to various parts of Africa. The genus *Olea* includes at least 30–35 species belonging to the family Oleaceae and subfamily Oleoideae (x = 23). The cultivated olive (*Olea europaea* L.) is an evergreen tree derived from tropical and subtropical species. Fossils from olive species have been found in Italy, France and in other countries.

The Mediterranean zone was, at one time, within the tropical zone, but drought and glaciers during the Pleistocene period constituted a means of natural selection for plants with the ability to avoid glaciers. The glaciers probably reduced the initial olive tree population, and only plants with the ability to survive at temperatures of between −5 and −12°C survived.

Olive culture was known from the year 4800 BC in Cyprus. The longevity of the olive tree may explain the great variability among its species. It is believed that genotypes that were crossed under various climatic conditions now constitute the species *O. europaea*. According to recent experimental results, significant genotypic variation exists among plants of the same cultivar; genetic diversity was studied by Angiolillo *et al.* (1999), and phylogenetic relations between *Olea* species by Ouazzani *et al.* (1993) and Baldoni *et al.* (2002). Varieties tolerant to cold are present in the northern areas of olive culture, although dry summers and mild winters are the most appropriate climatic conditions for olive trees.

The olive tree originating from the Eastern Mediterranean is one of the oldest cultures, belonging to the family Oleaceae with 30 genera, among which there are certain decorative plants. Most of the olive groves belong to the species *O. europaea*, with 2x = 46 chromosomes. The species *O. europaea* includes many groups and more than 2600 cultivars, many of which may be ecotypes.

Olea europaea does not seem to be a true species but one group of forms derived from hybridism and mutation. The tropical and subtropical Afro-Asian species, such as *O. chrysophilla* and *O. excelsa*, probably participated in the evolution of the culture.

Subspecies of olive are distributed in the Mediterranean countries and also in West Africa, Tanzania, the Canary Islands, the Azores, South Africa, etc. Olive trees have been introduced to the USA, Australia, South Africa and China in more recent decades.

The native olive *O. oleaster* (Breton *et al.*, 2006) and the cultivated olive *O. sativa* are the main olive species in the Mediterranean. A classification of the olive germplasm is presented by Ganino et al. (2006). The species *O. oleaster* has thorny shoots, small oval to spherical leaves and small black fruits, with low oil content. *O. sativa* has cylindrical shoots and big oval or elliptical fruits, with high oil content. The genus *O. oleaster* is native in a small number of areas. In some areas *O. europaea* is considered to be *O. oleaster* due to the juvenile stage of plants created by grazing. These plants enter the reproductive stage after appropriate treatments.

Countries with olive cultivation and the botanical species

Olive trees are cultivated in the following countries:

Europe: Spain, Italy, Portugal, France, Albania, Montenegro, Greece, Cyprus
Asia: Turkey, Syria, Lebanon, Jordan, Palestine, Iran, Iraq, Japan, China
Africa: Tunisia, Algeria, Morocco, Egypt, South Africa
America: USA, Mexico, Peru, Chile, Argentina, Uruguay
Oceania: Australia, New Zealand

Table 2.1 summarizes the current stage of olive cultivation in non-Mediterranean areas.

The botanical species of olive grown throughout the various continents are presented in Table 2.2.

Table 2.1. Olive cultivation in non-Mediterranean areas.

Continent	Countries with a tradition of olive culture	Countries with no tradition of olive culture	Countries that have started planting olives	Countries planting olives experimentally
Asia	Iran	–	Saudi Arabia	India
	Iraq	–	China	Japan
	–	–	–	Korea
Oceania	–	Australia	New Zealand	–
America	USA	Mexico	Brazil	–
	Argentina	Peru	–	–
	Chile	Uruguay	–	–
Africa	–	South Africa	–	–

Table 2.2. Botanical species of *Olea*, by continent.

Continent	Species	Countries/areas of cultivation
Oceania	O. apetala	New Zealand
	O. paniculata	Australia
	O. europaea	Australia, Java
America	O. floribunda	
	O. americana	USA: Florida, Georgia, Carolina, Virginia
	O. europaea	Chile, Peru, Mexico, Guadeloupe, Tahiti
	O. passiflora	
	O. maritima	
	O. microcarpa	
Asia	O. attenuata	Myanmar
	O. dentata	
	O. lindley	India: Calcutta
	O. salicifolia	India
	O. dioica	India
	O. cuspidata	Afghanistan, eastern India
	O. compacta	India
	O. roxburghiana	
	O. heyneata	
	O. glandulifera	
	O. acuminata	
	O. europaea	Asia Minor
Africa	O. chrysophylla	
	O. laurifolia	
	O. verrucosa	
	O. verrucosa brachybotris	
	O. capensis	
	O. foveolata	
	O. concolor	
	O. exaesprata	
	O. humilis	
	O. obtusifolia	
	O. lancea	
	O. europaea	
Europe	O. europaea	

CLASSICAL TAXONOMY OF THE CULTIVATED OLIVE

Kingdom: Green Plants
Subkingdom: Tracheobionata
Superdivision: Spermatophyta
Division: Magnoliophyta
Class: Magnoliopsida
Subclass: Asteridae

Order: Scrophulariales or Lamiales
Family: Oleaceae
Genus: *Olea*
Species: *europaea*

Other genera, except *Olea* (olive), within the family Oleaceae include the following: *Chionanthus* (Snow flower), *Forestiera*, *Forsythia* (Golden bell), *Fraxinus* (Ash), *Jasminum*, *Ligustrum* (Privet) and *Syringa* (Lilac). Some of the better known plants belonging to the Oleaceae family are listed below (common name(s) in parentheses).

Chionanthus retusus (Chinese fringe tree)
Chionanthus virginicus (Fringe tree, Snow flower)
Forestiera pubescens (New Mexico privet)
Forsythia intermedia (Border forsythia)
Forsythia mandschurica
Forsythia ovata (Korean forsythia)
Forsythia suspensa (Weeping forsythia)
Forsythia viridissima (Greenstem forsythia)
Fraxinus americana (White ash)
Fraxinus angustifolia (Narrow-leaved ash)
Fraxinus dipetala (Foothill ash)
Fraxinus excelsior (European ash)
Fraxinus latifolia (Oregon ash)
Fraxinus ornus (Flowering ash)
Fraxinus pensylvanica (Green ash)
Fraxinus quadrangulata (Blue ash)
Fraxinus velutina (Velvet ash)
Jasminum angulare (South African jasmine)
Jasminum floridum (Snowy jasmine)
Jasminum humile (Italian jasmine)
Jasminum le-ratii (Privet-leaved jasmine)
Jasminum mesnyi (Primrose jasmine)
Jasminum multiflorum (Star jasmine)
Jasminum multipartitum (African jasmine)
Jasminum nitidum (Shining jasmine)
Jasminum nudiflorum (Winter jasmine)
Jasminum officinale (Poets' jasmine)
Jasminum parketi (Dwarf jasmine)
Jasminum polyanthum (Winter jasmine)
Jasminum sambac (Arabian jasmine)
Ligustrum amurense (Amur privet)
Ligustrum ibodium (Privet)
Ligustrum japonicum (Japanese privet)
Ligustrum lucidum (Glossy privet)
Ligustrum ovalifolium (California privet)

3

MORPHOLOGY AND TAXONOMY OF THE OLIVE

The olive tree is a long-lived evergreen reaching 1000 years of age, or more. Olive trees have a titanic resistance that renders them almost immortal (see Fig. 3.1). In spite of cold winters and very hot and dry summers they continue to grow, bearing fruit that nourishes and heals.

HABIT

The olive tree is an evergreen with grey–green leaves and small, white, fragrant flowers in the spring that produce a lot of pollen (Martin, 1994a). A mature

Fig. 3.1. A very old olive tree growing on the Greek island of Crete. Its age is estimated to be over 1000 years, its trunk circumference is 15.6 m and its diameter at 60 cm above the soil level is 4.5 m. This tree is still producing fruit.

© CAB International 2009. *Olives* (I. Therios)

tree can reach a height of 25–30 feet (8–10 m) and live for many hundreds of years. An olive tree tends to grow densely, with thin branches. Cork cambium is present and secondary thickening develops from a conventional cambial ring.

The cultivated olive belongs to the family of Oleaceae, genus *Olea*; the scientific name is *Olea europaea*. In this family belong 30 genera and 180 species. The olive cultivars have 2x = 46 chromosomes. Excepting olive, in the Oleaceae family various genera are classified accordingly:

- Fleshy fruit   *Ligustrum*
  Chionanthus
- Dry fruit   *Fraxinus*
  Syringa
  Forsythia

According to De Candolle the genus *Olea* includes *Gymnelaea* and *Euelaea*, along with the species listed in Table 3.1 (Therios, 2005b).

Olea europaea includes various sub-species (see below). For paternity testing in olive progenies microsatellites are used (de la Rosa *et al.*, 2002, 2004).

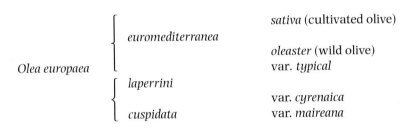

The olive tree is long-lived, with an average height of 15–20 m. However, under agricultural practices the height obtained following pruning is generally 4–5 m. The greatest proportion of roots grow at a soil depth of 60–70 cm, the thick roots growing in the top 20 cm. In dry areas the root system covers an area seven to eight times greater than the leaf area (Fernández *et al.*, 1990). When the soil is heavy the fibrous roots grow close to the soil surface. In sandy soils, and generally in light-textured soils, the root system is extensive. In Tunisia, a country characterized by a dry climate, the side-growth of roots is 12 m away from the trunk and the depth 6 m. With this deep root system the olive tree absorbs water and nutrient elements from soils with very low rainfall. The root system of olive trees (seedlings or asexual propagation) during the first 3–4 years grows vertically, and subsequently this vertical root system is replaced by a side-root system.

In the root system of olive trees, symbiosis with endotrophic mychorizzae is very common (Mancuso and Rinaldelli, 1996); this phenomenon of symbiosis

Table 3.1. Various species of olive, by continent.

Europe	Asia	Africa	America	Oceania
O. europaea	O. europaea	O. europaea	O. americana	O. apetala
	O. microcarpa	O. chrysophylla	O. floribunda	O. paniculata
	O. maritima	O. laurifolia		
	O. passiflora	O. verrucosa		
	O. heynana	O. branchybotris		
	O. attenuata	O. capensis		
	O. dentata	O. faveolata		
	O. lindley	O. concolor		
	O. salicifolia	O. exaesprata		
	O. dioica	O. humilis		
	O. cuspidata	O. obtusifolia		
	O. compacta	O. lancea		
	O. roxburghiana			
	O. glandulifera			
	O. acuminata			
	O. longifolia			

is greater in poor soils. Olive trees do not have a dominant taproot system, and the root system is restricted to the first 1 m of soil depth.

The tissue between the tree trunk and the root crown (xylopode) is characterized by ovoid hypertrophies, containing sphaeroblasts (Therios, 2005b). About 30% of the total carbon fixed by the plant is stored in this tissue, from which the roots are derived. This form of root system may allow more efficient water absorption after intermittent rainfall, in comparison with a deep root system. One drought-avoiding response achieved with a deep root system is to reach into moist soil. The ratio of the root size between irrigated and non-irrigated olive trees is 2.3:1. Irrigation reduces carbon transport to roots.

ANATOMY OF THE OLIVE TREE

Root system

The root system has the following functions:
- Anchorage of the tree.
- Water and nutrient absorption.
- Synthesis of various organic materials.
- Storage of nutrients.

The functions of the root system depend upon the rootstock, variety, soil conditions and cultivation practices.

Anchorage
The root is the most efficient anchorage mechanism for the tree when it is very deep and branched. Such a root system is found most commonly when the scion is vigourous and when the rootstock is a seedling and not originated from rooted cuttings. A very deep and branched root system does not grow in shallow soils or in soils having claypan, hardpan or a sand layer.

Absorption
Trees with a deep and well-branched root system are better adapted to water and nutrient absorption. Water absorption is a function of soil properties and the plant itself. The active absorption of nutrient elements requires utilization of metabolic energy in the form of adenosine triphosphate (ATP).

Synthesis
In the root, gibberellic acids (GAs) and cytokinins are produced in the root tip. Furthermore, in the root ethylene is synthesized, which at very low concentration induces root growth and branching (Abeles, 1973). When stress due to root damage is exerted a significant amount of ethylene is produced which, after its transport to the tree top, causes senescence and leaf abscission. Another growth regulator, abscisic acid (ABA), is produced in the root cap and, following its transport to leaves, is responsible for stomatal closure.

Storage
In the root system carbon and nitrogen compounds are stored. Carbon is stored in the form of starch and soluble carbohydrates, while nitrogen is stored as amino acids and proteins. The accumulation of certain compounds at the end of summer and early in autumn exerts a significant influence on both the induction of flowering buds and shoot growth the following spring. The storage role of the root system and the other functions depend on the availability of photosynthates. Every factor that reduces photosynthetic carbon fixation leads to a reduction in root growth. Under normal conditions, 50% of photosynthates are transported to the root, with 50% of these being utilized in respiration and growth and the remainder being stored.

Root growth may be affected by the following factors:

PRUNING   Severe pruning tends to reduce root growth during spring. This is due to the fact that shoot growth precedes root growth and all photosynthates are used for top and not for root growth.

SOIL CONDITIONS   Fungi, bacteria and nematodes reduce root growth. Adequate soil moisture is a prerequisite for optimal root growth. Furthermore, excess soil water reduces oxygen (O_2) and root growth. Also, accumulation of salts – and especially those containing sodium (Na) and chlorine (Cl) in the soil or in irrigation water – reduces root growth. Concentrations considered to be toxic are > 0.5% for Cl and > 0.2% for Na (d.w.). The root system could be a seedling or clonal rootstock.

The use of dwarfing rootstocks reduces the tree height and harvesting cost. Hence, seedlings of the cv. Ascolano reduce the height of the cvs Mission, Manzanillo and Sevillano. The rootstock *O. oblonga* reduces the height of Manzanillo less than that of Mission, while it promotes the growth of Sevillano. Also, the cv. Redding Picholine reduces the vigour of Manzanillo. Other genotypes tested as rootstocks, but without promising results, were *O. ferruginea*, *O. chrysophylla*, *Forestiera neomexicana*, *Fraxinus* and *Syringa vulgaris*.

Trunk

The olive trunk is cylindrical, with an uneven surface, bearing a lot of swellings. The wood is yellowish and darker towards the centre of the trunk. The graceful, billowing appearance of the olive tree can be rather attractive. In an all-green garden its greyish foliage serves as an interesting counterpoint. The attractive gnarled branching pattern is also quite distinctive. The trees are tenacious, easily sprouting back even when cut to the ground.

Main branches

The branches originate at a height of 1.2 m in the classical olive grove and at 20–40 cm in the modern dense olive plantings. The number of branches is three or more. The main branches give secondary and tertiary branching bearing the leaves, flowers and fruits. The small shoots are classified into four categories:

- *Vegetative shoots* bearing only vegetative buds and producing new shoots and leaves.
- *Fruit-bearing shoots* bearing flowering buds; their number is greater in the low-vigour trees.
- *Mixed shoots* bearing vegetative and flowering buds concurrently; the flowers and fruits are borne at the base of the mixed shoots.
- *Water sprouts* originating from the trunk, branches and the thick shoots; these are very vigourous, grow vertically and they should be removed, unless they are going to substitute for a low-vigour branch or stem.

Leaves

The leaves of olive trees are grey–green and are replaced at 2–3 year intervals during the spring after new growth appears. The olive's feather-shaped leaves grow opposite one another. Their skin is rich in tannins, giving the mature leaf its grey–green appearance. Leaves have stomata on their lower surface only (Fernández et al., 1997). Stomata are nestled in peltate trichomes (see Fig. 3.2), restricting water loss and protecting leaves against UV radiation (Karabourniotis et al., 1992, 1995). The leaves are covered by a layer of wax and cutin (cuticle). The deposition of cutin in the leaves takes place during leaf growth and stops when leaf growth terminates. The weight of the cuticle on the upper leaf surface is 1.4 mg/cm^2; the thickness of the cuticle on the upper leaf surface is 11.5 μm, and on the lower one 4.5 μm.

On both surfaces peltate trichomes exist and their concentration is 143/mm^2 on the lower surface but only 18/mm^2 on the upper. Stomates are present (470/mm^2) only on the lower surface (Martin, 1994a; Fernández et al., 1997). Leaf age affects stomatal conductance (Gucci et al., 1997b). Stomata play a significant role in sensing and driving environmental change (Hetherlington and Woodward, 2003). The large number of trichomes significantly increases the efficient leaf surface, which is three times greater on the lower surface in comparison with the upper. Both leaf surfaces are difficult to hydrate and the contact angle of distilled water is 106° for the upper and 125° for the lower leaf surface. Table 3.2 summarizes these characteristics.

The leaves may have juvenile or adult characteristics. The juvenile and adult phases can be differentiated by several morphological characteristics such as thorniness, leaf size and shape, pigment accumulation, phyllotaxy and ability to form adventitious roots (Hackett, 1985). Furthermore, in olive trees a difference is recorded between juvenile and adult leaves concerning lipid and protein composition (Garcia et al., 2000). Between the juvenile and mature

Fig. 3.2. Peltate trichomes on the lower surface of olive leaves (from Therios, 2005b).

Table 3.2. The main characteristics of olive leaves.

Characteristic	Upper surface	Lower surface
Stomates (n/mm^2)	0	470
Trichomes (n/mm^2)	18	143
Cuticle thickness (μm)	11.5	4.5
Contact angle (°)	106.0	125.4

stage there is a transition stage, which is followed by increasing internode length and leaf blade size. Furthermore, during the transition phase changes in composition take place. The changes (chemical, anatomical and compositional) that occur during the transition stage affect the optical properties of leaves. Hence, the electromagnetic spectrum could provide information easily and quickly about the developmental phase of the olive plant. The duration of the juvenile phase can be as long as 15–20 years in olive trees which creates problems for breeding and production of new olive genotypes through crossing of certain parents. During the juvenile period an olive plant cannot be induced to flower, but this is a drawback in breeding programmes (Leon and Downey, 2006). The previous analysis indicates that every effort should be exerted to shorten the length of the juvenile period (Lavee et al., 1996). The optical properties of leaves are a useful criterion in categorizing juvenile plants (Leon and Downey, 2006).

Buds

These are classed as either vegetative or flowering. The former are small and conical, while the latter are spherical and of greater size. Furthermore, latent buds grow following severe pruning or after frost damage. A proportion of the buds of the fruit-bearing shoot remains inactive.

Inflorescences and flowers

Inflorescences are born in the axil of each leaf (see Fig. 3.3). Each inflorescence contains 15–30 flowers, depending on both the prevailing conditions and the cultivar. Inflorescences originate from buds of the current season's growth. These buds are induced to become flowering ones after the winter's chilling effects. They then begin to grow, producing inflorescences. The olive flowers are small, creamy white and hidden within the thick leaves. Each flower consists of a four-segmented calyx, a tubular corolla with four lobes, two stamens and an ovary with two carpels and a short style (Martin, 1994a). The blossoms usually begin to appear in May. A wild seedling olive tree normally

Fig. 3.3. Olive inflorescence before and after blooming (cv. 'Chondrolia Chalkidikis') (from Therios, 2005b).

begins to flower and produce fruit at the age of 8 years. The olive tree produces both perfect and staminate flowers. The perfect flowers contain two stamens and a well-developed ovary (green in colour), while staminate flowers contain an aborted ovary (very small in size) and normal stamens. The percentage of perfect flowers depends on many factors such as the cultivar, the number of inflorescences per plant, the soil moisture and the leaf nitrogen content during the period of flowering differentiation. The phenomenon of flower absorption was mentioned by Theophrastus for olives in Italy as long ago as 350 BC.

Flower bud induction takes place in winter and 8 weeks before full-bloom flower formation can be seen under a microscope. During the following 8 weeks flower development proceeds rapidly and flowering occurs at the end of May, or 1–2 weeks earlier in more southerly areas. The pistil of perfect flowers contains two ovules from which only one is fertilized, giving one seed. Figure 3.4 illustrates the duration of flowering for a selection of cultivars over two separate growing periods.

Fruit

The olive fruit is a drupe, spherical or elliptic in shape and consists of the exocarp (skin), which contains stomata, the mesocarp (flesh), which is the edible portion of the fruit, and the endocarp (pit), including the seed. The fruit of the olive tree is purplish black when completely ripe, but a few cultivars are green when ripe and some olives develop the colour of coppery brown. The size of the olive fruit is variable, even on the same tree, and depends on cultivar, fruit load, soil fertility, available water and cultural practices (see Fig. 3.5).

CULTIVAR	MONTH (DATE)											
	MAY (1978)									JUNE (1980)		
	7	10	13	16	19	22	25	28	31	1	3	5	7
Kalamon			--------θ--------								——○——	
Amphissis		--------θ--------								——○——		
Chondrolia Ch.		--------θ--------						——○——			
Gordales		--------θ--------					——○——				
Manzanilla	--------θ--------						——○——					
S. Agostino		--------θ--------					——○——				
N. Blanco			--------θ--------						——○——		
Ladolia		--------θ--------						——○——			
Adramitini			--------θ--------					——○——			
Kolovi				--------θ--------						——○——	
Koroneiki		--------θ--------						——○——				

Fig. 3.4. The duration of flowering of 11 olive cultivars in two growing periods (from Therios, 2005b).

Fig. 3.5. Olive tree with fruit setting (from Therios, 2005b).

GROWTH, MORPHOLOGY AND PHYSIOLOGY OF THE OLIVE TREE

Olive trees may attain a significant size. However, the slow growth rate and the height are important characteristics of this species – olive trees require many years for full fruit bearing. Shoot extension of olive trees occurs in spring and is related to the rise in temperature. The growing tips are characterized by apical dominance, and this inhibits the growth of axillary buds from the tip towards the physiological base. Olive forms with both a physical spindle shape and strong apical dominance give the greatest yields.

The vegetative buds start sprouting at the end of March, a little bit later than the axillary flowering buds. The spring vegetative wave is the most important phase and lasts up to the middle of July. A second growth wave may occur in September–October after rain or irrigation. The length of annual shoots is affected by fruit load, since fruits antagonize shoot growth. The axillary buds induce or hinder growth from the middle of June to the middle of October. Fruit removal or abortion of seed before hardening of the endocarp (7–8 weeks after full bloom) increases the flowering percentage. Injection of GA_3 (gibberellin) into the branches of trees not bearing fruits, between May and November, inhibits flowering the following year.

Between the middle of October and the middle of November a greater quantity of RNA was recorded in the buds of trees that flowered. These changes lead to the conclusion that the induction of flower buds occurs up to the middle of November. Subsequently, the buds enter into dormancy. The lack of growth in the meristem of flowering buds from the middle of November up to March could be ascribed to two factors: (i) endodormancy; and (ii) ecodormancy.

Endodormancy leads to ecodormancy between the middle of January and the end of February, depending on the cultivar. Sprouting of flowering buds triggers inflorescence and flowering bud growth. This is a continuous process until flowering. However, certain periods are critical for productivity. Hence, the lack of water and nitrogen between bud sprouting and 6 weeks before flowering reduces the number of flowers per inflorescence and increases the percentage of ovarian absorption. Furthermore, the embryo sac growth, which lasts for a period of 20 days before flowering up to full flowering, may fail, due to genetic factors. Flowering determines the initiation of the main period of flower or fruit abscission and lasts for 6 weeks following full bloom.

Pollination and fertilization are followed by fruit set and growth. Of the four ovules only one is fertilized, while the other three degenerate. Ovarian growth triggers the abscission of the neighbouring, unfertilized, perfect flowers or the fruits with the smallest weight. The end of this period of abscission coincides with hardening of the pit, and the rapid growth of the embryonic fruit follows a double sigmoid curve, as a typical drupe. At the end of the abscission period fruit size is 20–25% of mature size. Fruit ripening starts in the middle of October, completion time depending on the cultivar. The maximum olive oil content is measured a little bit earlier than the stage of full fruit ripening. The periods of ripening and harvesting of 25 Greek and 36 worldwide cultivated cultivars are given, respectively, in Figs 3.6 and 3.7.

#	CULTIVAR	USE	RIPENING PERIOD														
			SEPT.			OCT.			NOV.			DEC.			JAN.			FEB.		
			1	2	3	1	2	3	1	2	3	1	2	3	1	2	3	1	2	3
1	Koroneiki	Olive oil								━	━	━	━	━	━	━	━	
2	Patrini	Olive oil								━	━	━	━					
3	Lianolia Kerkiras	Olive oil									━	━	━	━	━	━	━	━
4	Mastoides	Olive oil								━	━	━	━	━				
5	Thiaki	Olive oil								━	━	━	━					
6	Mavrelia Messinias	Olive oil				━	━	━	━	━								
7	Tragolia	Olive oil								━	━	━	━	━				
8	Smertolia Lakonias	Olive oil								━	━	━	━	━	━			
9	Aguromanako	Olive oil								━	━	━	━	━	━	━		
10	Adramyttini	Olive oil								━	━	━	━	━				
11	Kolovi	Olive oil									━	━	━	━	━	━		
12	Dafnelia	Olive oil								━	━							
13	Throubolia	Dual purpose							━	━	━	━	━					

#	CULTIVAR	USE	RIPENING PERIOD														
			SEPT.			OCT.			NOV.			DEC.			JAN.			FEB.		
			1	2	3	1	2	3	1	2	3	1	2	3	1	2	3	1	2	3
14	Kalokerida	Olive oil		━	━	━	━	━	━	━	━	━	━	━	━			
15	Kothreiki	Dual purpose					━	━	━	━	━	━	━	━	━			
16	Megaritiki	Dual purpose						━	━	━	━							
17	Amphissis	Table olive					━	━	━	━	━	━						
18	Kalamon	Table olive						━	━	━	━	━						
19	Gaidurelia	Table olive						━	━	━	━							
20	Vasilikada	Table olive						━	━	━	━							
21	Karolia	Table olive							━	━	━	━						
22	Mavrelia Lefkados	Olive oil						━	━	━	━							
23	Lefkolia Lefkados	Olive oil							━	━	━	━	━					
24	Hondrolia Chalk.	Dual purpose					━	━										
25	Galatsaniki	Dual purpose			━	━	━	━										

Fig. 3.6. Ripening periods of 25 selected Greek olive cultivars. Ripening period: 1, days 1–10 of the month; 2, days 11–20; 3, days 21–31.

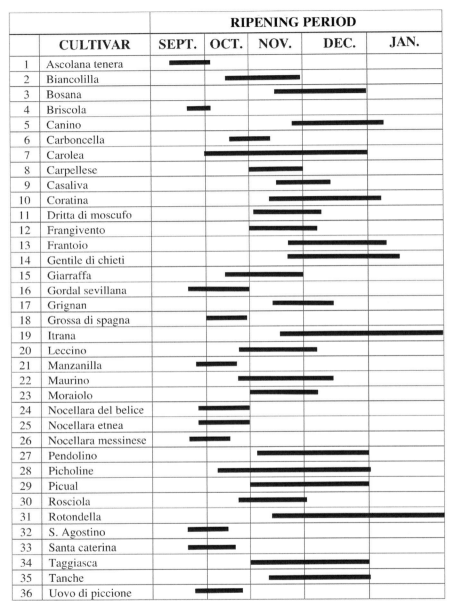

Fig. 3.7. Ripening periods of 36 selected worldwide olive cultivars.

STRUCTURE AND COMPOSITION OF THE OLIVE FRUIT

STRUCTURE

The olive fruit can be separated into three distinct anatomical parts: epicarp (skin), mesocarp (pulp or flesh) and woody endocarp (stone), the last containing the seed; these three fruit components follow distinctive growth periods (Hartmann, 1949; Duran and Izquierdo, 1964; Farinelli *et al.*, 2002).

Epicarp

This is the protective tissue that accounts for 1.0–3.0% of the drupe weight. The skin is covered by a layer of wax, accounting for 45–70% of the skin weight. The skin is green during the early stages of development due to chlorophyll, and later changes to straw yellow, pink, purple and black. This colour change is due to varying concentrations of chlorophyll, carotenoids and anthocyanins (Minguez-Mosquera and Garrido-Fernández, 1989). The epicarp plays a significant role in minimizing mechanical damage and pest attacks. Furthermore, the epicarp is impermeable to water, affecting the processing of table olives.

Mesocarp

This is the most important part of the olive fruit, being the edible portion and comprising 70–80% of the whole fruit. The water content of the flesh is 70–75% of its weight and the oil content ranges between 14 and 15% in green table olives and up to 30% in black mature olives. The table olives should have low oil content. However, in certain types of black olives lipid biosynthesis results in medium to high oil content (Donaire *et al.*, 1984).

Olives contain oxalic, succinic, malic and citric acids and high levels of free fatty acids. The sugars glucose, fructose, saccharose and mannitol – 3.5 to

6.0% of the flesh in total – are all present. The sugar content decreases with maturation. Protein content ranges between 1.5 and 2.2% of the fruit by weight. The pectic substances cement the cells and affect the texture of the olive flesh. These pectic substances during processing are hydrolysed by pectinolytic enzymes and the fruit texture becomes softer.

Endocarp (stone)

The endocarp represents 10–27% of the olive by weight, the seed comprising 2–4% of the weight. The seed contains 22–27% oil and the shell 1%. Some important characteristics of the endocarp include size, weight and ease of separation (cling or free stone).

COMPOSITION

Table olives are characterized by the following parameters: (i) shape and fruit size; (ii) the ratio of flesh to pit; (iii) flesh firmness; (iv) skin thickness; (v) stone size and shape; and (vi) taste of the final product. Varieties with large fruit size are used for table olive processing. The time of harvesting for table olive cultivars depends on the type of the final product and it is a function of climatic conditions. The determining factors include the following.

Colour

This depends on the intended use of the olive. Hence, for green olives the colour will be straw-yellow green, with a deep black skin for black olives due to anthocyanins, which could be evaluated in situ (Agati *et al.*, 2005). The amount of flesh should not be lower than 70%. Furthermore, the flesh should be firm enough for the processing to be performed without loss of fruit quality. The composition of the various parts of the olive fruit is presented in Tables 4.1, 4.2 and 4.3.

Table 4.1. Main composition of the olive fruit (from Marsilio, 2006).

Fruit			Weight (%)
Pericarp	Epicarp		
Pericarp/endocarp	Epicarp/mesocarp	Pulp	70–90
		Stone	10–27
		Seed	1–3

Table 4.2. Chemical composition of the olive fruit pulp, by weight.

Component	Weight (%)
Moisture	50–75
Lipids (oil)	10–30
Reducing sugars	2–6
Non-reducing sugars	0.1–0.3
Crude protein (N × 6.25)	1–2
Fibre	1–4
Phenolic compounds	1–3
Organic acids	0.5–1.0
Pectic substances	0.3–0.6
Minerals	0.6–1.0
Others	3–7

Table 4.3. Composition (percentage) of $CHCl_3$–EtOH-extractable classes of compounds in the epicarp, mesocarp, shell and seed of cv. 'Coratina'.

Compound	Epicarp	Mesocarp	Shell	Seed
Alkanes	8	trace	1.7	trace
Squalene	–	0.1	–	trace
Alkyl esters	2	trace	–	0.3
Methyl phenyl esters	2	–	–	–
Alcohols	10	trace	0.1	0.2
Aldehydes	2	–	–	–
Free fatty acids	1	5	7	8
Pentacyclic triterpene alcohols	6	trace	1.5	0.4
Pentacyclic triterpene acids	63	0.2	0.6	4
Free sterols	1	trace	trace	trace
Steryl esters	–	trace	1.1	2
Triacylglycerols	3	92	78	80

Alkanes

The major alkanes are found in both the epicarp and the woody part of the shell. They are a mixture of C23–C33 compounds; of those the odd chains C25, C27 and C29 dominate. The dominant component in green olives is C29, while in the black olive it is C27.

Alcohols, aldehydes and fatty acids

Alcohols are absent in the pulp, 0.1–0.2% in the stone and 10% in the skin. The major chains are C22–C28. Aldehydes are found in the skin in the even chains C26–C32. Fatty acids consist of two groups, C16–C18 and C22–C28. Alkyl esters are found in significant amounts in the skin lipids and at trace levels in the seed and pulp. Two groups are present, C40–C44 and C46–C56 chains. The dominant chains are C40 and C42 in the first group and C52 and C54 in the second one.

Triterpenoids

In the flesh and seed are found cycloartenol and methylene cycloartenol. Furthermore, pentacyclic triterpene acids are found in the epicarp, such as oleanolic and maslinic acids. Free sterols and terpenes are present in significant amounts in the cuticular lipids, and the stony endocarp contains steryl esters. Among the sterols the most important are β-sitosterol, stigmasterol and campesterol.

Phenols

Olive flesh contains 1–14% (d.w.) phenolic compounds, depending on the variety (Amiot *et al.*, 1986). The phenolic compounds are important in many aspects, such as protection of plants from bacteria, fungi and viruses (Hanbury, 1954) and fruit browning as maturation proceeds. Furthermore, phenolic compounds play significant roles in human nutrition (Bravo, 1999) and health (Christakis *et al.*, 1982; Manna *et al.*, 1999). The bitter taste of raw olives is due to phenols and, especially, oleuropein, which is water soluble (Gutiérrez *et al.*, 1992). Other phenols include ligstroside, verbascoside, 4-hydroxytyrosol, tyrosol and glucosides or aglycones, 3,4-dihydroxyphenyl glycol and flavonoids (Manna *et al.*, 1999). The concentration of phenols is a function of fruit maturity and can be as high as 14% (d.w.) in young fruits, though it may be close to zero in black-type fruits.

The size of the fruit, which is a characteristic of the cultivar and irrigation method employed (Costagli *et al.*, 2003), is related to phenol concentration. Hence, varieties with small-sized fruits have a high oleuropein content compared with large-fruited varieties. As olive fruits mature, oleuropein concentration is reduced (Amiot *et al.*, 1989) while the levels of oleoside-11-methylester and dimethyloleuropein start to increase, reaching maximum levels when the fruit is black and mature. Therefore dimethyloleuropein is the major component of black olives. Furthermore, as maturation procceds, verbascoside concentration increases.

Another phenolic compound, ligustroside, is abundant in the early stage, but levels decrease during fruit development. The oleuropein content depends on the variety and its origin. Hence, the Greek and Italian cultivars contain greater amounts of oleuropein in comparison with the Spanish and Portuguese.

Oleuropein

Oleuropein is present in all parts of the olive tree and especially in the leaves, at concentrations of 60–90 mg/g dry weight (Soler-Rivas et al., 2000). *Olea europaea* contains, besides oleuropein, other phenolic glucosides such as verbascoside (ester of caffeic acid and hydroxytyrosol), ligustroside, dimethyloleuropein (acid derivative of oleuropein) and cornoside. Phenolic composition differs between the various cultivars. Among the phenolics the three most important are oleuropein, dimethyloleuropein and verbascoside.

Of all the olive varieties, the small-fruited ones contain high levels of oleuropein and low amounts of verbascoside, while the opposite is true for the large-fruited varieties. Other phenolics common in olive cultivars are tyrosol, hydroxytyrosol and tyrosol glucoside. Furthermore, olive leaves and buds contain flavonoids such as quercetin, kaempferol and hesperitin. Also found in the olive fruit are the flavonol glycosides such as luteolin 7-glucoside, rutin and anthocyanin (cyanidin, delphinidin).

BIOSYNTHESIS   Oleuropein belongs to the secoiridoids, found very commonly in the Oleaceae (Damtoft *et al.*, 1993). These compounds are produced from the secondary metabolism of terpenes. Oleuropein is an ester of hydroxytyrosol. Oleuropein biosynthesis in the olive commences during secondary metabolism, via a branching of mevalonic acid. This produces oleosides, which finally lead to formation of secoiridoids. The carbon skeleton is derived from mevalonic acid. The precursors for oleuropein in olives are both epoxides of secologanin and secoxyloganin.

DEGRADATION   Oleuropein is abundant in the young olive fruits and comprises around 14% of dry matter. Also, ligustroside and cornoside are abundant. While the fruits mature and their colour changes from green to black, the emerging compounds of oleuropein are elenolic acid, glucoside and dimethyloleuropein and glucosylated derivatives of oleuropein. At the green mature stage and when the fruit turns black, dimethyloleuropein is the major substance produced by the activity of esterases. The olive fruit accumulates glucosylated derivatives of oleuropein. Another transformation of oleuropein includes dihydroxytyrosol, which is found in leaves.

Other oleocides, which are reduced as oleuropein declines, include ligustroside, while verbascoside levels are increased. Furthermore, tyrosol and hydroxytyrosol are present in olive fruits during ripening.

OLEUROPEIN AS A PROTEIN DENATURANT AND AS A DEFENCE MECHANISM AGAINST HERBIVORES   Oleuropein is a secoiridoid glycoside, which is common in the family Oleaceae – such as olives and *Ligustrum obtusifolium*. Oleuropein is stable and is retained in a compartment separate from the activating enzymes. When herbivores digest the leaves, enzymes present in the organelles activate oleuropein and transform it into a strong protein denaturant, which promotes protein cross-linking and decreases lysine levels. These changes have adverse effects on herbivores by decreasing the nutritive value of the dietary protein. The enzyme β-glucosidase in organelles converts oleuropein to a glutaraldehyde-like substance. Therefore, oleuropein activated by β-glucosidase has very strong protein-denaturing, protein-cross-linking and lysine-alkylating activities, and *L. obtusifolium* has developed an effective defence mechanism, with oleuropein as a protein cross-linker.

Antioxidant activity of olive oil

Among the phenolic compounds found in extra virgin oils, gallic, caffeic, vanillic, p-coumaric, syringic, ferulic, homovanillic, hydroxybenzoic, protocatechuic acids, tyrosol and hydroxytyrosol are the most important. During cooking under domestic conditions the oil is heated to temperatures of up to 190°C. Polyphenols of extra virgin oil are stabilizers of α-tocopherol during heating (Andrikopoulos, 1989). Therefore, they contribute to the nutritional value of cooked foods, and polyphenols prevent the decrease in antioxidant activity in olive oil during heating.

Olea europaea extracts containing oleuropein and hydroxytyrosol were found to be much more effective than vitamin E in their antioxidative activities. These extracts owe their antioxidative properties both to their high oleuropein content (19% w/w) and to a lower amount of flavonoids (1.8% w/w). Tyrosol showed neither antioxidant nor pro-oxidant activity (Lerutour and Guedon, 1992).

Phenolic biosynthesis

In olive plants the aromatic amino acids, phenylalanine and tyrosine, are produced via the shikimate pathway. Furthermore, carbohydrates supply carbon skeletons necessary for the biosynthesis of acetate, shikimic acid and aliphatic amino acids. The initial reactants are phosphoenolpyruvate and erythrose-4 phosphate, which are produced via the non-oxidative glycolysis of glucose. Phenylalanine is the precursor of most phenolic compounds. Phenylalanine ammonia lyase (PAL) is the enzyme involved in phenolic biosynthesis.

Hydroxytyrosol in olive fruits

Hydroxytyrosol is a natural phenolic compound that is present in olive fruits, table olives, olive oil pomace and waste water produced during processing of olives. When olives were treated with NaOH, the major phenolic compounds present were hydroxytyrosol, tyrosol, caffeic acid and p-coumaric acid. Hydroxytyrosol glucoside is a polar substance and it is logical to expect its presence in the waste water and not in the olive oil.

Phenolic compounds in various olive varieties

Phenolic compounds in olive fruits are a very important factor in the evaluation of the quality of virgin olive oil, since they are responsible for its antioxidation stability. Oleuropein is the main substance giving bitterness to olives. The phenolic content of virgin olive oil is affected by the variety, location, degree of ripeness and the type of machinery used, e.g. three-phase or two-phase decanter (Di Giovacchino et al., 1994; Servili et al., 2006).

Phenolic content of waste water and its antioxidant activities

Olive oil is extracted mechanically by pressure and by a three- or two-phase decanter. The three-phase centrifugation needs 0–6 m^3 H_2O/t of olives and is the most extensively used method for virgin olive oil production. The two-phase decanter separates virgin olive oil by recycling the waste water produced from processed olives. The two-phase extract has the highest antiradical activity, which is twice the value obtained by the three-phase extract. The compound exerting most antioxidant activity is hydroxytyrosol (Galli and Visioli, 1999).

5

HECTAREAGE, NUMBER OF TREES AND PRODUCTION OF OLIVE OIL AND TABLE OLIVES

OVERVIEW

Most olive trees, on a worldwide basis, are cultivated in the Mediterranean region. More than 75% of the global olive production occurs in Europe, which cultivates 500 million olive trees. Europe is followed by Asia (13%), Africa (8%) and America (3%). Spain is the leader in olive culture, followed by Italy and Greece.

The hectareage and number of olive trees worldwide are given in Table 5.1, the average annual total world production of olives in Fig. 5.1, the global olive oil production in Table 5.2, the olive oil production in the Mediterranean in Table 5.3, the EU production of olive oil in Table 5.4 and the world production of olive oil (percentage of total) in various countries in Fig. 5.2.

In recent decades olive oil production has had periods of growth followed by stagnation. At the beginning of the 1980s world production was about

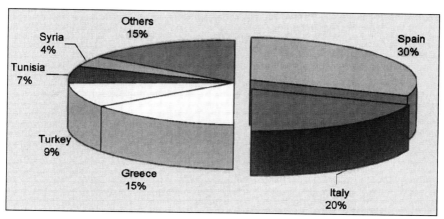

Fig. 5.1. The average annual total world production of olives during the period 1998–2001 (15,090,620 t) (from FAOSTAT, 2003).

© CAB International 2009. *Olives* (I. Therios)

Table 5.1. Hectareage and number of olive trees worldwide (2005). Data elaborated using the International Olive Oil Council (IOOC) statistics.

Country	Hectareage (ha × 1000)	Trees (n × 1000)
Spain	2,340	200,000
Italy	2,250	185,000
Greece	670	133,000
Cyprus	12	2,450
Rest of Europe	1,203	62,150
Total, Europe	6,475	7,000
Argentina	70	582,600
USA	44	4,500
Mexico	15	1,540
Rest of America	11	1,570
Total, Americas	140	14,610
Turkey	723	72,000
Syria/Lebanon	168	24,160
Palestine	66	12,130
Rest of Asia	2	240
Total, Asia	959	108,530
Tunisia	1,240	52,000
Morocco	222	22,000
Algeria	166	16,000
Rest of Africa	117	4,550
Total, Africa	1,742	94,000
Others areas	681	12,550
Total, World	10,000	800,000

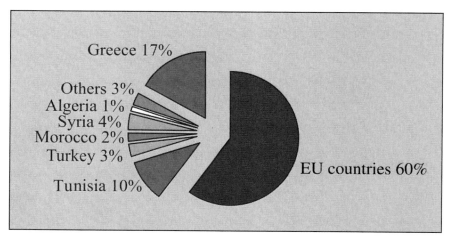

Fig. 5.2. World production of olive oil (percentage of total, 1999–2000; total 2,033,500 t) (from IOOC).

Table 5.2. World production of olive oil (1000 t), 1998–2000 (from IOOC).

	1998/2000			1999/2000		
Country	Production	Consumption	Export	Production	Consumption	Export
EU[1]	1,698.5	1,660.0	235.0	1,563.0	1,679.0	257.0
Spain	789.2	500.0	75.0	550.0	500.0	80.0
France	3.4	78.8	1.0	3.0	80.0	1.5
Greece	473.0	245.0	6.0	426.0	245.0	7.0
Italy	397.0	705.0	140.0	620.0	710.0	150.0
Portugal	36.0	67.0	12.5	40.0	68.0	17.5
Tunisia	215.0	49.0	175.0	200.0	65.0	120.0
Turkey	170.0	97.0	60.0	70.0	60.0	30.0
Morocco	65.0	55.0	20.0	40.0	50.0	10.0
Syria	115.0	88.0	4.0	80.0	85.0	5.0
Algeria	39.5	35.0	0.0	25.0	29.0	0.0
Other countries	70.6	392.2	14.6	55.5	391.9	9.4
Total	2,373.6	2,385.2	508.6	2,033.5	2,359.9	431.4

[1]EU, European Union; incorporates several of the countries in this table.

Table 5.3. Production of olive oil in the Mediterranean area (percentage of total, by country) (from IOOC).

Country	Production
Spain	33
Italy	23
Greece	17
Portugal	8
Tunisia	8
Turkey	5

1.8 million t, 40% above the value recorded in the mid-1960s. After a relatively stable period production again increased in the second half of the 1990s, to reach 2.5 million t. Average world production for the most recent years is about 2.7 million t.

The EU is the dominant producer on the olive oil market. Until 1981 its 425,000 t accounted for only one-third of world production. EU production after the accession of Greece, Spain and Portugal rose sharply, averaging about 80% of world production.

During the 1990s a rapid rise in production in the EU occurred due to the doubling of Spain's production and an increase of 16–18% from Greece and Italy. The next planned expansion of the EU will have limited impact on EU olive

Table 5.4. Olive oil production in the European Union (1000 t, 1992–2003) (from IOOC).

Year	Spain	Italy	Greece	Portugal	France	Total
1992/1993	623.1	435.0	310	22.0	1.6	1391.7
1993/1994	550.9	520.0	254.0	32.1	2.3	1359.3
1994/1995	538.8	448.0	350.0	32.2	2.0	1371.0
1995/1996	337.6	620.0	400.0	43.7	2.3	1403.6
Average	512.6	505.75	328.5	32.5	2.1	1381.4
1996/1997	947.3	370.0	390.0	44.8	2.5	1754.6
1997/1998	1077.0	620.0	375.0	42.0	2.7	2116.7
1998/1999	791.9	403.5	473.0	35.1	3.4	1706.9
1999/2000	669.1	735.0	420.0	50.2	4.1	1878.4
Average	871.3	532.1	414.5	43.0	3.3	1864.2
2000/2001	973.7	509.0	430.0	24.6	3.2	1940.5
2001/2002	1411.4	656.7	358.3	33.7	3.6	2463.7
2002/2003	865.0	590.0	375.0	29.0	4.7	1863.7
Average	1083.4	585.2	387.8	29.1	3.8	2089.3

oil production, since only three of the new members are producers (6000 t for Cyprus, 400 t for Slovenia and 150 t for Malta, annually). These three members represent 0.4% of the national guaranteed quantities of the other member states. The global production of table olives is presented in Table 5.5 and Fig. 5.3.

Olive oil accounts for only about 3% of the global market of edible oil. World consumption of olive oil has been progressing steadily (see Table 5.6),

Table 5.5. Global production of table olives (1000 t, 1999–2000) (from IOOC).

Country	Production
EU[1]	563.5
Spain	380.0
France	2.0
Greece	92.0
Italy	80.0
Portugal	9.5
Turkey	100.0
Morocco	85.0
Syria	75.0
Argentina	103.5
Egypt	40.0
Other countries	142.5
Total	1109.5

[1]EU, European Union

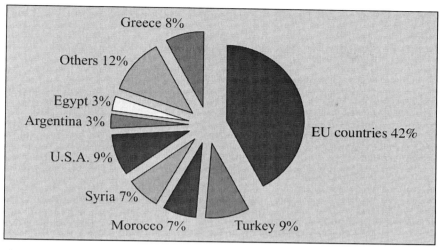

Fig. 5.3. World production of table olives (percentage of total, 1999–2000; total 1,149,500 t) (from IOOC).

Table 5.6. World consumption of olive oil (1000 t, 1995–2003) (from IOOC).

Year	EU	USA	Japan	Australia	Canada	Other	EU (%)
1995/96	1402	105	17	17	14	374	72.7
1996/97	1687	144	26	22	19	473	71.9
1997/98	1841	152	34	18	18	485	72.2
1998/99	1824	159	29	24	19	501	71.4
1999/00	1844	174	28	24	20	480	71.8
2000/01	1918	212	30	31	25	497	70.7
2001/02	1994	221	32	28	24	461	72.2
2002/03	2028	225	33	29	26	490	71.6

the EU being the world's largest consumer. Other minor consumers in the Mediterranean basin are Syria (100,000 t), Turkey (70,000 t), Morocco (50,000 t) and Tunisia (40,000 t). Of the non-European countries the USA is the world's second biggest consumer of olive oil. Appreciable rises were also recorded in Australia, Japan, Canada and Brazil. The EU is the world's leading consumer of olive oil, averaging 1.8 million t, representing 71.6% of world consumption in 2002/2003. The main consumers in the EU are Spain, Italy and Greece, together accounting for more than 85% of the union's total consumption.

Through the 1990s, olive oil consumption showed an increase of 3.3% in the EU as a whole, but growth was smaller in Greece, Spain and Italy

(2.2–2.6%). The consumption rate rose much faster in Portugal (9.7%) and France (10.8%) in the 1990s. The per capita increase in consumption of olive oil in the EU over that decade is presented in Table 5.7, while EU exports for the period 1992–2003 are shown in Table 5.8.

Greek exports consist mainly of extra virgin olive oil (73%), while those of Italy, Spain and Portugal represent 45, 44 and 21%, respectively. Most of the EU's exports are directed to the USA, Australia, Japan, Canada and Brazil. Other major exporters are Turkey (to Canada, the USA, Australia), Tunisia (to the USA) and Argentina (to Brazil). The average prices of exported olive oil are given in Table 5.9.

The ratio of the prices of olive oil to other edible oils is 4–5:1. The difference in quantities of olive oil produced from year to year results in major variation in the prices paid to producers; consumer prices tend to be more stable. In Italy and Greece producer prices for olive oil followed a downward trend in the 1990s. The producer price (€/t) for extra virgin olive oil in Italy fell from an average of 3015.8 to 2318.7 towards the end of the decade. For Greece this figure dropped from 2708.6 to 1905.2 and in Spain from 2775.9 to 1826.6.

Table 5.7. Apparent consumption of olive oil in the EU (kg per capita) (from IOOC).

Year	Greece	Spain	Italy	Portugal	France	Other	Whole EU
1990	20.2	10.1	9.5	2.7	0.5	0.1	3.3
2000	25.0	12.6	12.3	6.9	1.4	0.5	4.6
Increase (%)	2.2	2.2	2.6	9.7	10.8	16.1	3.3

Table 5.8. EU olive oil exports (× 1000 t, 1992–2003) (from IOOC).

Year	Spain	Italy	Greece	Portugal	France	Other	Total
1992/93	51.6	90.8	10.3	7.5	0.9	0.3	161.4
1993/94	54.6	104.8	9.2	10.5	1.1	2.5	182.7
1994/95	54.0	105.8	5.5	13.1	1.0	3.1	182.5
1995/96	48.8	90.5	11.0	11.8	1.1	1.6	164.8
Average	52.3	98.0	9.0	10.7	1.0	1.9	172.9
1996/97	66.7	129.5	5.2	17.0	1.1	0.7	220.2
1997/98	76.2	123.5	8.0	17.4	1.1	1.0	227.2
1998/99	63.6	125.3	5.4	12.4	1.0	0.9	208.5
1999/00	87.7	182.7	8.2	17.5	1.4	1.0	298.5
Average	73.6	140.3	6.7	16.1	1.2	0.9	238.6
2000/01	88.3	173.0	10.0	17.3	1.3	1.1	291.0
2001/02	112.5	182.9	10.0	16.2	1.0	1.7	324.3
2002/03	115.0	200.0	15.0	16.0	1.1	0.8	347.9
Average	105.3	185.3	11.7	16.5	1.1	1.2	321.1

Table 5.9. Average annual prices (€/t) of exported extra virgin and lampante[1] olive oil, 1992–2002 (from IOOC).

Year	Spain		Italy		Greece	
	Extra virgin	Lampante	Extra virgin	Lampante	Extra virgin	Lampante
1992/93	2244.5	2151.3	2586.2	2128.1	2265.6	1799.4
1993/94	2424.9	2322.1	2772.8	2287.8	2419.0	1960.6
1994/95	2770.4	2622.4	2855.3	2538.1	2557.8	2118.0
1995/96	3663.9	3473.0	3849.0	3385.8	3592.0	2864.8
Average	2775.9	2642.2	3015.8	2585.0	2708.6	2185.7
1996/97	2467.0	2045.4	3564.6	2170.5	2765.1	1986.4
1997/98	1812.2	1564.2	2507.7	1545.1	2011.9	1364.3
1998/99	2325.2	2164.7	2591.1	2122.7	2221.2	1781.8
Average	2201.5	1924.8	2887.8	1946.1	2332.7	1710.8
1999/00	1919.8	1782.5	2268.8	1734.7	1841.4	1514.1
2000/01	1712.9	1598.0	2250.0	1547.6	1752.2	1391.4
2001/02	1847.2	1733.3	2437.2	1708.7	2122.1	1482.2
Average	1826.6	1704.6	2318.7	1663.7	1905.2	1462.6

[1]Lampante oil is of a grade not intended for human consumption; the word derives from the ancient use of olive oil as a fuel for lamps.

TABLE OLIVES

World production in 2003 was about 1.6 million t (see Table 5.10). There is a variety of table olives (whole, pitted, sliced, stuffed) and, as observed in the case of olive oil, the production of table olives fluctuates from year to year, due to both weather conditions and the alternate bearing habit of olive trees. When the price of olives for the production of table olives is low, part of the yield is used for oil production. The EU is the largest table olive producer (40%), others being Turkey (13%), the USA (10%), Morocco (8%), Syria (7%) and Egypt (4%).

Table 5.10. Global production of table olives (1000 t, 1995–2003) (from IOOC).

Year	EU	Turkey	Syria	USA	Morocco	Egypt	Total
1995/96	369	120	75	66	85	60	947
1996/97	370	185	90	144	100	25	1063
1997/98	486	124	60	91	85	50	1094
1998/99	500	178	85	78	95	41	1201
1999/00	621	150	93	129	80	85	1351
2000/01	576	224	142	60	80	70	1343
2001/02	765	75	80	120	90	135	1464
2002/03	614	165	170	81	80	300	1647

The average production from the EU in the period 2000–2003 was 651,000 t (see Table 5.11): Spain, 478,000 t (73.4%), Greece, 100,000 t (15.3%), Italy, 62,000 t (9.5%), Portugal 10,000 t (1.5%) and France, 2000 t (0.3%).

The average world consumption of olives in 2003 was about 1.6 million t (see Table 5.12). The EU is the greatest consumer (33%), followed by the USA (13%), Egypt (10%), Turkey (9%), Syria (7%) and Brazil (3%).

EU consumption (see Table 5.13) comprises 39.8% for Spain, 29% for Italy, 7.8% for France, 5.4% for Greece, 2.6% for Portugal, 6.6% for Germany and 2.6% for the UK.

Table 5.11. Table olive production from the EU (1000 t, 1992–2003) (from IOOC).

Year	Spain	Italy	Greece	Portugal	France	Total
1992/93	224	70	60	17	2	372
1993/94	205	77	60	16	2	360
1994/95	236	60	60	10	2	368
1995/96	203	86	70	9	1	369
Average	217	73	63	13	2	367
1996/97	244	55	60	9	2	370
1997/98	310	80	85	9	2	486
1998/99	360	45	85	9	2	500
1999/00	431	75	100	13	2	621
Average	336	64	83	10	2	494
2000/01	416	65	85	9	2	576
2001/02	575	60	115	12	2	764
2002/03	442	60	100	10	2	614
Average	478	62	100	10	2	651

Table 5.12. Global consumption of table olives (1000 t, 1995–2003) (from IOOC).

Year	EU	USA	Syria	Brazil	Egypt	Turkey	Total
1995/96	352	149	71	47	48	129	1043
1996/97	319	173	80	48	29	132	1066
1997/98	341	179	65	50	33	127	1075
1998/99	392	177	70	48	44	149	1186
1999/00	430	184	91	51	77	130	1250
2000/01	454	185	110	45	57	125	1297
2001/02	525	205	74	51	75	100	1380
2002/03	520	205	111	51	160	135	1575

Table 5.13. EU consumption of table olives (1000 t, 1992–2003) (from IOOC).

Year	Spain	Italy	Greece	Portugal	France	Other	Total
1992–1996 average/year	120	125	27	17	28	33	348
1996–2000 average/year	135	115	22	12	33	55	372
2000–2003 average/year	199	145	27	13	39	76	500

WEB SITE

International Olive Oil Council (IOOC): http://www.internationaloliveoil.org

6

ROOTSTOCKS

The planting material for olive orchards is originated from either budded seedlings or leafy cuttings. Budded plants are preferable, since the rootstock affords plants a better adaptability to stress conditions. *Olea europaea* seedlings – and especially those of wild olive – are often used with budding or grafting. However, composite plants show significant variation concerning tree vigour and productivity.

Among olives more suitable as rootstocks, with no variation in tree vigour, are included the rooted cuttings of a vigourous variety. Under orchard conditions the olive tree has to adjust to certain problems, such as cold stress (Charlet, 1965), water stress, salinity and, in moist soils, *Verticillium* wilt and other pathogen problems (European and Mediterranean Plant Protection Organization, 2006). One such rootstock is *O. oblonga* (Hartmann *et al.*, 1971). Other *Olea* species do not provide satisfactory rootstocks for olive trees. Genera and species tested as rootstocks for olives include *Phyllirea, Ligustrum, Syringa, Chionanthus, Forsythia, Fraxinus forestiera, O. verrucosa* and *O. chrysophylla*. Their compatibility with olives is very limited. In addition to poor growth characteristics, incompatibility is most evident through unsatisfactory union. Therefore, a defective union is produced, leading to a defective union structure. In addition, rooted leafy olive cuttings from low-vigour cultivars can be used as dwarfing rootstocks.

The root system of rooted olive cuttings differs from that of olive seedlings, at least during the first years of life, as rooted cuttings have all their roots originating at the base of the cutting and growing at the same soil level. In contrast, in plants grafted to seedlings the roots originate along a 40 cm axis and produce an angle greater than 90°. Therefore, such roots grow more deeply into the soil. Such differences result in roots more tolerant to water stress and cold.

© CAB International 2009. *Olives* (I. Therios)

ROLES OF ROOTSTOCKS

The rootstocks of olives affect the size of the tree, its mineral nutrition (Chatzissavvidis and Therios, 2003), the fruit size and shape and the season of ripening.

Tree size

New systems of olive planting, such as super-high density, have started to expand all over the world. Therefore, the reduction of tree vigour is very important in olive culture. Today dwarfed rootstocks, as in the case of apples, are not available. Very dwarfed rootstocks are not the most appropriate. Certain rootstock–scion combinations achieve the dwarfing effect (Troncoso et al., 1990; Lavee and Schachtel, 1999). These combinations are exemplified by the following:

- 'Mission', 'Sevillano' (Hartmann and Whisler, 1970) and 'Manzanillo' grafted on to 'Ascolano' seedlings.
- Self-rooted 'Sevillano' gives relatively small trees appropriate for high-density planting.
- 'Manzanillo' grafted on to 'Oblonga'.
- Self-rooted 'Mission' and 'Manzanillo' give more vigourous trees than grafted-on rootstocks.

Fruit size, shape, ripening time and mineral content of olive fruits

No consistent trend between scion and olive fruit ripening season was recorded. Fruit shape could be affected or not by the kind of rootstock (Connell and Catlin, 1994). For example, the length:width ratio of the fruits of the 'Mission' cultivar is greater in self-rooted plants in comparison with grafted ones on various rootstocks. Furthermore, olive fruit size and/or productivity are affected positively or negatively by grafting, depending on the cultivar (Caballero and del Rio, 1990). Moreover, the mineral content of olive leaves is a function of the rootstock used.

Salt tolerance

Ion toxicity is one way in which salinity can adversely affect olives. Olive trees are considered moderately tolerant of salinity, though certain cultivars are

DESCRIPTION OF SELECTED OLIVE ROOTSTOCKS

Olea oblonga

This is a seedling first found in California in 1940. This rootstock is *Verticillium dahliae Kleb.* resistant but not immune to *Verticillium*. 'Oblonga' is easily propagated under mist or fog using small leafy cuttings. Grafting of sensitive cultivars to *V. dahliae* resistant rootstocks reduces their susceptibility (Porras *et al.*, 2003). The 'Oblonga' rootstock has a significant dwarfing influence on 'Manzanillo', less dwarfing on 'Mission' and an invigorating effect on 'Sevillano'. 'Sevillano' scions grafted on to this rootstock produced large and heavy-yielding olive trees (Connell and Catlin, 1994).

'Allegra'

This is also a rootstock resistant to *Verticillium* wilt. When 'Allegra' is used as rootstock the *Verticillium* is transmitted though the rootstock to the susceptible scion.

Wild olive (Oleaster)

Oleaster olives differ from the cultivated clones by the presence of spinescent juvenile shoots, smaller fruits with less mesocarp, lower oil content and furthermore by a long juvenile stage, long-lasting in some seedlings. The genetic variations of *Oleaster* can be analysed by using RAPD markers, RAPDs and allozyme polymorphism.

Olive oil and table olive cultivars

'Mission'
'Mission' seedling rootstocks are variable, and this variation may account for variation in commercial olive orchards grafted on to 'Mission' seedling rootstock. Cultivars grafted on to 'Mission' rootstock include 'Manzanillo' and 'Sevillano'.

'Nevadillo'

This is an oil-producing cultivar and is recommended as rootstock for 'Sevillano', 'Mission' and 'Manzanillo' cultivars.

'Redding Picholine'

'Redding Picholine' is a large-fruited variety and is used as a rootstock for 'Sevillano' and 'Mission'.

'Megaritiki', 'Chondrolia Chalkidikis', 'Amphissis', 'Koroneiki', 'Agiou Orous' and 'Matolia'

In one experiment own-rooted cvs 'Megaritiki', 'Chondrolia Chalkidikis', 'Amphissis', 'Kalamon', 'Koroneiki', 'Agiou Orous' and wild olives – as well as scion × rootstock combinations – were grown in a greenhouse (Chatzissavvidis, 2002). The lowest boron (B) concentration in leaves and roots was found in 'Kalamon' and wild olives, respectively. 'Megaritiki' had higher leaf B concentration when grafted on to 'Megaritiki' or 'Chondrolia Chalkidikis' compared with own-rooted plants. The same cultivar as own-rooted plants had higher root B concentration than a rootstock of the other tested cultivars. Budding reduced stem potash (P) concentration of 'Megaritiki' and 'Chondrolia Chalkidikis' grafted on to 'Chondrolia Chalkidikis' and 'Amphissis', respectively. 'Chondrolia Chalkidikis', as a rootstock, increased stem potassium (K) in 'Megaritiki'. Similarly, 'Amphissis' and 'Megaritiki', as rootstocks of 'Amphissis', reduced stem K concentration. Budding of 'Chondrolia Chalkidikis' on to 'Chondrolia Chalkidikis' or 'Megaritiki', and 'Amphissis' to 'Megaritiki', both reduced stem calcium (Ca) concentration.

In Table 6.1 the concentration of B in four rootstocks (*O. oblonga*, 'Chondrolia Chalkidikis', 'Matolia' and wild olive) at six B concentrations in the nutrient solution is presented and indicates that the maximum increase of B in leaves was recorded at a treatment rate of 20 mg B/l in the cultivar 'Matolia' (897%) (Chatzissavvidis, 2002).

The accumulation of B increased very rapidly in all the treatments and in a relatively short time. In the following Table (6.2) is presented the correlation between B concentration in the nutrient solution and mineral concentration in the leaves of the four rootstocks (*O. oblonga*, 'Chondrolia Chalkidikis', 'Matolia' and wild olive).

CLASSIFICATION OF ROOTSTOCK VIGOUR

The various degrees of vigour of olive plants exhibited by common rootstocks are as follows:

- Rootstocks inducing increased growth: 'Acebuche', 'Moriscade', 'Badajor', 'Lechin de Sevilla', 'Real Sevillana' and 'Cornezuelo'.

Table 6.1. Boron (B) concentration at 30, 60 and 90 days after initiation of the experiment and percentage change in B concentration at 90 days in comparison with that at 30 days (from Chatzissavvidis, 2002).

Variety	Treatment (B, mg/l)	Days			Change (%, 30–90 days)
		30	60	90	
Olea oblonga	0.27	26	28	21	−17
	1.00	20	26	27	+34
	2.50	24	34	49	+104
	5.00	30	53	89	+193
	10.00	50	97	189	+279
	20.00	83	147	239	+187
'Chondrolia Chalkidikis'	0.27	21	27	22	+6
	1.00	22	34	32	+49
	2.50	25	29	37	+47
	5.00	31	45	87	+180
	10.00	37	54	143	+283
	20.00	55	153	126	+129
Wild olive	0.27	23	35	15	−34
	1.00	27	39	37	+38
	2.50	25	38	–	+52
	5.00	36	62	118	+224
	10.00	54	110	259	+381
	20.00	111	114	255	+130
'Matolia'	0.27	38	27	24	−35
	2.50	36	49	59	+63
	20.00	23	56	233	+897

In wild olives (2.50 mg B/l) the per cent change represents a period of 60 days.

- Rootstocks inducing normal growth: 'Gordal', 'Blanqueta', 'Tempranilla de la Sierra', 'Changlot', 'Manzanillo de Jaén'.
- Rootstocks inducing little growth: 'Redondilla de Logrono', 'Picual', 'Hojiblanca' and 'Habichuelero'.

Listed below are other methods of reducing growth.

- Use of growth regulators; the GA_3 inhibitor uniconazole had too drastic an effect on tree size. The increased yield due to uniconazole treatment is due to increased fruit setting; similar growth inhibition was achieved with paclobutrazol.
- Selection of a suitable cultivar for high-density systems.

The best cultivars for super-high-density planting were 'Askal', 'Arbequina', 'Arbosana' and 'Koroneiki'.

Table 6.2. The correlation between boron (B) concentration in the nutrient solution and mineral concentrations in the leaves of *Olea oblonga*, 'Chondrolia Chalkidikis', 'Matolia' and wild olive, used as rootstocks (Chatzissavvidis, 2002).

	Olea oblonga		'Chondrolia Chalkidikis'		'Matolia'		Wild olive
	Base	Top	Base	Top	Base	Top	Base + top
B	0.934**	0.993**	0.570**	0.586**	0.843**	0.707**	0.960**
N	−0.723**	−0.698**	−0.845**	−0.802**	NS	NS	−0.761**
P	NS	NS	−0.440*	−0.563**	NS	NS	NS
K	−0.756**	−0.737**	−0.832**	−0.598**	−0.585*	−0.849**	−0.532**
Na	NS	NS	NS	NS	NS	−0.563*	NS
Ca	NS	−0.707**	−0.537**	−0.422*	−0.877**	−0.808**	−0.841**
Mg	NS	NS	NS	NS	NS	NS	NS
Fe	−0.877**	−0.635**	−0.530**	−0.549**	NS	−0.808**	−0.593**
Mn	NS	NS	NS	NS	NS	NS	−0.456*
Zn	0.679**	NS	NS	NS	NS	−0.777**	NS

N, nitrogen; P, potash; K, potassium; Na, sodium; Ca, calcium; Mg, magnesium; Fe, iron; Mn, manganese; Zn, zinc.
Statistically significant differences for $P \leq 0.05$ (*), ≤ 0.01 (**) and < 0.001 (***).

Major Trends in Olive Farming Systems

The total area under olive cultivation within the EU has evidenced significant fluctuation within its various member states and areas since the 1970s. Throughout the 1970s, in all the EU countries, a decrease in the number of olive groves was recorded. This could be ascribed to EU programmes aimed at digging out old trees. During the following decade olive cultivation began to expand in certain Mediterranean countries such as Greece and Spain and, to a lesser extent, Italy and Portugal.

Significant changes in cultivation techniques have occurred since the 1970s, involving such as the use of chemical fertilizers, herbicides and insecticides and modern cultivation and irrigation systems. The expansion of new plantation systems and the abandonment of the old, traditional ones have produced significant environmental consequences. We shall now examine what changes and progress in olive culture have been at work in the three main EU olive-producing countries, i.e. Spain, Italy and Greece.

SPAIN

During the mid-1960s olive groves in Spain covered 2.4 million ha, while during the following decade many old plantations were cleared out and replaced by arable crops. In 1986, Spain joined the EU and, with the support of the Common Agricultural Policy (CAP), olive cultivation increased both in terms of production and the total area of olive groves, partly through the introduction of new plantations, these new olive plantations replacing arable crops and grassland areas. In the year 1995 alone 67,000 ha of olives were planted in Spain and, from 1990 to 1998, about 150,000 ha of new olive groves were developed. In other words, the average rate of planting from 1990 was around 20,000 ha/year and the total olive area in Spain is now over 2.4 million ha. Plantations developed after 1998, according to the EU support policy, cannot receive EU aid.

It was in the 1980s that olive growing became more intensified, with biological production and chemical pest control being practised in many areas – with the exception of marginal ones. In Spain the use of fertilizers is increasing, and likewise irrigation in new plantations. Another new technique introduced to Spain is mechanical harvesting for both new, dense planting systems and traditional plantations. Biological olive oil production is increasing, but still represents only a small proportion of the total olive grove area. Furthermore, integrated pest management systems have been developed in olive groves in Spain but they represent only a small percentage of the total area.

ITALY

The latest data from EU surveys indicate that the total olive-growing area is 1.4 million ha, which is greater by about 400,000 ha when compared with the area in existence in the 1990s. There are significant differences in planting trends in various regions. Hence, in some areas old olive trees were grubbed out and some new and more efficient cultivars planted. Furthermore, many traditional orchards were transformed into modern, densely planted orchards. A new trend in Italy is the planting of biological olive groves, and the area of those in 1997 was 15,200 ha, which still represented a small proportion. However, their area is now increasing due to financial support from the EU (Regulation 2078/92).

GREECE

The area of olive orchards has increased significantly since the late 1970s. This is due to the planting of new olive orchards in the high-density system (6–7 m between trees and 6–7 m between rows). Furthermore, olive oil and table olive production have both increased over the same period. This enhanced productivity is due to various factors: (i) mechanization; (ii) land levelling; and (iii) irrigation, wherever there is available water. Other factors responsible include: (i) the high level of EU support; (ii) high prices for olive oil; (iii) improvements in olive cultivation and mechanization; (iv) establishment of new, intensive plantations; (v) biological olive oil production; and (vi) efficient water use by means of drip irrigation (Metzidakis and Koubouris, 2006).

On many small Greek islands, and also in marginal, semi-mountainous areas, olive production is generally not intensive due to difficult soil/climatic conditions (lack of irrigation water, shallow soil, soil inclination and difficulties with mechanization, use of labour in touristic areas, etc.). Under such circumstances olive production yields only a low income.

8

CLIMATIC AND SOIL CONDITIONS

CLIMATIC CONDITIONS: TEMPERATURE REQUIREMENTS

The olive tree is cultivated under various climatic and soil conditions and its cultivation is feasible throughout the entire temperate and subtropical zone, i.e. between 30° and 45° (Bongi and Palliotti, 1995). The olive tree could also grow under tropical conditions without producing fruits, unless it is cultivated either: (i) at adequate elevation in order to fulfil its needs of low temperatures; (ii) if the cultivars have low chilling requirements; or (iii) by supplemental pollination (Denney et al., 1985; Ayerza and Sibbett, 2001; Ayerza and Coates, 2004). The greatest olive tree accumulation is recorded in the Mediterranean countries with mild winters and hot, dry summers.

The areas of olive culture have a mean annual temperature 15–20°C, with a minimum of 4°C and a maximum of 40°C. The minimum temperature should not drop below −7°C, otherwise damage to trees occurs. However, this limit is only an estimation, since the tolerance of olive trees depends on various factors such as the duration of very low temperatures, atmospheric humidity, cultivar, etc. Damage also varies from one tree to another. If the drop in temperature is gradual the olive tree can stand temperatures down to −12°C (Bongi and Palliotti, 1995). The sensitivity of olive trees to frost reduces the extension of olive cultivation northwards; in the marginal northern limit of olive cultivation olive culture is confined to coastal areas.

Due to the late blooming of olives, spring frosts are not a determining factor in olive culture. Olive trees require low temperatures for flowering bud differentiation and the requirement of vernalization explains why the tree does not produce in tropical regions. Some varieties such as 'Sevillano' and 'Ascolano' require about 2000 h below 7°C. Not only low but also high temperatures play a significant role, especially during the summer, since these instigate fruit drop.

Areas of high altitude are not appropriate for olive culture, due to frost danger and to the shorter vegetative period. Commonly, olive trees are not planted at altitudes greater than 800 m in Mediterranean countries. High relative humidity is responsible for disease problems. Hail damages olive trees by damaging the fruits and by increasing the danger of *Bacterium savastanoi* infection.

Although olive trees are tolerant to wind, wind-affected areas should not be used for olive culture. Cold, moist and hot winds during spring reduce flower fertilization and fruit growth. Furthermore, hot winds during the summer instigate fruit drop.

Concerning soils, olive trees can grow well even in poor, dry, calcareous and gravelly soils. However, the best soils for annual bearing are the deep, sandy-loam soils adequately supplied with nitrogen (N), P, K and water. Among the fruit trees olives are considered to be very tolerant to water stress. Root density is positively related to tolerance to water stress and water absorption. Hence, in olives the ratio of roots:leaves is 74:1, while in *Actinidia chinensis* (kiwi fruit) it is 9–10:1. The root length of olive plants having the same age as kiwi fruit plants is four times greater in comparison with *A. chinensis*.

The available soil volume/tree is significant. Clay soils that have high moisture content and immobilize K and P are not suitable for olives. Also, soils with a hardpan close to the soil surface are not appropriate for olives. Another factor is that the sodium chloride (NaCl) content of soils should be less than 1g/l.

Olive trees grow and produce in soils with both medium acid and medium alkaline pH; however, pH values greater than 8.5 reduce growth significantly. Furthermore, olives grow with no toxicity problems in soils with a relatively high boron content (Chatzissavvidis, 2002).

Heat stress and optimum temperature range for olive trees

Adequate knowledge of the optimal thermal range is essential in quantifying thermal stress and in assessing the efficiency of the tolerance or resistance mechanisms in plants (Mahan *et al.*, 1995). High temperature stress has negative effects on olive plants and, as a consequence, olive yields are reduced (Denney *et al.*, 1985; Mancuso and Azzarello, 2002). This imposes limitations on the expansion of olive culture in heat-stressed environments. The determination of the optimal thermal range is useful in order to predict the ability of the plant to tolerate heat stress; when we say heat stress tolerance we mean the ability of the plant to maintain its productivity when exposed to thermal stress.

Temperatures below and above the optimal thermal range determine the low and high points of temperature stress. The heat load of the olive plant and the energy exchange within the olive tree canopy are a function of many factors such as leaf size, shape and leaf orientation. The architecture and the pruning system of olive trees can modify the temperature of the olive leaf canopy, thus affecting water consumption and heat stress. Other important factors for increasing the tolerance of olive plants include the following:

- Root:leaf ratio.
- Leaf orientation, size, shape and characteristics of the surface. The existence of pubescence in leaves reduces heat stress problems.
- Leaf thickness, stomata size and distribution.
- Optimal root system, which absorbs sufficient water in order to maintain olive canopy temperature.

Molecular techniques can increase the thermal stress tolerance by incorporation of genes giving heat tolerance. The incorporation of enzymes with different heat tolerances into the olive plant will widen the range of optimum temperatures.

Thermal requirements of the olive

Temperature is the important climatic factor that geographically limits olive culture. Olive trees do not survive temperatures less than −12°C. However, olives require a period of low temperatures (0–7°C) for flowering bud differentiation. Therefore, temperature limits olive cultivation to the range of 30–45° latitude. Olive oil quality is improved in areas with cold winters, and this explains why 10–15% of olive oil production is located in such areas (Bongi and Palliotti, 1995). The olive yield/m^2 of leaves at 34.8° latitude is 50% greater in comparison with that at 45° latitude, due to both growth decrease during winter and to a high degree of photoinhibition. The total number of low-temperature hours should be at least 10 weeks with temperatures <12.2°C for flower bud differentiation (Hartmann, 1953; Hartmann and Porlingis, 1957). The particular cultivar of the olive tree is the determining factor in chilling requirements.

Constant temperatures of <7°C or >15°C can inhibit flowering bud induction, and winter temperatures of >20°C for 2–3 weeks inhibit sprouting of flowering buds. The hormone balance of olive trees controls chilling requirements; therefore, a late harvesting period could increase the chilling requirements (Lavee and Harshemesh, 1990). The differentiation of flowers needs higher temperatures in comparison with those required for flowering bud induction. Furthermore, very high temperatures in spring may inhibit flower development. Another factor is that high temperatures in spring reduce the duration of flowering and inhibit pollen tube growth (Galán *et al.*, 2005). In contrast, low spring temperatures lengthen the period of flowering by up to 3 weeks and can also postpone flowering or inhibit cross-pollination.

With regard to the germination of olive stones, the optimum temperature is 10°C for 1 month before sowing (Voyiatzis and Porlingis, 1987). When the temperature during sowing is 25°C or higher it reverses the effect of chilling for a period of up to 5 weeks. The optimum temperature for photosynthesis depends on the olive variety. Furthermore, temperature affects the accumulation of various metabolites in the different parts of olive plants. For example, starch and mannitol accumulation in the leaves is reduced during summer (Drossopoulos and Niavis, 1988). During winter, starch concentration is reduced in the leaves, while that of soluble sugars is increased. Also, the transport of mannitol from bark to wood is reduced with low temperatures. Furthermore, low soil temperatures induce a water deficit (Pavel and Fereres, 1998).

The southern limit of olive cultivation is determined by winter temperatures, which should be <14°C for 2 weeks for flowering bud induction. From the known cultivars, 'Koroneiki', 'Arpa' and 'Rubra' have the lowest chilling requirements (Hartmann and Porlingis, 1957) and can be used for expanding the southern limit of olive cultivation. High temperatures reduce olive productivity, since these destroy protein structure. On average, leaves and shoots are injured at temperatures of around 48 and 50°C, respectively (Mancuso and Azzarello, 2002). When temperatures reach >35°C photosynthesis is inhibited, and the leaf cuticle is irreversibly damaged at temperatures >55°C. High temperatures are associated with a high vapour pressure deficit (VPD) between leaves and the surrounding air. The same applies to fruit, where high temperatures may cause fruit drop.

Low temperature stress

When winter temperatures drop below 5°C, the acclimation and hardening processes of the olive plants begin (Burke et al., 1976; Levitt, 1980; Alberdi and Corcuera, 1991; Ruiz et al., 2006). With low temperatures sucrose accumulates (Guy et al., 1992), starch is redistributed and is accumulated in organs protected from frost, such as roots. Furthermore, changes in biochemical parameters of olive leaves occur during the vegetative season, correlated with frost resistance (Bartolozzi et al., 1999). Temperatures < −12°C damage the leaf canopy, shoots and branches. Buds and inflorescences are also very sensitive to spring frosts, and temperatures <10°C during flowering reduce the percentage of pollination. When water freezes ice is produced initially in the leaf apoplast, which has 100 times less density than the symplast. Before harvesting, frosts cause shrinkage of olive fruits, wood and cambium browning, splitting of shoots and necrosis of the whole tree. The matric potential (Ψ_m) forces of the cell wall inhibit ice formation in the apoplast at temperatures of down to −10°C (Bongi and Palliotti, 1995). When an ice crystal is formed in the apoplast, the cell maintains its permeability to water and a significant difference in water potential (Ψ) causes water loss from the symplast (Palta and Weiss, 1993). A high degree of supercooling is observed in olives, which is a mechanism for avoidance of ice formation (Fiorino and Mancuso, 2000). Leaves in which ice has formed damp downwards following melting of the ice due to difficulty in reversion of plasmolysis. Furthermore, due to low winter temperatures olive fruit will shrivel (see Fig. 8.1), shoots dry out (see Fig. 8.2), bark splits (see Fig. 8.3) and wood and cambium become brown in colour (see Fig. 8.4).

Fig. 8.1. Olive fruit shrivelling due to low winter temperatures prior to harvesting.

Fig. 8.2. Shoots of cv. 'Koroneiki' drying due to winter frosts.

Fig. 8.3. Bark splitting on an olive tree after a severe frost.

Fig. 8.4. Wood and cambium browning on olive trees after severe winter frosts.

Low temperature sensing in olive trees and the role of calcium

The capacity of a species to survive at freezing temperatures is a function of the cultivar and its ability to acclimatize to cold (Alberdi and Corcuera, 1991; Palta and Weiss, 1993). Cold acclimation is followed by accumulation of cryoprotective substances, which influence membrane structure and anatomy. The leaves perceive the cold acclimation stimulus and also produce the necessary chemicals for cold acclimation (Weiser, 1970). Cold acclimation requires expression of specific genes (Palya, 1993). However, in olives genetic cold tolerance has not been studied in depth and only a few publications exist (D'Angeli et al., 2003; D'Angeli and Altamura, 2007). Nevertheless, based on both experience and macroscopic appearance, olive varieties can be classified as cold tolerant, cold semi-tolerant or cold sensitive (Roselli et al., 1989).

Olive is an evergreen tree with low tolerance to frost, limiting its cultivation in cold regions, where the slow post-maturation process improves oil quality (Palliotti and Bongi, 1996). Therefore, finding a way for tree survival in the colder climate is of great significance. However, a transgenic approach for improving cold resistance in olives has not yet been carried out (D'Angeli and Altamura, 2007).

Calcium plays a very important role in freezing tolerance. Hence, in many fruit trees exogenous application of Ca^{2+}, in the form of sprays or fertilizers, increases fruit and leaf Ca levels and thereby increases cold tolerance (Raese, 1987). From many experiments with olives it was found that young leaves are more sensitive to cold than are older leaves, which have a greater Ca^{2+} concentration. The results of D'Angeli et al. (2003) indicate that the non-acclimated olive leaf protoplasts counter a rapid fall in temperature by increasing their Ca^{2+} concentration. This includes an influx of Ca^{2+} via the plasmalemma and an efflux of the Ca^{2+} from the organelles. A decrease of $0.075°C/s$ is adequate to create an increase in Ca^{2+}. An increase in Ca^{2+} is genotype dependent and is also essential for the development of cold tolerance. Therefore, Ca is involved as a signalling agent in cold resistance of olives, and Ca^{2+} signalling could be an early marker for selection of cultivars for cold resistance.

In order for olive trees to become cold tolerant the process of cold acclimation is necessary, in which osmotin is involved. Osmotin is a pathogenesis-related protein having cryoprotective functions. Plant cold acclimation involves the acquisition of freezing tolerance by exposure to non-freezing low temperatures. This process requires accumulation of cryoprotectants and anatomical changes in specific tissues enabling the plant to become tolerant to extracellular freezing. Like other species not displaying dormancy, the olive tree needs several weeks to exhibit cold damage symptoms at the level of plant anatomy. Such symptoms include necrosis of cambium, tylosis, suberization and cell wall lignification. During adaptation to cold stress an increase in calcium concentration occurs. In *Olea europaea* osmotin, which is a pathogenesis-related protein, is positively involved in the induction of programmed cell death (PCD). This protein is produced during the cold acclimation period of the tree and involves more than one of the processes leading to complete acclimation. Osmotin is also involved in arresting cold-induced Ca signalling. The effect of long exposure to cold is presented in Fig. 8.5 (D'Angeli and Altamura, 2007).

Assessment of frost tolerance in the olive: methods

Various methods are used in order to assess frost tolerance in olive varieties (Fiorino and Mancuso, 2000). These include the following: (i) visual observations; (ii) electrical resistance changes during exposure to low

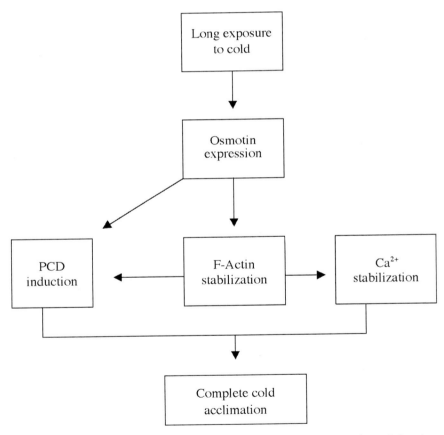

Fig. 8.5. The role of osmotin during cold acclimation in the olive. The stabilization of F-actin and Ca^{2+} and the induction of programmed cell death (PCD) are necessary for complete acclimation (from D'Angeli and Altamura, 2007).

temperature; (iii) stomatal size and density; (iv) electrolytic conductivity; (v) differential thermal analysis; and (vi) tissue staining.

Visual observations
This is the traditional method of screening chilling and freezing tolerance. However, this method is not objective and is influenced by various factors including wind, relative humidity, and nutritional and water stress of the plants. With this method scientists record their observations only of the leaf canopy and not of the root system. At $-12°C$ trees are damaged, i.e. leaves die, leaf abscission occurs and twigs dry up.

Chlorophyll fluorescence quenching analysis
This method does not have adequate sensitivity and is too labour intensive.

Electrical resistance
With this method we measure electrical resistance changes within different plant organs (Mancuso and Rinaldelli, 1996) and, based on the electrical resistance changes, we determine the critical and freezing temperatures (Mancuso, 2000). Both critical and freezing temperatures are lower for the more chill-tolerant genotypes, and the absolute critical temperatures are lower for the chill-tolerant varieties (8.8°C for 'Ascolana') and higher for the chill-sensitive ones (13.6°C for 'Coratina'). The sensitivity to chilling of various organs in the olive follows the order: roots > leaves > shoots > vegetative buds.

Nevertheless, individual components do not represent the cold hardiness of the whole plant. Therefore, for an estimation of frost damage it is better to use the entire tree instead of merely parts of the tree.

Measurement of the changes in electrical resistance of olive tissues at low temperatures consists of a simple, easy, accurate and non-destructive method to screen those olive varieties having cold tolerance. The absolute critical temperatures (Mancuso, 2000) for the cultivars 'Ascolano', 'Leccino', 'Frantoio' and 'Coratina' are 8.8, 10.6, 12.6 and 13.6°C, respectively.

Stomatal size and density
Stomatal density is used as a criterion for selection of cold-tolerant varieties (Roselli *et al.*, 1989; Roselli and Venora, 1990). From a study of stomatal density in some Greek olive varieties we can classify them as follows:

- Some varieties considered tolerant to cold had smaller stomatal density in comparison with sensitive ones (see Fig. 8.6 and 8.7).
- Varieties with the greatest stomatal density are: 'Pierias' (551 stom/mm^2), 'Megaritiki' (548 stom/mm^2) and 'Maronias' (547 stom/mm^2).
- The varieties with the smallest stomatal density are: 'Mavrelia Serron' (441 stom/mm^2), 'Koroneiki' (441 stom/mm^2) and 'Galatistas' (355 stom/mm^2).

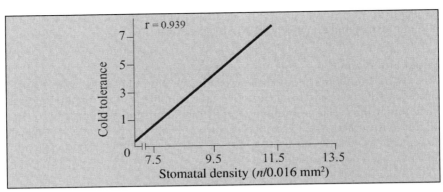

Fig. 8.6. Linear regression between stomatal density and cold tolerance of ten olive cultivars. Cold tolerance: 1 = hardy, 7 = very susceptible (from Roselli and Venora, 1990).

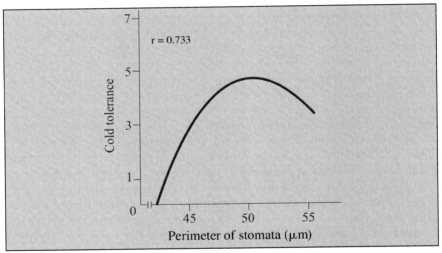

Fig. 8.7. Parabolic correlation between stomatal perimeter and cold tolerance of olive trees. Cold tolerance: 1 = hardy, 7 = very susceptible (from Roselli and Venora, 1990).

In comparison with deciduous fruit trees, olive trees have greater stomatal density. Figure 8.6 shows the linear correlation between stomatal density and cold tolerance of ten varieties, while Fig. 8.7 shows the parabolic relationship between stomatal perimeter and cold tolerance.

Electrolytic conductivity

Electrolytic conductivity is a sensitive method and is utilized to determine differences between olive cultivars with regard to frost resistance (Murray *et al.*,

1989). Leaves were exposed in a cold room for 4 h, with a controlled drop in temperature of 7°C/h. Determination of the cold tolerance of olive leaves is achieved by measuring the conductivity of leaf disk extract. Thirty leaf disks, 1 cm in diameter, were placed in a plastic bag and 10 ml of distilled water at 25°C added for 6 h. Another 20 disks were incubated for 1 week at room temperature in order to determine the degree of cold damage. Recording of damage was conducted with image analysis, being measured by both image analysis and measurement of electrical conductivity of the leaf disk extract. The relative release of electrolytes (REL) is calculated from the equation:

$$REL = [(L_1/L_2) \times 100]$$

where L_1 is the first measurement of electrical conductivity. This represents the liberation of ions caused by cold damage but also includes ion release under normal conditions; L_2 is the second measurement of electrical conductivity and represents the conductivity in the dead tissue due to cold. This represents the total quantity of the electrolyte in the leaf disk. Cold tolerance can be expressed as LT_{50} (temperature at which 50% of tissues are damaged).

Differential thermal analysis (DTA)

Olive trees exposed to low temperatures survive either by avoidance of ice formation in their tissues (supercooling) (Bongi and Palliotti, 1995) or by increasing their frost hardening and resistance. Supercooling is generally studied by the differential thermal analysis (DTA) technique (Fiorino and Mancuso, 2000).

Olive plants survive winter temperatures by deep supercooling of tissue solutions to temperatures close to those of homogenous nucleation of the aqueous solution (−40°C). When this fraction of the solution freezes, an exotherm can be observed by DTA (Fiorino and Mancuso, 2000). In order to conduct DTA, excised olive tissues are placed on one side of a thermopile plate and a 'dried sample of tissue' on the other as a control in order to measure the differential temperature changes between the two samples during the process of freezing. The samples are placed in a freezing cabinet and cooled down to −30°C at 5°C/h. During this process we measure the signals from the thermopiles.

Tissue staining

Five cross-sections of two to three cells' width were cut with a razor blade from the middle of leaves. Cutting was performed under demineralized water and maintained under various temperatures for 20 min in a 0.02% solution of neutral red [3-amino-7-dimethylamino-2-methyl-phenazin (HCl)]. The samples were stored at 5°C for 24 h in order to absorb the dye. For each different temperature 100 cells per section were examined under a light microscope (200 × magnification). Only living cells absorb the dye and exhibit intense red coloration. The percentage dye absorption of the control is measured at temperatures from 0 to −20°C (Fiorino and Mancuso, 2000).

Dye absorption was reduced depending on cultivar, by: 14–32% at −10°C; 21–64% at −14°C; and 82–100% at −16°C. Finally, at −20°C all tissues were dead (see Table 8.1).

Comparison of the methods – electrolytic conductivity (REL), visual observation and DTA gave comparable results. However, significant differences in cold tolerance exist between olive varieties.

Frost tolerance of olive varieties and methods of protecting olive trees against frost

Frost affects plant cell membranes and these become less permeable and even porous, allowing leakage of electrolytes. Certain cultivation techniques can reduce the danger of cold damage; such techniques include: (i) reduction of irrigation and N fertilization; (ii) maintenance of the soil in a bare condition; and (iii) late pruning and use of cold-hardy varieties, since cold tolerance is a genetic characteristic. The high K:Ca ratio in the plant increases tolerance to cold and drought. Spraying olive trees with copper (Cu)-containing chemicals (such as Kocide) hardens olive trees and protects them from frost. Spraying was conducted during winter. The Cu-containing sprays protect against both frost and bacterial infection (Antognozzi and Catalano, 1985; Antognozzi et al., 1990c, 1994; Martin et al., 1993; La Porta et al., 1994; Bartolozzi and Fontanazza, 1995, 1999; Mahan et al., 1995; Barranco et al., 2005).

Foliar application of mefluidine (Palliotti and Bongi, 1996), a plant growth regulator, at 1000 ppm concentration to cultivars sensitive to frosts, induced a hardening effect and increased freezing resistance by reducing the leakage of K^+, Na^+ and Ca^{2+} through the various biological membranes (Palliotti and Bongi, 1996). The maintenance of bare soil increases its temperature and a bare soil is warmer during a frost in comparison with that covered by vegetation. Furthermore, stony soils have higher temperature and waterlogged soils are colder than well-drained soils.

Table 8.1. Neutral red absorption at temperatures below freezing for four olive cultivars.

Variety	Neutral red absorption (percentage of control at 0°C)						
	0°C	−5°C	−10°C	−12°C	−14°C	−16°C	−20°C
'Ascolano'	100	100	86	82	79	18	0
'Leccino'	100	100	76	66	51	12	0
'Frantoio'	100	100	73	53	35	0	0
'Coratina'	100	100	68	51	36	0	0

Olive varieties resistant to cold include the following (Barranco et al., 2005). The frost-hardiest cultivars are 'Cornicabra', 'Arbequina' and 'Picual'; other frost-hardy cultivars include 'Mission', 'Leccino', 'Carolea', 'Chemlali', 'Moraiolo' and 'Picholine'. Conversely, the most frost-susceptible variety is 'Empeltre'. The cultivar 'Hojiblanca' is generally considered cold hardy (Barranco et al., 2000) and is classed together with 'Frantoio' in the intermediate cold-resistance group.

From the Greek varieties those considered most resistant include the following: 'Galatistas', 'Arvanitolia Serron' and 'Mavrelia Messinias'. 'Kalamon' has medium tolerance and 'Koroneiki', 'Amphissis' and 'Chondrolia Chalkidikis' are sensitive to cold.

SOIL CONDITIONS

Important characteristics of soils

Soil is a complex system consisting of varying proportions of four components, i.e. mineral particles, non-living organic matter, the soil solution and air occupying the pore space. Furthermore, soil gives shelter to many living organisms such as bacteria, fungi, algae, protozoa, insects and small animals.

Mineral particles comprise the main part of most soils and the most stable component. These are derived by weathering of mother rocks or deposition by water or wind. Organic matter constitutes less than 2% of soil volume, with the exception of organic soils. The term organic matter includes the various organic residues such as roots, litter and decomposition products. The solid matrix forms a pore space which constitutes 30–60% of soil volume. The pore space at field level is filled with soil solution at 40–60%, while the remainder is filled with air. In saturated soils the entire porosity is filled with water, whilst in dry soils most of the pore space is filled with air. Excepting organic matter, water and air, soil provides shelter for microorganisms that help in the decay of organic matter and release of mineral nutrients. Small animals, such as earthworms and insects, drill tunnels in the soil, which improve infiltration and water movement.

The properties of a soil depend on soil texture, i.e. size and distribution of mineral particles, their arrangement and the type of clay minerals. The clay of soils consists of four minerals, i.e. kaolinite, montmorillonite, vermiculite and illite. Kaolinite has a 1:1 proportion of silica to aluminum and its crystals have the least tendency to shrink or swell upon hydration changes. Montmorillonite and illite have a 2:1 silica:alumina ratio and soils containing these minerals swell and shrink markedly upon hydration or drought, and crack significantly upon drying. The significant characteristic of clay minerals is that, during replacement of silicon or aluminum with other cations, they develop a negative charge, which determines their efficiency in holding cations or their cation exchange capacity (CEC). Furthermore, the organic matter content contributes to CEC. The CEC of clay minerals and organic matter is as follows: organic matter (100–300 meq/100 g dry soil) – illite, 3, kaolinite, 10, montmorillonite, 100 and vermiculite, 160.

Soil texture

The relative contents of sand, silt and clay in the soil determine soil texture, including sand, loam, silt, clay and various intermediate classes such as sandy loam, silt loam or clay loam. The diameter of solid particles in coarse sand is 2.00–0.20 mm, in fine sand 0.20–0.02 mm, in silt 0.02–0.002 mm and in clay < 0.002 mm.

A sandy loam soil contains 66.6% coarse sand, 17.8% fine sand, 5–6% silt and 8.5% clay. A loamy soil contains 27.1% coarse sand, 30.3% fine sand, 20.2% silt and 19.3% clay. Finally, a heavy clay soil contains only 0.9% coarse sand, 7.1% fine sand, 21.4% silt and 65.8% clay.

Sandy soils have a large volume of non-capillary pore spaces and are characterized by adequate aeration and drainage. Furthermore, sandy soils have a low CEC and low water-holding capacity; they are easy to cultivate and chemically inert. Clay soils have a high surface area and greater water- and nutrient-holding capacity. The loamy soils contain about equal amounts of sand, silt and clay and they have intermediate behaviour between clay and sand. Moreover, such soils are more appropriate for olive cultivation, have good water- and nutrient-holding capacity and good aeration.

Soil structure

Soil particles are united as aggregates in the form of crumbs, giving rise to the particular soil texture. Clay particles are flocculated when there is good availability of cations in the soil solution, which neutralize the negative charge of the clay particles. Hence, neighbouring particles stop repelling each other and form aggregates in the form of crumbs. The role of iron and/or aluminum oxides in organic colloids is important in cementing these particles together. Wetting can destroy aggregation. Furthermore, the predominant ions may affect soil structure and permeability. Soil saturated with sodium has a poor structure: it is easily dispersed and does not drain easily.

Soil porosity

The pore space is that fraction of soil filled with air or with soil solution, and this constitutes at least 50% of the soil volume. The pore space consists of two types of pore, i.e. capillary and non-capillary. Capillary pores are filled with water after drainage of the soil, while the non-capillary do not hold water and drain easily after irrigation. Therefore, such pores are filled with air. Sandy soils have good aeration and drainage, due to the non-capillary pores. However, such soils have low water-holding capacity. An ideal soil for plant growth has 50% non-capillary and 50% capillary pores.

Soil horizons

Olive groves can be found in areas with a uniform soil texture to a depth of 1.0–1.5 m. Older, undisturbed soils have definite horizons with different properties. These horizons are categorized as below, starting from the surface of the soil:

- AOO: under-composed organic debris.
- AO: partly decomposed organic debris.
- A1: high content of organic matter plus mineral material.
- A2: light-coloured horizon with maximum leaching.
- A3: transition zone to horizon B.
- B1: similar to A3.
- B2: zone of maximum illuviation.
- B3: transition zone to mother rocks.
- C: mother rocks.

Classification and retention of soil water

Soil water content is defined as the amount of water per unit of dry soil or the volume of water per unit volume of bulk soil. However, this signifies little with regard to water availability for plant growth. Water is retained by different mechanisms in various soils, such as pressure or tension applied to remove water and water retention in the presence of osmotically active salts, which lower the vapour pressure of the soil water. The chemical potential of soil water is termed the soil water potential (Ψ_{soil}).

Soil water potential
The forces that participate in the soil water potential are associated with the soil, the osmotic pressure of the soil solution, etc. The forces associated with the soil matrix are termed *matric forces* and constitute the *matric potential* (Ψ_m). The forces associated with soil solution constitute the osmotic potential (Ψ_s), and any applied pressure gives rise to the pressure potential (Ψ_p). Therefore, the water potential of soil water is the sum of three components:

$$\Psi_{soil} = \Psi_m + \Psi_p + \Psi_s$$

In the case of when another force, such as soil gravity, is applied, the previous equation is transformed into the following:

$$\Psi_{soil} = \Psi_m + \Psi_p + \Psi_s + \Psi_g$$

Soil water can also be described by two other terms: *field capacity* and *permanent wilting percentage*.

Field capacity
This is the water content after drainage 1–3 days after the soil has been thoroughly wetted by rain or irrigation. Field capacity is a condition in which water moves slowly and the water content does not change significantly between measurements. The deeper the soil the more time is required for the surface layer to reach field capacity. Furthermore, field capacity depends on soil characteristics and the condition under which it is measured. Because field capacity is a function of soil structure, laboratory measurements are not a very good estimation of its value under field conditions.

The *moisture equivalent* is an estimation of field capacity, which is the water content of a wetted and sieved soil sample having been drained by centrifugation at a gravitational acceleration of 1000 g. In order to measure moisture equivalent, soil samples are exposed on a pressure plate to a 0.3 bar pressure and the soil moisture content is an approximation of the moisture equivalent.

Permanent wilting percentage
Permanent wilting percentage (PWP) is the soil water content at which plants remain permanently wilted. PWP is the lowest limit of soil water and is measured by cultivating seedlings in well irrigated containers until adequate growth is achieved. Subsequently, the soil surface is sealed with plastic or aluminum foil and the plants are allowed to deplete the available soil water, leading to wilting. When no recovery of plants is recorded in a moist area then the soil water content measured is termed PWP. The soil water potential at wilting has a mean value of about -15 bars. Therefore, the percentage of water content at -15 bars is used as an estimation of PWP. In soils PWP is determined on soil samples by using a pressure membrane. The appearance of wilting in an olive tree may be due to the inability of water supply from roots to meet the transpirational demands of the tree. Other factors responsible for wilting in the orchard include depth and soil volume where the root system grows, soil water conductivity, osmotic stresses and other factors.

Readily available soil water
With the term *readily available water* we mean the amount of water retained in a soil between field capacity and the PWP. The available water range has a greater value in fine-textured soils in comparison with that in coarse-textured. The value of available water is a function of many variables, such as high rooting depth or restricted root growth and rate of infiltration of water.

Infiltration of water and its movement into the soil

For maintenance of olive growth, soil moisture content should be restored by rain and/or irrigation. When water is applied to the soil surface it moves downward and creates five zones, i.e. the saturation zone, a transition zone, the main transition zone, a wetting zone and the wetting front. Water infiltration is affected by the initial water content, soil organic matter, colloids and surface permeability, soil temperature and duration and intensity of the rainfall. Existence of claypan or hardpan significantly decreases soil infiltration.

Surface permeability and hence infiltration can be increased by the addition of organic matter and its incorporation into the soil surface. Water within the soil moves via the soil pore spaces. The driving forces include gravity and the influence of surface tension forces. Also, water could be diffused in the form of vapour via the air space of soil. The direction of movement is towards the lower water potential. In waterlogged soils (saturated flow) the driving force is gravity, which controls the water potential gradient. In dry soils the water potential gradient is controlled by the water potential and water moves in the form of films. As a soil becomes drier, the matric potential Ψ_m is the main force. According to hydraulic conductivity laws we can classify soils as poorly drained when the k value (the property of a material that indicates its ability to conduct heat) is less than 0.25 cm/h, and as well drained when k is greater than 25 cm/h. Soils with a very high k value are not appropriate for olive cultivation. The smaller the pore size the less the soil permeability. Another means of water movement is in the form of vapour, when the soil water potential is about −15 bars.

Soil temperature shows seasonal fluctuation; the soil surface temperature increases during the summer and becomes colder during winter. This helps upward movement during summer, in the form of vapour.

Measurement of soil water content

Soil water content significantly affects olive tree growth. Various methods of measuring soil water content have been devised. Due to the variability of soil texture and soil water distribution it is not easy to obtain an accurate picture of the soil water content throughout the entire root zone. Therefore, soil sampling should be repeated in order to give an accurate determination of soil water content. The methods for determination of soil moisture may be direct or indirect.

Direct methods of measuring soil water content
Soil samples of known weight or volume are collected and the water content is expressed in either g H_2O/g oven-dry soil or /cm^3 of oven-dry soil. The samples are collected in a specific type of container of known volume. The soil sample is

oven dried at 105°C to constant weight. The amount of water in the sample is determined as the difference in the weight of the soil sample before and after drying. The water content is determined as the percentage of dry weight of the soil sample. Also, this can be expressed as water content per unit of volume.

Indirect methods of determining soil water content

NEUTRON PROBE   This is the most commonly used indirect method of measuring soil water. The principle of this method is that hydrogen (H_2) atoms can slow down and scatter fast neutrons. Therefore, counting of low neutrons near a source of fast neutrons gives an estimation of H_2 content of the soil. Since the main source of H_2 in a soil is H_2O, this technique provides an estimation of soil water content. However, in organic soils (organic material contains H_2) this may modify the estimation of water content by this method.

ELECTRICAL CONDUCTANCE   Water has a significantly higher dielectric constant than dry soil. Therefore, changes in water content change electrical capacitance, which varies with temperature and is less affected by the soil solution salt concentration. Since it is difficult to obtain uniform electrical contact with the soil, capacitance measurements are neither accurate nor reliable.

THERMAL CONDUCTIVITY   As the water content of the soil decreases, heat conduction via the soil is decreasing. Thermal conductivity can be used to measure the water content of the soil, especially in moist and sandy soils. This method uses a heating element buried in the soil; the element is heated by electricity and the rate of heat transfer is measured.

ELECTRICAL RESISTANCE BLOCKS   This method is based on the measurement of electrical resistance as the soil moisture changes. The electrodes are embedded in small blocks of plaster of paris (gypsum); the blocks are buried in the soil and connected to a resistance bridge. The water content of the gypsum changes with the water content of the soil and results in changes in electrical conductivity of the soil solution between the electrodes. The range of matric potential where gypsum blocks work is -0.5 to -15 bars. Therefore, gypsum blocks are more appropriate in dry soils for soil water measurement.

Soil water potential measurement

Tensiometry
In salty soils the water potential is almost equal to matric potential, and tensiometry provides measurements closely related to total water potential of greater than -0.8 bar. However, in salty soils, the osmotic potential has a significant value and matric potential is not a good approximation of the total water potential.

Freezing point depression
With cryoscopic methods the freezing point depression of soil water is measured. Afterwards, the solute potential is calculated from the relationship between chemical potential and freezing point depression.

Thermocouple psychrometry
The relative vapour pressure is measured by a thermocouple psychrometer. This measurement is based on the relationship between the chemical potential of water and the depression of vapour pressure.

Measurement of osmotic potential
Osmotic potential is measured within the soil solution following soil saturation, its extraction from the soil and filtering. With this technique the saturation extract is produced. The osmotic potential can be determined by either cryoscopy or vapour pressure psychrometry. Furthermore, the electrical conductivity of the soil solution is measured since there is close agreement between osmotic potential and electrical conductivity.

Achievement of certain levels of soil water stress
Plants are the best indicator of the need for irrigation, and plant growth is controlled by the water potential value of its tissue. The colour of certain crops turns dark blue-green as water stress develops; this colour change is a good irrigation criterion. Furthermore, as water stress develops the olive leaf angle changes and leaf rolling is observed. Also, guard cells are very sensitive to water stress and closure of stomata has been used as an indicator of the need for irrigation. *Indicator plants* can be used to determine the need for irrigation. The rate of growth of fruits, leaves and trunks (dendrometers) has been used as an indicator of the needs of the olive for irrigation.

Waterlogging

The olive is a species sensitive to waterlogging or root hypoxia, and its sensitivity is a function of olive cultivar (Hassan and Seif, 1990). Of the tested cultivars cv. 'Mission' died after a 30-day duration of waterlogging, while the cv. 'Kalamon' survived for a period of 60 days after waterlogging was imposed. The mechanism of tolerance to waterlogging is based on the production of adventitious roots near the soil surface. Some rootstocks, such as *O. oblonga*, are tolerant to waterlogging. Waterlogging alters certain metabolic functions of roots and results in reduced absorption and transport of water and nutrients (Larson *et al.*, 1991).

GASEOUS POLLUTANTS

Significant modifications of the gas composition of the atmosphere have been observed in recent years for sulphur dioxide (SO_2), carbon dioxide (CO_2) and ozone (O_3). These pollutants are present at high concentrations, and for several months in the Mediterranean region, where the olive plant is mostly cultivated, reduced growth was observed in annual plants and forest trees (Agrawal and Agrawal, 2000). However, the data concerning pollutant effect on olive growth and metabolism are scarce (Minnocci et al., 1995; Vitagliano et al., 1999).

Sulphur dioxide

Experiments have indicated that differences in SO_2 sensitivity were observed between two olive varieties, 'Moraiolo' and 'Frantoio' (Giorgelli et al., 1994). Due to EU regulations for the protection of the environment, emissions of SO_2 in more recent years have been significantly reduced.

Carbon dioxide

The CO_2 concentration of the atmosphere has increased from 280 to 360 μmol/mol in recent times and it is expected to reach 650–700 μmol/mol by the year 2075 (Carbon Dioxide Information Analysis Center). Carbon dioxide is responsible for most of the global greenhouse effect. Exposure of olive plants ('Frantoio' and 'Moraiolo') to elevated CO_2 enhanced net photosynthesis and decreased stomatal conductance (Tognetti et al., 2001, 2002; Sebastiani et al., 2002a). Chlorophyll concentration decreased in an elevated CO_2 environment in the leaves of only 'Frantoio'. Stomatal density and leaf nutrients did not differ between treatments.

Ozone

Ozone in the stratosphere protects the living world from exposure to UVB radiation. In the Mediterranean region O_3 concentration is around 40–100 ppb and is an atmospheric pollutant. Ozone pollution effects on the varieties 'Frantoio' and 'Moraiolo' were studied (Vitagliano et al., 1999). Of these two varieties 'Moraiolo' showed greater visible injury to O_3 than did 'Frantoio'. Therefore, in Mediterranean olive culture, O_3 pollution may decrease olive growth and productivity (Minnocci et al., 1999; Sebastiani et al., 2002b).

LIGHT AND PHOTOSYNTHESIS

Light interception and photosynthesis

Olive orchards include both intensive and extensive olive cultivation. The light interception in extensive orchards is 20–30% of the solar radiation, while in the new, intensive orchards light interception can be as high as 70%. Photosynthetic active radiation (PAR) reaching the leaves is the critical factor for dry matter accumulation. However, PAR is a function of light intensity, canopy morphology (Norman and Welles, 1983) and its optical characteristics (Ross, 1981).

The arrangement of olive leaves within the whole plantation is a function of the planting system, canopy design and orientation of tree lines in the orchard. Therefore, the high and super-high density plantings are two-dimensional systems, while the classical olive orchards are a three-dimensional system. Other factors controlling PAR are the inclination of leaves and light transmittance and reflectance (Norman and Welles, 1983).

Effect of leaf development, senescence and abscission on photosynthesis

Net assimilation of olive leaves increases with leaf area expansion and obtains maximum Pn value when the leaves attain 40% of their final size. Also, total carbohydrate and chlorophyll content and leaf dry matter are a linear function of leaf area. The leaf initially develops as a sink for carbohydrates and gradually becomes a source (Foyer and Galtier, 1996). Therefore, initially the growing leaf imports carbon from neighbouring organs and later becomes a carbon exporter. The onset of carbon export depends on the position of each leaf along the shoot. Olive leaves persist for 2–3 years and their age increases from the tip to the base of the shoot and affects photosynthesis (Bongi et al., 1987). From the total number of leaves a certain percentage abscises and is replaced by new ones. This means that leaves of all ages are present on one tree at any one time. During leaf senescence a change from anabolism to catabolism happens and the net photosynthetic rate (Pn), stomatal conductance (Gs), respiration, chlorophyll content, dry matter, starch and N decrease. As a leaf ages, 50% of the total N is exported to other organs and parts of a tree before abscission; senescing olive leaves contribute to carbon and N balance. Therefore, olive trees should be trained appropriately to allow adequate light penetration of the tree canopy.

Effect of olive cultivars and season on photosynthesis

The photosynthetic rate is high during the period from spring to autumn. The principal factors affecting Pn are: (i) radiation; (ii) water stress (Diaz-Espejo *et al.*, 2006); (iii) salt stress (Bongi and Loreto, 1989); (iv) temperature (Bongi *et al.*, 1987); (v) adequate nutrients; (vi) disease control; (vii) leaf area and anatomy; (viii) leaf inclination; and (ix) the concentration of the carbon-fixation enzyme, RuBisCo. However, olive productivity is a function of both Pn and respiration of the various organs. The photosynthetic rate (Pn) is affected by both the time of day and the season (Proietti and Famiani, 2002). Therefore, the highest Pn values were recorded in October and the lowest ones in August and December. The explanation is simple, since the temperature is favourable during October for photosynthesis and unfavourable during August and December. Both high and low temperatures affect chloroplast integrity, light scattering and light reflectance. Measurement of Pn in five Greek olive cultivars (Hagidimitriou and Pontikis, 2005) indicated that 'Koroneiki', a drought-tolerant cultivar, had the highest Pn and Gs values (21 $\mu mol/m^2/s$), followed by 'Megaritiki' (18 $\mu mol/m^2/s$), 'Conservolia' (16 $\mu mol/m^2/s$), 'Lianolia Kerkiras' and 'Kalamon' (both 13–14 $\mu mol/m^2/s$).

The Pn is significantly higher in 1-year old leaves than current-season leaves early in spring, while later current-season leaves can also supply equal amounts of assimilates for plant and fruit growth (Bongi *et al.*, 1987). Olive leaves have a pronounced seasonal variation in Pn, and this is affected by the cultivars, date and ambient conditions (Hagidimitriou and Pontikis, 2005).

Gas exchange of olive fruits: effect of fruit load

The dry matter of olive fruit increases gradually from fruit set to the end of October. The oil content starts to increase 40–50 days after full bloom up to the period of initiation of fruit maturation. Dark respiration (R_D) and stomatal conductance (Gs) were both high after fruit set, and then decreased. Also, Pn and chlorophyll were high in the first 3 weeks after fruit set and progressively decreased thereafter (Proietti *et al.*, 1999). For a significant part of the fruit-growing period the CO_2 intake by fruit permitted the reassimilation of 40–80% of the CO_2 produced by dark respiration. As the olive fruit grows, Pn decreases due to the reduction in the ratio between fruit surface area and volume.

Fruit load and the presence or absence of fruit on the shoot does not significantly influence certain parameters such as Pn, Gs, internal CO_2 (Ci), transpiration rate (E) and respiration rate (RD). Furthermore, chlorophyll, sugar content and water content were significantly influenced by fruit load. As the fruit grows it becomes a great sink, which does not improve the leaf's photosynthetic efficiency.

Girdling and photosynthesis

An immediate reduction of 80% in photosynthesis was observed following girdling (incising the bark through 360° to enhance fruiting). After healing of the girdling area, photosynthetic activity was restored. The presence of fruits on the shoot resulted in such a decrease not being observed. Girdling resulted in a decrease in Gs. The decrease in photosynthesis due to girdling could be ascribed to assimilate accumulation, implying that Pn should be controlled by assimilate demand. Therefore, transport of assimilates from source to sink is the controlling factor of sink growth and the Pn of the source plant. When the production of photosynthates exceeds their utilization the accumulative sugars in the leaves become a type of stress factor. Other effects of girdling include the increase in both percentage of oil in the fruit and flower bud differentiation.

Elevated carbon dioxide and photosynthesis

Rising CO_2 has the potential directly to alter plant growth via its effects on gaseous exchange (Kirschbaum, 2004). Assimilation rates are enhanced (Centritto, 1998) and Gs is reduced for plants grown in high CO_2. Rising CO_2 resulted in a persistent increase in Pn, as happens with the majority of the C3 plants (see above) exposed to elevated CO_2. The decline in Gs between elevated and ambient CO_2-grown plants was a result of short-term stomatal response to CO_2. Furthermore, as a result of changes in Pn and Gs, water use efficiency (WUE) increased under elevated CO_2. Water use efficiency can be defined as either the ratio of CO_2 assimilation to water lost via transpiration or the ratio of net CO_2 assimilation rate to transpiration.

Water stress and photosynthesis

Maximum values of Pn rates are reached early in the morning in both control and water-stressed plants, and subsequently declined gradually. This photosynthetic inactivation was accompanied by changes in the fluorescence characteristics of the upper leaf surface. In the water-stressed plants the maximum fluorescence (F_p) and the ratio $F_v:F_p$ (variable fluorescence (F_v) decreased by midday. However, the initial fluorescence (F_0) rose to a maximum value at midday and declined in the afternoon. In the control plants the F_p and the ratio $F_v:F_p$ increased again in the afternoon and had recovered completely by 20.00; in the stressed plants the ratio $F_v:F_p$ declined gradually as water stress developed.

CLIMATE CHANGE AND ITS EFFECT ON OLIVE CULTURE

General

The fundamentals of world climate include:
- The strength of the incident radiation from the sun, which controls the Earth's overall temperature.

- The greenhouse effect of water vapour and other gases.
- The energy from the sun which the Earth receives is 1370 W/m², and, since the Earth is spherical, each m² receives about 342 W. The majority of this incoming solar radiation is absorbed in equatorial regions, while considerable quantities of radiation are emitted from the polar regions. The oceans cover 71% of the Earth's surface and play a significant role in heat redistribution, and its heat capacity in comparison with the atmosphere is huge.

The greenhouse effect

Atmospheric air is constituted by O_2, N and a number of minor gases such as water vapour, CO_2, O_3 and nitrous oxide (N_2O). Oxygen and N are transparent to both solar radiation and infrared radiation, the latter being emitted from the Earth's surface. Furthermore, the minor atmospheric gases are transparent to solar radiation and absorb the infrared radiation that the soil surface emits. The main components identified with the greenhouse effect are illustrated in Fig. 8.8.

Water vapour content varies from < 0.01% by volume to > 3.00%. Other gases include CO_2, methane, N_2O, O_3 and halocarbon compounds such as O_3-depleting chlorofluorocarbons and hydrofluorocarbons. The previously mentioned gases absorb radiation and re-emit some radiation towards the soil surface, leading to its warming. In the absence of the greenhouse effect the Earth would maintain a temperature of –18°C, while with the presence of greenhouse gases the average temperature is 14°C. Most of this 32°C difference

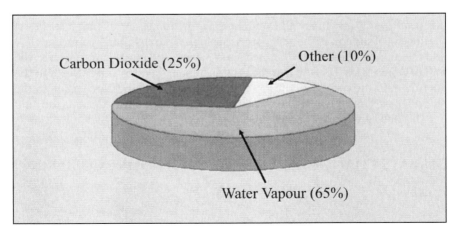

Fig. 8.8. Percentage participation of CO_2, water vapour and other factors involved in the greenhouse effect.

in temperature is due to water vapour (21°C) and CO_2 (7°C), while O_3 contributes 2°C. The presence of greenhouse gases in the atmosphere is essential to life on Earth, allowing it to maintain an average surface temperature between the boiling and freezing points of water.

Carbon dioxide

Carbon dioxide is released into the atmosphere through both natural and human processes. Carbon dioxide production and absorption take place through the carbon cycle in the biosphere. Human processes such as fuel burning increase the CO_2 concentration in the atmosphere; this has increased by 30% from the late 18th century up to modern times, and is now at a level of about 370 ppm, while it was recorded at 280 ppm before the Industrial Revolution. The stomata of leaves therefore play a role in sensing and driving environmental change (Hetherlington and Woodward, 2003).

Methane

Methane concentration has increased by about 150% since 1750. Methane is a greenhouse gas produced from both natural and human activities. Anthropogenic factors result in greater emissions of methane. The level of atmospheric methane increased from about 1610 ppb in 1983 to 1745 ppb in 1998.

Why climate change is happening

Climate change is happening fast and it is caused by the human input of greenhouse gases into the atmosphere. Climate changes in the past were slow, allowing plants and animals to adapt.

During the 20th and 21st centuries the average global temperature has increased to levels not existing in the previous 650,000 years of Earth's history. The global temperature rose by 0.7°C during the last century and has already risen by another 1.6°C, due to increasing concentrations of greenhouse gases in the atmosphere. However, olive varieties have certain heat requirements in order to flower, and various classical models could be used in order to evaluate their heat requirements (Ribeiro et al., 2006). The CO_2 concentration has been constant to 280 ppm for the last 1000 years but, due to industrial growth, has risen to 370 ppm today. Besides CO_2, other gases are increasing in concentration. Methane emissions constitute 15% of greenhouse gases today and are going to increase to 50% by the year 2100. This intensifies the warming effect, being equivalent to 430 ppm of CO_2. This increase in atmospheric greenhouse gases is due to human activity, mainly through accumulation of greenhouse gases due to the burning of fossil fuels to cover human needs. Other significant factors are deforestation, farming methods and poor waste management. Over the previous 140,000 years the climate has been harsh and fluctuating; however, over the last 8000 years the climate of our planet has been stable, leading to agricultural growth and therefore human civilization as we know it.

Side effects of climate change

At the present time the climate is 0.7°C warmer than before industrialization and has caused the following significant changes in our planet (IPCC, 2001).

1. Melting of polar ice and glaciers, leading to a sea level rise of 0.2 m.
2. Unpredictable and violent weather changes (see Fig. 8.9).
3. Heatwaves (see Fig. 8.10) and extended drought and flooding (Rodriguez Diaz et al., 2007). It is predicted that CO_2 concentration in the atmosphere is going to rise to 550–700 ppm by the year 2050 and to 650–1200 ppm by the year 2200. This will lead to global warming of 2.5°C or even more by the year 2050. The polar regions will have more severe levels of warming, which will lead to sea levels rising by > 0.5 m, creating problems for millions of people. Such climate changes will create water shortages for 4 billion people, and melting glaciers will create flood risk and displacement from their areas for 200 million people.

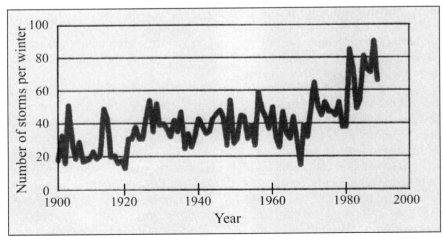

Fig. 8.9. Storm frequency in the northern hemisphere due to climatic changes over the last 100 years (from Lambert, 1996).

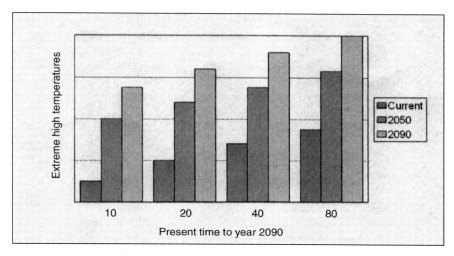

Fig. 8.10. Expected change in extreme high temperatures, from the present day to the year 2090 (from Hangeveld, 2000).

4. The decreased water supply, together with rises in temperature, will decrease food supplies.
5. Ocean acidification with increased CO_2 concentration has negative effects on and consequences for fish populations and marine ecosystems.
6. Increase in deaths due to heat stress, lack of food, malaria, etc.
7. Increase in the frequency of very intense hurricanes, typhoons and tropical storms.
8. Loss of an increasing percentage of biodiversity, since various species fail to adapt to harsh weather conditions. A temperature rise of 2°C will lead to extinction of 15–40% of species that are a source of food and medicines.
9. Changes in sea level. The rate of global mean sea level rise during the 20th century was in the range of 1.0–2.0 mm/year. The most rapid rise in global sea level was between 15,000 and 6,000 years ago, with an average rate of 10 mm/year.

IMPACTS OF CLIMATE CHANGE ON OLIVE TREES

Cycle length

Higher temperatures in late winter and early spring will promote the developmental stages of olives. With respect to olives no experimental results are known, but it can be expected that the flowering season will be promoted

(Frenguelli *et al.*, 1989; Gioulekas *et al.*, 1991; Chuine *et al.*, 1998; Galán *et al.*, 2005). However, higher temperatures could be catastrophic to flowering quality (perfect flowers) and therefore it could reduce olive production.

Higher temperatures increase biomass production. Furthermore, in olives and in other woody species, the influence of changing climate will depend on how long the increased CO_2 concentrations promote growth and production and also how the influence of CO_2, temperature and non-availability of water affects olive fruit and quality.

Weeds, diseases and pests

Greater CO_2 concentrations promote weed growth. The competition for olive plants from weeds will depend on their respective reactions to climate change. However, the nature of damaging effects depends on the weed species.

Increasing temperature in the northern hemisphere leads to increased air humidity and disease growth, such as the fungus *Cycloconium oleaginum*, which can cause considerable commercial damage to plantations. The risk of olive damage from pests and diseases has increased under warming of the climate; more generations of pests and diseases can be expected. This implies an increased requirement for pest and disease control, with negative effects on the environment due to agrochemical pollution. Furthermore, O_3 concentrations are increasing in warmer temperatures, and high O_3 concentration may lead to yield losses. Pesticide use is projected to increase by 10–20% for olives under climate warming. This increase in pesticide use leads to reduced economic profit.

Modification of culture areas

The olive tree is a Mediterranean crop. However, climatic changes will create desertification of the Mediterranean region and will render the temperatures of more northerly countries more beneficial (Parry, 1992). Therefore it is now possible to plant an olive grove in the UK, where already Mark Diacono has planted his grove of 120 olive trees along the banks of the River Otter and is expected to produce the UK's first olive oil within a few years. Furthermore, the seasons will change and summer and winter will be extended (Sparks and Menzel, 2002).

Organic olive farming impacts on climate change

Many people do not realize that agriculture is a significant contributor to atmospheric CO_2. Furthermore, the various types of agriculture produce varying amounts of CO_2. Organic olive culture controls weeds by ploughing, which doubles CO_2 emissions from the soil compared with non-tillage culture. Conventional olive culture relies on synthetic fertilizers while organic olive culture uses primarily manure. The use of manure fertilizers increases soil respiration rates, and therefore CO_2 emissions, by two- to threefold. Some researchers suggest a complete conversion to organic agriculture. However, this conversion requires a tremendous increase in the numbers of agricultural animals to produce manure fertilizer, which would result in a significantly greater CO_2 production in comparison with conventional olive culture.

Today, the worldwide tendency is to plant more olive trees, even in countries not traditionally cultivating the crop. This greater number of olive trees helps to restore the productivity and richness of the soil. Olive trees also absorb CO_2 from the environment and fight the symptoms of global warming, which has now become the Earth's biggest threat in the 21st century.

Climatic impact on olive irrigation

The most recent estimate of climate change in Europe suggests an increase in temperature of up to 0.7°C per decade during summer and, furthermore, a reduction and change in the distribution of precipitation during the 21st century (Moreno, 2005). The available irrigation water is going to decrease while water demand is going to increase. The reduction in available water will be greater in the southern Mediterranean areas. Also, the frequency of days with temperature extremes, droughts or storms is going to be significantly increased.

With regard to CO_2, its concentration is increasing, leading to greater productivity (see Fig. 8.11). This positive effect is counterbalanced by the higher temperatures and lower rainfall that increase the olive's water requirements. The irrigation seasons are also now longer, due to lower rainfall. Climate change requires that irrigation systems will have to be designed for longer and higher peaks.

Climatic change and olive tree phenology

Olive phenology is a sensitive indicator of climate change and especially of the future climatic warming of areas where olives are grown (Osborne *et al.*, 2000). By phenology we mean the study of the timing of periodic biological and/or natural events, and especially during the months of spring.

Rising spring temperatures during the previous century have advanced the time of emergence of flowers on olive trees. Several models based on thermal time and chilling have been developed for predicting the flowering date of olives.

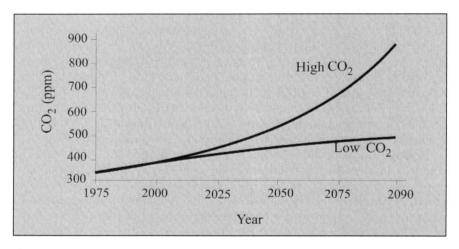

Fig. 8.11. Actual plus estimated increase in atmospheric CO_2 concentration over a period of 115 years (1975–2090).

Their analysis shows that the best predictions are from thermal time models based only on spring temperatures (Chuine *et al.*, 1998). The annual flowering date for an individual site may vary by 29–30 days within a single decade.

Phenology is the most responsive aspect of nature to warming, and is an ideal means of demonstrating that warming may already have an influence on the natural world. Olive flowering and deciduous tree leafing are amongst the easiest processes to record. Furthermore, in deciduous trees any increase in the mean temperature leads to advancement of harvesting date. Planetary warming and higher spring temperatures affect the timing of bird migration and breeding periods.

Adaptability to climate change

Adaptational measures are likely to be critical and include the following:

- Breeding for high CO_2 concentration and selecting for CO_2 response.
- Breeding for tolerance to water and heat stress. Certain olive varieties have proved to be more tolerant to high temperatures or to water stress conditions.
- Selection for tolerance to diseases and insects, since an increase in the use of pesticides and herbicides could threaten drinking water quality.

WEB SITE

Carbon Dioxide Information Analysis Center: http://cdiac.ornl.gov/

FLOWER BUD INDUCTION AND DIFFERENTIATION

INTRODUCTION

Flower bud inflorescences are borne in the axil of each leaf. Usually the bud is formed on the current season's growth and begins growth the following season. Lombardo *et al.* (2006) report observations regarding the floral biology of 150 Italian olive cultivars. Buds may stay in a dormant condition for more than 1 year and subsequently begin growth, giving inflorescences bearing flowers. Each inflorescence contains 15–30 flowers and this depends on the cultivar and the prevailing conditions for development.

The flowers are small, yellowish white and inconspicuous and contain a four-segmented calyx and a tubed corolla with four petals (lobes). The flower also has two stamens and a short style with a two-loculed ovary. The flowers are divided between two categories: *perfect*, having stamen and pistil, and *staminate* (male) flowers, where the pistil is aborted while the two stamens are functional. In the perfect flower the pistil is large, green in colour and fills the space in the floral tube. Staminate flowers are very small and do not fill the floral tube; the style is greenish white and small.

Flower initiation takes place very early in November and the flower parts develop during March–April, as has been shown by histochemistry. Induction in the olive may occur about 6 weeks after full bloom, while the initiation is not visible until 8 months later (February). Various authors report on the time of floral induction (Hartmann, 1951; Fernández-Escobar *et al.*, 1992; Cuevas *et al.*, 1999).

For flower induction, chilling is a prerequisite. Some olive cultivars ('Koroneiki', 'Mastoidis') in southern Greece require very little chilling, while other cultivars ('Kalamon', 'Amphissis' and cultivars from California and Spain) require adequate chilling. Experiments in Thessaloniki indicated that the cultivars 'Chondrolia Chalkidikis' and 'Amphissis' need longer periods of chilling than do 'Megaritiki' and 'Koroneiki' to satisfy their chilling requirements for flower induction (Hartmann and Porlingis, 1957; Hartmann and Whisler, 1975; Porlingis and Therios, 1979). The first appearance of sepal primordia occurs during the first 12 days of April, the beginning of carpel differentiation was observed during the last 10 days of April and full bloom took place during the last few days of May or the first days of June (Porlingis and Dogras, 1969).

Olive inflorescences appear during spring and originate from buds produced during the previous year's vegetative period. Initially, these buds are not differentiated to flowers. However, after exposure to low winter temperatures, internal changes occur in endogenous inhibitors and promoters, leading to their differentiation to flowering buds and to inflorescences (Hartmann *et al.*, 1967). Exposure of buds to relatively low temperatures is the necessary condition for flowering, and both the level of low temperatures and the cultivar determine to a great extent the percentages of buds that will produce inflorescences.

STAGES OF FLOWERING BUD INDUCTION AND DIFFERENTIATION

Observations with the cultivar 'Chondrolia Chalkidikis' under the climatic conditions pertaining in the area of Thessaloniki, Greece and also of other papers (Hackett and Hartmann, 1963, 1967) indicate that the flowering process of the olive can be divided into the following stages (Porlingis and Dogras, 1969).

1. The stage of *flowering bud induction*, which starts in October and lasts up to the end of February. The vegetative buds are subjected to winter chilling, which causes physiological changes necessary for flowering bud induction. However, these buds remain unaltered morphologically.

2. The stage of *morphological changes*, which leads to the development of a central axis and side branches of the inflorescence and the meristems destined to develop into flowers. The duration of this stage for the climatic conditions of Thessaloniki is about 40 days. For this stage low temperatures exert a beneficial influence.

3. The stage of *differentiation of flowers*. Initially the induction of sepals takes place and is followed by petals, stamens and carpels. The induction takes place the first 15 days of April, while that of carpels after 16–20 days.

4. The stage of *completion of growth* of the various parts of flowers. The full bloom happens from the end of May to the first days of June. The duration of the 3rd and 4th stages is 45–60 days (see Figs 9.1, 9.2 and 9.3). Figure 9.4 gives the biennial cycle of vegetative and reproductive processes in the olive (Rallo *et al.*, 1994).

The olive inflorescence is a panicle. The number of flowers per inflorescence is a function of the cultivar, soil moisture during its development (Hartmann and Panetsos, 1962) and N content of leaves. Prolonged and very cold weather during April and May, when the olive flower buds should be developing rapidly,

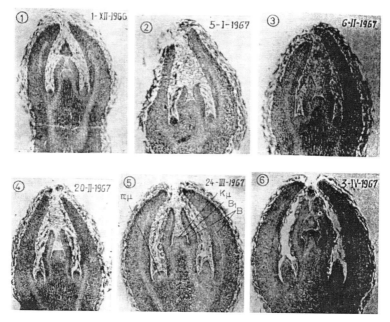

Fig. 9.1. First stages of floral differentiation in olive buds (cv. 'Chondrolia Chalkidikis'). At the upper right of each microphotograph the date of bud collection is given. 1, vegetative bud; 2, 3, no change is observed; 4–6, the apical meristem becomes broader and a new pair of bracts is formed (B1). Kμ, apical stem; πμ, lateral stem; B, bracts (magnification × 30) (from Porlingis and Dogras, 1969 [in Greek]).

has a detrimental influence on flowering, pollination and fruit set. From the olive flowers optimum yield can be achieved when 1 or 2% of these flowers remain as developing fruit 14 days after full bloom. Most of the flowers not setting fruit have abscised.

Olives are polygamous–monoecious since, on the same tree, we may find both perfect flowers with weak, well-developed ovaries and staminate (imperfect flowers with well-developed stamens and underdeveloped ovary) (Levin and Lavee, 2005).

The percentage of perfect flowers on a tree depends on: (i) the cultivar; (ii) shoot vigour; and (iii) environmental conditions, such as the lack of soil moisture during spring, when the inflorescences develop. Furthermore, the leaf:bud ratio affects the percentage of perfect flowers. When the leaf:bud ratio is increased by disbudding, the production of perfect flowers is promoted.

Constant temperature (12.5°C) or ideal temperature variations during the 24-h period (12.5°C for 20 h and 21°C for 4 h) promote the development of imperfect flowers, whereas low temperatures – or when the temperature varies (7°C for 20 h and 26°C for 4 h) – increase the percentage of perfect flowers.

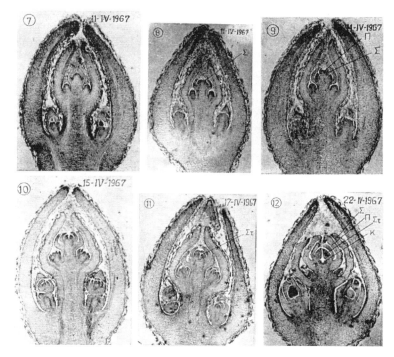

Fig. 9.2. Further stages of floral differentiation in the olive bud (cv. 'Chondrolia Chalkidikis'). 7, 8, broadening of the apical meristem and appearance of the primordial sepal (Σ); 9, sepal and petal (Π) primordia in the apical meristem; 10, 11, stamen differentiation (Στ); 12, carpel differentiation (K) (magnification × 20) (from Porlingis and Dogras, 1969 [in Greek]).

FACTORS AFFECTING FLOWERING BUD INDUCTION

Light

Light is an important factor in flower induction; shading reduces flower bud differentiation (Tombesi and Cartechini, 1986). Shading following flower bud differentiation does not affect flowering, however, it can cause morphological sterility or abortion of the ovary. Shading or leaf removal may postpone flowering.

The light intensity and quality are reduced as light passes through the top of olive trees and varies according to the shape of the trees and the density of their leaves. The index LAS (leaf area:soil area) is 2.5 (Bongi and Palliotti, 1995). The maximum net assimilation of CO_2 under optimum conditions is less than that in other C3 (see Chapter 8) plants. The Pn of olive leaves is 18 μmol $CO_2/m^2/s$. 'Shotberries' are parthenocarpic fruits that occur

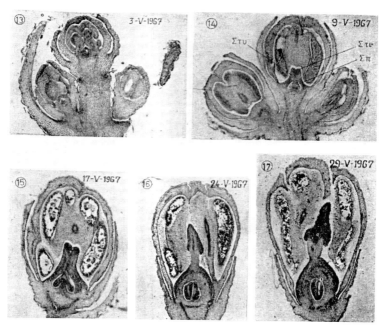

Fig. 9.3. Pistil development in the apical flower of the olive inflorescence. 13, pistil formation; 14–17, formation and development of ovary, ovules (Σπ), style (Στυ) and stigma (Στι) (13, 14 and 15: magnification × 20; 16, 17: magnification × 15) (from Porlingis and Dogras, 1969 [in Greek]).

randomly in the form of clusters on each inflorescence; these mature much earlier than normal fruits.

Juvenility

Juvenility is generally defined as the period during which a plant cannot be induced to flower. The duration of the juvenile phase in woody plants is quite variable, and can be quite lengthy (Meilan, 1997). Phase change describes the period during which a plant undergoes the transition from juvenility to maturity. This transition is a gradual and continuous process. Phase change has occurred if the olive plant flowers independently of the cause of the induction of flowering.

The juvenile and mature growth phases are often distinguishable by other morphological characteristics, which in olives can include leaf shape and size, phyllotaxy and the ability to form adventitious roots and/or buds. The term maturation describes the transition from juvenility to maturity, while ageing means loss of vigour and increased complexity. In assessing the ability to flower,

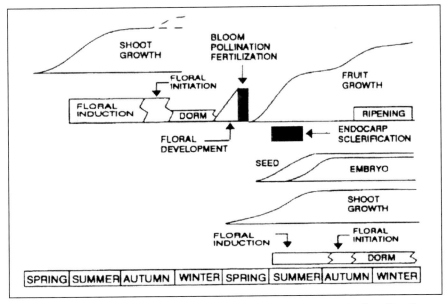

Fig. 9.4. Biennial cycle of vegetative and reproductive processes in the olive (from Rallo *et al.*, 1994). Dorm, dormancy.

it is necessary to use treatments that are known to induce flowering but which do not promote maturation.

Temperature

Winter chilling

Several climatic factors are apparently critical for olive performance, with respect to flowering, fruit growth and olive oil production and quality. Some winter chilling is required for floral initiation, but this requirement varies quantitatively between varieties (Hartmann and Porlingis, 1957; Hackett and Hartmann, 1963, 1964, 1967; Navarro *et al.*, 1990; Piney and Polito, 1990; Ferrara *et al.*, 1991; Rallo and Martin, 1991). Varieties originating from Italy, Spain, northern Greece and California require more winter chilling than those grown in southern Greece (Crete), Israel or Egypt (Martin *et al.*, 1994a), where the chilling periods range from 86 to 1400 h (Hartmann and Whisler, 1975). Temperature fluctuation affects flowering: fluctuation between 2 and 15°C for 80 days produced adequate flowering, while fluctuation between 7 and 18°C produced smaller numbers of inflorescences.

A chilling period is required prior to flowering. Afterwards, the determination of heat requirements in the first developing phases of plants has been expressed as temperature accumulation by means of parameters such as Heat Units, which are usually expressed as Growing Degree Days (GDD). Their determination is useful for forecasting when flowering will occur. The most suitable threshold temperatures were carried out in a range of 7–15°C. The causes responsible for varying threshold temperatures recorded in areas of the same latitude include different climatic conditions and the various cultivars present within olive groves on the same latitude.

Variation of temperature and cultivars
Due to the late blooming of olives, flower damage due to late spring frosts is rare. On the contrary, hot and dry winds during flowering reduce fruit set. Strong winds in June can increase the percentage of fruit drop.

In an experiment, olive plants grown in containers of the varieties 'Koroneiki', 'Kolovi' and 'Chondrolia Chalkidikis' were exposed continuously to temperatures of 10°C, 14°C, outdoors, outdoors + 10°C and outdoors + 14°C (Porlingis and Therios, 1979). After temperature exposure measurements were made of the number of inflorescences, the percentage flower bud differentiation, the number of flowers per inflorescence, the percentage of perfect flowers and the number of fruits/100 perfect flowers. These data are presented in Tables 9.1 and 9.2.

The data indicate that low temperatures exerted a favourable effect during both the first and the second stages of flowering bud differentiation in 'Koroneiki', but in 'Chondrolia Chalkidikis' only during the first stage. These observations indicate the existing relationship between temperature and number of flowers per inflorescence. The fact that the variety 'Kolovi' has a constant number of flowers per inflorescence at all temperature treatments indicates that the response to low temperatures is cultivar dependent. The effect of temperature on the percentage of perfect flowers is not clear. Therefore, the level of temperature during the first stages of flower bud differentiation affects, at least in some cultivars, the percentage of flower buds and number of flowers per inflorescence, with no effect either on pistil development or the percentage of perfect flowers. In some olive cultivars, such as 'Arbequina', flowering and fruiting can be achieved in the absence of chilling temperatures ($\leq 7°C$) (Malik and Bradford, 2005, 2006a). The absence of flowering in subtropical climates is not due to lack of chilling but due to high temperatures during winter days. Hence, temperature during the day $\leq 23°C$ could produce adverse effects on flowering.

Growth regulators

The levels of ABA, IAA and GA_3 in leaves, nodes and fruits during the induction, initiation and differentiation periods in the *on years* (those with a high percentage of flowering and fruit set) were lower than those in the *off years* (those with a low percentage of flowering and fruit set in the alternate bearing

Table 9.1. Number of inflorescences per plant, percentage of flower buds and number of flowers per inflorescence of three olive varieties maintained at various temperatures (from Porlingis and Therios, 1979).

Treatment	Variety		
	'Koroneiki'	'Kolovi'	'Chondrolia Chalkidikis'
Inflorescences/plant (n)			
Outdoors	1343	228	1150
10°C	1174	24	200
Outdoors + 10°C	1524	171	765
14°C	924	11	0
Outdoors + 14°C	1302	99	0
Flower buds/inflorescence (%)			
Outdoors	76.6^a	15.8^a	77.5^a
10°C	75.7^a	2.0^c	12.7^b
Outdoors + 10°C	84.9^a	12.0^a	48.8^a
14°C	56.2^b	0.9^d	0.0^c
Outdoors + 14°C	73.4^a	4.5^b	0.0^c
Flowers/inflorescence (n)			
Outdoors	18.7^a	7.6^a	12.9^a
10°C	18.1^a	7.0^a	8.6^b
Outdoors + 10°C	18.3^a	6.9^a	12.4^a
14°C	14.6^c	7.6^a	–
Outdoors + 14°C	15.8^b	8.3^a	–

Different superscript letters within the same column indicate statistically significant differences for $P < 0.05$ (Duncan's multiple range test).

cycle). However, GA_3 levels during these periods were higher in the on years. The fact that GA_3 decreased and GA_4 levels increased during the induction and initiation periods in the off years suggests that these affect flower bud formation. Zeatin increase during the induction period in the off year suggests that an increase in cytokinin during the induction period possibly has a positive effect on floral formation (Ülger et al., 1999, 2004).

The ABA content of the inflorescences was consistently > 40 nmol/100 g fresh weight (FW) from early anthesis until full bloom, reaching a maximum 1 week before full bloom. The ABA concentration changes are thought to be due to the shedding of bracts and imperfect flowers during the period of inflorescence development prior to full bloom. After full bloom, endogenous ABA concentrations fell to < 20 nmol/100 g FW. The ABA content of bark tissues was about 30 nmol/100 g FW before full bloom. Shedding of old leaves led to an increase in ABA concentration in bark tissues in late May. Exogenous application of GA_3, asparagine and glutamine affected the induction of flowering buds (Proietti and Tombesi, 1996a).

Table 9.2. Number of perfect flowers and the number of fruits per plant and per 100 perfect flowers of three olive varieties maintained at various temperatures (from Porlingis and Therios, 1979).

Treatment	Variety		
	'Koroneiki'	'Kolovi'	'Chondrolia Chalkidikis'
Perfect flowers/plant (%)			
Outdoors	77.3^a	90.3^a	84^a
10°C	31.9^b	87.9^a	90.3^a
Outdoors + 10°C	35.6^b	90.0^a	69.7^a
14°C	62.7^a	98.7^a	–
Outdoors + 14°C	43.5^b	96.5^a	–
Perfect flowers/plant (n)			
Outdoors	1759	60	301
10°C	1597	147	66
Outdoors + 10°C	1935	186	211
14°C	1580	16	0
Outdoors + 14°C	1732	29	0
Fruit (n)/100 perfect flowers			
Outdoors	9.1	3.8	2.4
10°C	23.6	11.4	4.5
Outdoors + 10°C	19.5	17.5	3.2
14°C	18.6	19.0	–
Outdoors + 14°C	19.4	3.7	–

Different superscript letters within the same column indicate statistically significant differences for $P < 0.05$ (Duncan's multiple range test).

Sugars and nutrient elements

Differences in concentration of any of the sugars, with the exception of fructose, were not significant in the on and off years (Ülger et al., 2004). The highest macro- and micro-element concentrations were found to be Ca and iron (Fe), respectively. The results suggest that carbohydrates and mineral nutrients may not have a direct effect on inducement of flower initiation (Ülger et al., 2004).

Oleuropein levels

Oleuropein levels sharply decreased during the transition from vegetative to flower buds (Malik and Bradford, 2006b), while levels rapidly increased with the expansion of fertilized pistils and then sharply declined with fruit maturation. Exogenous application of oleuropein inhibited flowering in

Kalanchoe blossfeldiana (Bongi, 1986). Therefore, oleuropein could play a role in flowering in olives. Drastic reduction in oleuropein levels was observed in floral buds, as compared with vegetative buds.

After floral bud differentiation, oleuropein levels progressively increased from the flowering to fruiting stage and then declined as the fruit started to mature, and were negligible in the fully mature black fruits (see Table 9.3).

Date of harvesting

The date of harvesting affects significantly the percentage of flower bud differentiation. Late harvesting depletes the olive tree and results in a lower percentage of flower bud differentiation (see Table 9.4 and Fig. 9.5).

Fruit thinning

Olive fruit thinning increases the percentage of flowering bud differentiation and flowering. Use of NAA (the auxin, naphthalene acetic acid) at 150 mg/l increased flowering bud differentiation by 60%.

Table 9.3. Sugar, hormone and mineral nutrient concentrations of 'Memecik' olive at the three floral stages (Ülger *et al.*, 2004).

Sugar	Differentiation	Induction	Initiation
Fructose (%)	0.18^c	0.34^a	0.26^b
Glucose (%)	3.08^a	3.26^a	3.12^a
Sucrose (%)	0.23^b	0.36^a	0.37^a
Total sugar (%)	3.48^b	3.99^a	3.81^{ab}
IAA (μg/g)	0.72^b	0.70^b	1.75^a
Zeatin (μg/g)	18.09^a	5.42^c	8.63^b
ABA (μg/g)	0.47^a	0.21^b	0.15^b
GA_3 (μg/g)	4.12^a	1.53^b	3.88^a
GA_4 (μg/g)	40.62^a	23.70^b	41.36^a
N (%)	1.04^a	0.97^b	0.61^c
P (%)	0.11^c	0.12^b	0.14^a
K (%)	0.59^c	0.91^a	0.73^b
Ca (%)	2.06^a	1.30^b	1.98^a
Mg (%)	0.17^a	0.13^a	0.17^a
Fe (μg/g)	93.77^a	63.98^b	60.09^b
Mn (μg/g)	33.82^a	22.73^b	19.20^b
Zn (μg/g)	21.61^a	16.95^b	11.21^c
Cu (μg/g)	29.13^a	17.77^b	16.04^b

Different superscript letters within the same column indicate statistically significant differences for $P < 0.05$ (Duncan's multiple range test).

Effect of defoliation

The percentage of flower bud differentiation decreased after defoliation due to diseases such as *Cycloconium* and nutrient and water stress.

Table 9.4. Flower bud differentiation (%) as affected by the date of harvesting and fertilizer applied.

Date of harvesting	Fertilizer (urea)	Fertilizer (10–10–10)	Control
30 November	6.88	3.48	3.33
15 December	4.05	1.31	1.63
25 January	0.85	0.29	0.81
14 March	0.46	0.20	0.09

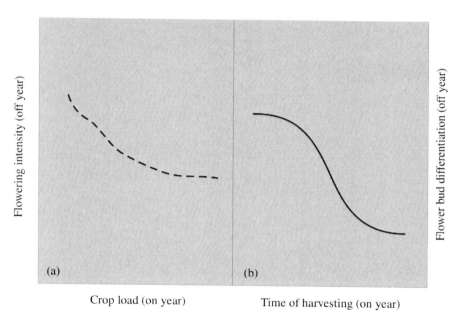

Fig. 9.5. The relationships between (a), olive production (kg) and flowering intensity the following year; (b) harvesting date and flower bud differentiation the following year (from Gucci and Cantini, 2004a).

10

FLOWERING, POLLINATION, FERTILIZATION AND FRUITING

OLIVE FLOWERING AS AFFECTED BY GLOBAL CLIMATIC CHANGE

Olive *flower phenology* is characterized by an annual cycle including bud formation during the previous summer, dormancy during the cold period, bud burst in late winter and flower structure development from bud burst to flowering in spring. Among the factors affecting plant development, temperature and photoperiod exert the strongest effect on flowering. Furthermore, variations in temperature and photoperiod due to altitude and latitude produce geographically differing growth phases.

Olive phenology is an indicator of global climatic change, especially during spring months (Osborne *et al.*, 2000). The rising of spring temperatures during the past and current centuries has advanced the timing of leafing and flowering in many species at high, northern latitudes. After chilling and breaking of bud dormancy the subsequent rate of bud development depends on the duration and warmth, usually known as 'heat sum' or 'thermal time'. In olives, flowering occurs between April and June in the Mediterranean region. The flowering date in olives may thus be a direct, simple and sensitive biological indicator of climatic warming. Model simulations indicate that the flowering date for olives in the Western Mediterranean could become significantly earlier by the year 2099 (Osborne *et al.*, 2000). The 4.5°C rise in temperature produced by greenhouse gases advanced flowering at a rate of nearly 30 days, or 6.2 days/°C. Furthermore, future monitoring of airborne olive pollen may therefore provide an early biological indicator of climatic warming in the Mediterranean. This model predicts that phenology will advance more strongly in large parts of southern France, Algeria, Tunisia, Morocco, southern Spain and Portugal than on the islands of the Mediterranean.

MODELS FOR THE PREDICTION OF FLOWERING DATE

Many previous studies on the prediction of flowering have used data from only one location (Frenguelli *et al.*, 1989; Alcalá and Barranco, 1992) and/or used a simple thermal time approach (De Melo-Abreu *et al.*, 2004). Model 1 is a *chill-heating model*, which is a generalization of the Utah model. Model 2 is a < 7°C chill-hours model and model 3 relies on a thermal time approach after 1 February. Model 1 is more appropriate for predicting the time of flowering in the olive and it is the only model considered in this study that is physiologically meaningful. Models 2 and 3 are also good predictors of the time of flowering but should not be used in warm areas, where some varieties do not satisfy their chilling requirement. Figure 10.1 gives the phenological stages of olive trees from inflorescence development to the third stage of fruit growth.

POLLEN AND ITS CYTOPLASMIC STERILITY

Two types of male sterility are distinguished, according to their mode of inheritance: (i) nuclear or genic male sterility (gMS); and (ii) cytoplasmic male sterility (CMS) (Besnard *et al.*, 2000). In most cases gMS is determined by a single locus and is due to a recessive allele. The olive is an allogamous, hermaphrodite and usually a wind-pollinated plant. Most of its cultivars are self-incompatible, a trait descended from *oleaster*, but some such as 'Oliviere', 'Chemlali' and 'Lucques' are totally male-sterile. This implies the necessity of using another variety as pollinator (Besnard *et al.*, 2000).

Compatibility between cultivars could reduce the problems of self-sterility that have been observed in olives. Emission of pollen at differing times could lead to lower reproductive success, since self-sterile cultivars may not be receptive when pollen is either released from other cultivars or pistils have passed their optimum receptivity, leading to reduced fertilization. Greater homogeneity could result in pollen emission throughout pistil receptivity and may result in more frequent cross-pollination between different self-sterile cultivars, resulting in earlier and higher levels of fertilization and an increase in total production. Farinelli *et al.* (2006) in their work have proposed the most appropriate pollinators for 21 olive cultivars (see Table 10.1).

Differences between various olive cultivars were observed regarding the onset and the start and length of flowering period. These differences are strictly correlated to differences in micro- and megagametogenesis and are cultivar-specific characteristics.

ANTIOXIDATIVE ENZYMES IN POLLEN

The mature pollen grains are released into the environment and are exposed to a number of agents such as high or low temperatures, pathogens, air pollutants

Fig. 10.1. Phenological stages of the olive tree. (a) Inflorescence development; (b) flower enlargement (the flowers become spherical); (c) full bloom; (d) petal fall and imperfect flowers; (e) fruit set and 1st stage of growth; (f) 2nd stage of fruit growth (hardening of endocarp begins); (g) 3rd stage of fruit growth (cell enlargement).

and UV radiation. The enzyme superoxide dismutase (SOD) has been found in olive pollen and catalyses the disproportionation of superoxide radicals (O_2^-), which is a defensive mechanism against oxidative stress (Alohé et al., 1998). SOD occurs in three forms, containing manganese (Mn), Fe or Cu.

Table 10.1. The most appropriate pollinators for 21 olive cultivars (Farinelli et al., 2006).

Main cultivar	Most appropriate pollinator(s)
'Arbequina'	'Carolea', 'Kalamon'
'Ascolana semitenera'	'Carolea', 'Itrana', 'Kalamon', 'Picholine'
'Ascolana tenera'	'Maurino'
'Bella di Spagna'	'Nocellara etnea'
'Carolea'	'Ascolana semitenera', 'Bella di Spagna', 'Leccino', 'Maurino', 'Moresca'
'Giarraffa'	'Ascolana tenera', 'Leccino', 'Nocellara etnea'
'Gordal sevillana'	'Ascolana tenera', 'Manzanillo'
'Grossa di Spagna'	'Giarraffa', 'Itrana', 'Santa Caterina'
'Itrana'	'Ascolana tenera', 'Carolea', 'Manzanillo'
'Kalamon'	'Arbequina', 'Ascolana semitenera', 'Carolea', 'Giarraffa', 'Koroneiki', 'Leccino', 'Manzanillo', 'Maurino'
'Koroneiki'	'Kalamon' (or self-fruitful)
'Manzanilla'	'Gordal', 'Leccino', 'Maurino', 'Nocellara etnea', 'Picholine', 'Santa Caterina'
'Maurino'	'Ascolana tenera', 'Leccino', 'Manzanillo', 'Picholine', 'Picual'
'Moresca'	'Ascolana semitenera', 'Carolea', 'Leccino', 'Picholine', 'Sorani'
'Nocellara etnea'	'Bella di Spagna'. 'Giarraffa', 'Leccino'
'Pendolino'	'Ascolana tenera'
'Picholine'	'Ascolana semitenera', 'Ascolana tenera', 'Manzanillo', 'Maurino', 'Moresca'
'Picual'	'Leccino', 'Maurino'
'Santa Caterina'	None
'Sorani'	'Gordal', 'Marocaine', 'Moresca', 'Picholine'
'Taggiasca'	'Ascolana tenera', 'Carolea', 'Leccino'

STAMINATE FLOWERS AND POLLEN

The presence of imperfect or staminate flowers in olive grants certain advantages to olive trees (see Fig. 10.2): (i) they can enhance pollen donation to fertilize other flowers; and (ii) they increase the attractiveness to plant pollinators. Hence, olive flowers can receive enough pollen. Therefore, a correlation exists between the quantity of pollen released and olive productivity. Furthermore, staminate flowers are not a non-functional byproduct, but they perform a vital role for olive trees (Cuevas and Polito, 2004). The dry weight of staminate flowers is 9% less than that of hermaphrodite flowers. Furthermore, concerning the amount of pollen per anther and pollen viability, no significant difference was recorded between staminate and pistillate flowers.

Fig. 10.2. (a) Olive inflorescence during the period of blooming; the flower parts (petals, stamens and pistil) are easily distinguishable; (b) olive inflorescences with imperfect flowers (male) (cv. 'Chondrolia Chalkidikis').

LOCATION OF FLOWERS WITHIN THE INFLORESCENCE AND POLLEN

The distribution of the flower (Lavee *et al.*, 1996) within the inflorescence is correlated with the time of anthesis and the percentage of imperfect flowers. Most hermaphrodite flowers are located at the apex and the primary pedicels, while flowers in secondary pedicels are mainly staminate and flower later.

FLOWERING AND POLLINATION

In the olive, full blooming is recorded in Mediterranean areas from late May to first days of June. The flowering period varies between cultivars (see Fig. 10.3), and this should be taken into account when selecting the most appropriate pollinators. The blooming period lasts 5–6 days in any particular cultivar, though a little more when the weather is cold. The amount of pollen and its germination capacity is a function of the cultivar, and ranges from low levels to 40–50%. Various factors influence pollination, such as temperature (Galán

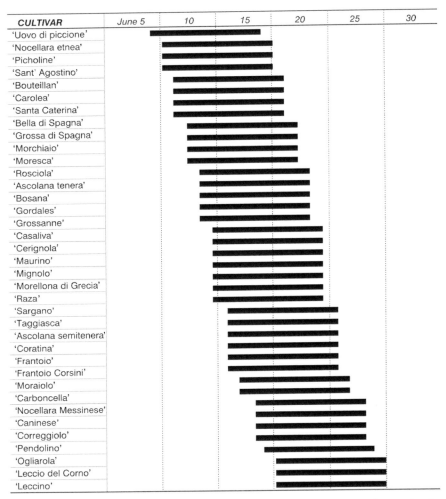

Fig. 10.3. Flowering periods of 37 olive cultivars in the Umbria hills of central Italy (from Antognossi et al., 1975).

et al., 2001), relative humidity, wind and rain. The temperature of the air should be < 30°C. However, very low temperatures reduce the speed of pollen tube growth, while high temperatures dry out the stigma and do not allow pollen germination. Wind is important for dissemination of pollen (Griggs et al., 1975). However, dry and hot winds destroy pollen grains and nullify pollen germination and pollen tube growth. Rain during the period of blooming also has a negative influence on pollen germination. When pollinizers are planted, their distance from the main cultivar should not be > 30 m, in order to attain an adequate fruit-setting percentage.

Each flower consists of a small calyx, four petals, two stamens and filaments with large anthers and a green pistil with a short style and a large stigma. Perfect flowers have a large green pistil, while imperfect ones are staminate and do not have a pistil or their pistil is rudimentary (Brooks, 1948).

Pollination in olives occurs through either self-pollination – as in the case of 'Koroneiki' – or cross-pollination (Cuevas and Rallo, 1990; Cuevas *et al.*, 2001), as in 'Kalamon', and the cultivars are characterized as either self-fruitful (Androulakis and Loupassaki, 1990) or non-self-fruitful (Diaz *et al.*, 2006b) (see Fig. 10.4). In the case of self-pollination, the pollen of a flower pollinates the pistil of the same flower. The pollen is transferred by gravity, bees or wind. In cross-pollinated cultivars, the pollen is transferred by wind. Olive pollen is poor in sugars and therefore bees do not commonly visit olive flowers.

Fig. 10.4. The cultivar 'Chondrolia Chalkidikis' following (a) self-pollination and (b) cross-pollination.

When pollen reaches the stigma it germinates, producing the pollen tube, which conveys the male germ cells (sperm) to the female germ cell (egg). The pollen tube can be stained for observation by microscopy (Cuevas *et al.*, 1994b). The pollen tube grows for a few days, reaching the egg cell in the embryo sac. One sperm cell fuses with the egg, producing the zygote which, after division, produces the embryo, which is destined to become the olive fruit. Cold weather reduces the speed of pollen tube growth; in this case, therefore, it reaches the embryosac late, after deterioration of the latter. After pollination and fertilization the ovary is transformed into fruit. Each ovary has two cavities, the locules, each one containing two ovules, i.e. four ovules per ovary. Only one developed ovule is required for seed formation. The presence of well-developed embryo sacs in the ovules and ovule longevity are determining factors in their capacity for fertilization.

FRUIT SET AND FACTORS AFFECTING IT

Many factors determine fruit set, such as: (i) inflorescence load and distribution (Lavee *et al.*, 1999); (ii) size of and number of florets per inflorescence; (iii) percentage of hermaphrodite flowers; (iv) ovarian size; and (v) physiological condition of the ovules and their longevity.

After fertilization, petals and stamens are abscised and the ovary starts to grow and enlarge (Rapoport and Rallo, 1991). The average fruit set is about 2%. Factors responsible for inadequate fertilization and subsequent fruitlet drop include: (i) incompatible pollinating cultivars; (ii) lack of pollinator trees; (iii) pest and disease problems; (iv) strong winds; (v) water stress and nutrient stress; and (vi) unfavourable climatic conditions (Cuevas *et al.*, 1995). Usually one fruit is maintained per inflorescence (see Fig. 10.5). Sometimes shotberries, i.e. fruits which are parthenocarpic (lacking seed), are produced. The reason why they are produced is unknown. Shotberries mature earlier than normal fruits.

Number of florets per inflorescence

The number of florets is affected by cultivar, the position of the inflorescence on the bearing shoot and the environmental and soil conditions, such as N levels and water stress.

Perfect and pistillate flowers

Olive flowers are distinguished by two types, hermaphrodite (*perfect*) and staminate (*imperfect*). Perfect flowers have both male and female organs, the

Fig. 10.5. High percentage of fruit setting after free pollination of 'Chondrolia Chalkidikis'.

ovary being relatively large and dark green in color. Imperfect flowers have only stamens and a rudimentary pistil and do not set fruits. The formation of imperfect flowers is due to arrested pistil growth as a consequence of water stress or N deficiency during the period of flowering bud differentiation. A small percentage of setting (1–2%) is sufficient for a good crop.

Ovarian size

The size of the ovary is affected by water and N availability during the period of flowering bud differentiation. Subsequently, through the process of cell division, expansion and/or differentiation, ovarian size increases 50-fold (Rapoport *et al.*, 2004) to produce the olive drupe.

Number and quality of ovules

The olive ovary has two carpels, each one containing two ovules. For efficient fertilization all the ovules should develop completely. However, of the four ovules contained in an embryo sac, only one is fertilized and the other three degenerate and shrink. In olives, degeneration of ovules could occur through various stress conditions, such as water stress and heat stress. Also, the type of cultivar affects embryo sac degeneration. When olive trees are not well supplied with N, ovule viability declines rapidly. Nitrogen concentrations in leaves of < 1.3% markedly reduce ovule longevity.

The olive tree produces an abundant number of flowers, but only 1–2% of these develop fruits that reach maturity. The level of normal fruit set in most cases is based on heavy blooming of olive trees during the 'on' year. Furthermore, in the 'off' year with limited flowering, flower quality is good and fruit set percentage is increased (Cuevas et al., 1994a; Martins et al., 2006; Rapoport and Martins, 2006).

Temperature and fruit set

Fruit set is completely inhibited at a constant temperature of 30°C, which significantly reduces pollen germination but does not prevent pollen tube growth. The most favourable temperature is 25°C, at which faster tube growth and higher fruit set were both observed. Most fruits are formed at the periphery of the tree, and very few fruits set in the interior of the olive canopy.

Olive trees are tolerant to high temperatures. However, very high temperatures during flowering may create problems. When the cv. 'Chondrolia Chalkidikis' is exposed to 35°C during flowering and 37°C before fruit set, fruit set is reduced (Porlingis and Voyiatzis, 1999). The most favourable temperatures for flowering, pollination and fertilization are between 18 and 22°C. The effect of high temperature could be alleviated by the application of paclobutrazol.

Pollination and fruit set in areas of mild winters

Today, olive cultivation is expanded not only in areas with a Mediterranean climate with mild, rainy winters and long, warm, dry summers, but also in other places with very different environmental conditions, e.g. Argentina, Mexico (Malik and Bradford, 2006c). Early in their life cycle, these trees show problems with fruit set and yield, with plant growth and metabolism being affected by the very different environmental conditions found in those regions. Cross-pollination improves fertilization and fruit set under hot environments, as well as in years of poor flower quality. The need for cross-pollination exists when temperatures during olive bloom exceed 30°C (Rallo, 1997). Another problem in warm, dry areas is a very short vernalization period (Ayerza and Sibbett, 2001). This increases the incompatibility of self-pollination for many cultivars, including 'Manzanillo', but it does not affect open pollination (Lavee et al., 2002).

High temperatures during olive flowering increase pollen incompatibility. Pollen tubes of the same cultivar frequently become blocked between the stigma and embryo sac in this condition. Furthermore, under hot, arid conditions the distance for effective pollination is reduced and is < 30 m radius, which is at variance with normal conditions. Increasing the number of pollinator trees in a grove could solve some of these problems, although it introduces the problem of establishing monocultivar plantings.

FRUIT GROWTH

The olive fruit is a drupe consisting of the pericarp and endocarp. The pericarp consists of the skin (exocarp) and the flesh (mesocarp), the latter producing the oil. The endocarp consists of a lignified shell which encloses the seed; each seed consists of the seed coat and the endosperm. The seed also contains an embryo, two cotyledons, a radicle and a plumule. The shape and size of the olive fruit depends on the particular cultivar: there are small-, medium- and large-fruited cultivars.

Olive fruit development follows a double sigmoid curve, very common in other fruits, having three stages. In the first stage, growth is fast; the growth is exponential and is characterized by cell division, and the pit is still growing. In the second stage the growth slows down or stops. However, at this stage the pit hardens and attains its final size. Finally, in the third stage the fruit grows by cell enlargement.

FURTHER READING

Acebedo, M.M., Cañete, M.L. and Cuevas, J. (2000) Processes affecting fruit distribution and its quality in the canopy of olive trees. *Advances in Horticultural Science* 14(4), 169–175.
Batanero, E., Ledesma, A., Villalba, M. and Rondriguez, R. (1997) Purification, amino acid sequence and characterization of Ole e 6, a cystein-enriched allergen from olive tree pollen. *FEBS Letters* 410, 293–296.
Bernier, G., Kinet, J.M. and Sachs, R. (1981) *The Physiology of Flowering. Volume II. Transition to Reproductive Growth.* CRC Press, Boca Raton, Florida, pp. 157–159.
Chaari-Rkhis, A., Maalej, M., Messaoud, S.O. and Drira, N. (2006) In vitro vegetative growth and flowering of olive tree in response to GA_3 treatment. *African Journal of Biotechnology* 5(22), 2097–2302.
Cimato, A., Cantini, C. and Sillari, B. (1990) A method of pruning for the recovery of olive productivity. *Acta Horticulturae* 286, 251–254.
Cuevas, J. and Polito, V.S. (1997) Compatibility relationships in 'Manzanillo' olive. *Journal of Horticultural Science* 32, 1056–1058.
Fernández-Escobar, R., Sanchez-Zamora, M.A., Uceda, M. and Beltran, G. (2002) The effect of nitrogen overfertilization on olive tree growth and oil quality. *Acta Journal of Horticulturae* 586, 429–431.
Fernández-Escobar, R., Benlloch, M., Herrera, E. and Garcia-Novelo, J.M. (2004) Effect of traditional and slow-release N fertilizers on growth of olive nursery plants and N losses by leaching. *Scientia Horticulturae* 101, 39–49.

Gioulekas, D., Chatzigeorgiou, G., Lykogiannis, S., Papakosta, D., Mpalafoutis, C. and Spieksma, F.T.M. (1991) *Olea europaea* 3-year pollen record in the area of Thessaloniki, Greece, and its sensitizing significance. *Aerobiologia* 7, 57–61.

Kitsaki, C.K., Drossopoulos, J.B. and Terzis, S. (1995) Endogenous free abscisic acid in floral, bark and leaf tissues of olive during anthesis and early fruit development. *Scientia Horticulturae* 64, 95–102.

Marquez, J.A., Benlloch, M. and Rallo, L. (1990) Seasonal changes of glucose, potassium and rubidium in 'Gordal Sevillana' olive in relation to fruitfulness. *Acta Horticulturae* 286, 191–194.

Martin, G., Nishijima, C. and Early, J. (1993) Sources of variation in olive flower and fruit populations. *HortScience* 28, 697–698.

Navarro, C., Benlloch, M. and Fernández-Escobar, R. (1990) Biochemical and morphological changes related to flowering in the olive. In: *XXIII International Horticulture Congress*, Florence, Italy, Abstract 3196.

Stutte, G.W. and Martin, G.C. (1986) Effect of light and carbohydrate reserves on flowering in olive. *Journal of the American Society of Horticultural Science* 111, 27–31.

Vitagliano, C., Minocci, A., Sebastiani, L., Panicucci, A. and Lorenzini, G. (1999) Physiological response of two olive genotypes to gaseous pollutants. *Acta Horticulturae* 474, 431–433.

11

ALTERNATE BEARING

INTRODUCTION

The productivity of each cultivar is a function of alternate-bearing habit and fruit abscission before ripening. Alternate bearing is defined as the characteristic of an olive tree to produce olives every 2 or more years. The year of maximum production is called the 'on' year, while the year with minimum or no production is called the 'off' year.

In olive trees we observe two growth flushes, the first one in spring and the second in autumn (October), under adequate soil moisture and temperature. Alternate bearing is directed by various factors such as water stress, warm soil, climatic factors and the cultivar. Significant genotypic variation was reported concerning cultivars, with respect to alternate bearing. The main causal factor of alternate bearing is the inhibition of flower bud induction by the seed of growing fruits, and this is due to competition for nutrients.

The two main factors that can reduce the intensity of alternate bearing are pruning and productivity. Furthermore, the irrigation schedule may reduce alternate bearing. Mineral nutrition is also a determining factor. Nutrient deficiency, especially of N during the period of flower bud differentiation, affects vegetative vigour, the size of fruits and the flowering of the following spring (Fernández-Escobar *et al.*, 1999, 2004b).

THE EFFECT OF FRUIT LOAD

The inhibition of floral induction by seeded developing fruits is the major factor in the olive's biennial behaviour. Environmental and other factors – such as cultivation practices that affect aspects other than floral induction – may amplify the alternate-bearing habit. Such factors include irrigation, pruning, fertilization, etc. Therefore, the control of biennial bearing should be achieved by (i) selection for breeding cultivars that produce every year; and (ii) cultural practices to stabilize the floral induction percentage.

GIRDLING

Many horticultural techniques such as girdling can increase flower initiation and fruit set in the off years (Hartmann, 1950; Lavee *et al.*, 1983; Ben-Tal and Lavee, 1984; Li *et al.*, 2003). Girdling increases soluble sugar and starch content in leaves and shoot bark above the girdle during the off year. Trees during the on year do not accumulate carbohydrates above the girdle. Furthermore, girdling reduces soluble sugar and starch concentrations below the girdle in both on and off trees. Girdling affects carbohydrate-related gene expression (Weiss and Goldschmidt, 2003).

ENDOGENOUS HORMONES

Endogenous plant hormones play a significant role in alternate bearing (Lavee, 1989; Ülger *et al.*, 1999). ABA and GA_3 have a direct role in flower bud initiation, while IAA and IAA-like compounds have an indirect role in the initiation of flower buds. Furthermore, exogenous application of growth regulators (Ben-Tal and Lavee, 1976; Akilioglu, 1991; Eriş and Barut, 1991), spermine or spermidine (Pritsa and Voyiatzis, 2005) or paclobutrazol (Lavee and Haskal, 1993) may alleviate the problem.

PROTEIN CONTENT

The total extractable protein content in the leaves of off trees is significantly lower than in the leaves of on trees (Lavee and Avidan, 1994). This was observed in the cultivars 'Koroneiki', 'Uovo de piccione', 'Manzanillo' and 'Barnea'. The least difference was recorded in the leaves of 'Koroneiki', which demonstrates the least alternating capacity.

Chlorogenic acid (CHA) is associated with alternate bearing and levels are high in the alternate years and low in the off years. Injection of exogenous CHA reduces flower bud differentiation when applied during the secondary induction period in the autumn and early winter.

In most cases proteins of 28 and 32 kD were present in samples of the off year leaves, and 22 and 14 kD were in higher concentration in the leaves of the on trees.

PHENOLIC ACIDS

Endogenous control of alternate-bearing includes the possible involvement of phenolic acids, with the end product of caffeoylquinic acid (see Fig. 11.1) (Lavee and Avidan, 1982; Lavee *et al.*, 1986).

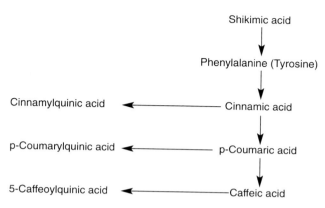

Fig. 11.1. Pathway for the production of 5-caffeoylquinic acid via cinnamylquinic acid (from Lavee *et al.*, 1986).

FURTHER READING

Martin, G.C., Ferguson, L. and Polito, V.S. (1994) Flowering, pollination, fruiting, alternate bearing and abscission. In: Ferguson, L., Sibbett, G.S. and Martin, G.C. (eds) *Olive Production Manual*. University of California, Division of Agriculture and Natural Resources, Berkeley, California, Publication 3353, 19–21.

Rallo, L., Martin, G.C. and Lavee, S. (1981) Relationship between abnormal embryo sac development and fruitfulness in olive. *Journal of the American Society of Horticultural Science* 106, 813–817.

Rallo, L., Torreño, P., Vargas, A. and Alvarado, J. (1994) Dormancy and alternate bearing in olive. *Acta Horticulturae* 356, 127–136.

12

Fruit Thinning

INTRODUCTION

Fruit size is the determining factor in the price offered per kg of raw table olives to growers. Therefore, the study of factors that tend to increase fruit size is fundamental. The 'first fruit drop' takes place after petal fall. During this process a heavy shedding of undifferentiated or poorly developed olives is recorded due to fruit competition (Suárez et al., 1984). After fruit set the olive fruit is small, and various factors can influence its final size.

The growth of olive fruit is performed initially by cell division and, later, by cell enlargement. Together with cell enlargement, air spaces increase to a maximum. The combination of cell division, cell enlargement and air space appearance results in the olive growth curve. The olive fruit size, especially in table olives, is increased by the following factors:

- Adequate soil moisture.
- Excess N fertilization.
- High leaf:fruit ratio.
- Light flowering and fruit set.
- Appropriate pruning.
- Early thinning, stimulating cell division.

OBJECTIVES OF OLIVE FRUIT THINNING

Olive fruit thinning aims at reducing limb breakage, increasing fruit size, improving fruit quality and stimulating flower bud initiation for the following year's production.

LEAF:FRUIT RATIO AND ASSIMILATE TRANSPORT

Thinning modifies the leaf:fruit ratio; therefore, it reduces fruit competition in olive (Rallo and Fernández-Escobar, 1985). A change in the leaf:fruit ratio causes significant differences in fruit growth. Under high levels of assimilate availability in fruits, as happens following fruit thinning, the fruit weight increases. In contrast, leaf removal reduces olive fruit size. Growth and fruit size are independent of each other, and neighbouring shoots can be considered an independent unit, since this will not be affected by conditions occurring in adjacent shoots. (Proietti and Tombesi, 1996b). The nutrients for olive fruit growth and development are supplied mainly by the leaves of the same shoot. Therefore, in shaded areas where carbohydrate availability is low, olive fruits cannot obtain optimum size although the availability for the whole tree may be high. Fruit thinning increases carbohydrate availability to remaining fruits and results in an increase in size. The factors, which increase or decrease thinning, are listed in Table 12.1.

THINNING METHODS

Generally, olive thinning can be carried out using the following methods: hand-, mechanical and chemical thinning.

Hand-thinning

Hand-thinning is the removal of flowers or fruits manually. Our method is to leave six to seven fruits per shoot (30 cm). This method incurs a heavy labour

Table 12.1. Factors which increase or decrease fruit thinning (from Westwood, 1978).

Increased thinning	Decreased thinning
Young trees	Mature trees
Moist weather	Dry weather
High relative humidity	Low relative humidity
High maximum temperature	Temperature lower than maximum
Softly sprayed water	Heavily sprayed water
Slow drying conditions	Fast drying conditions
High concentration of chemicals	Low concentration of chemicals
Trees of low vigour	Vigorous trees
Close spacing of trees	Wide planting distances
Light pruning	Severe pruning
Heavy flowering	Light flowering percentage
Poor pollination	Adequate pollination
Previous heavy crop	Previous light crop
Addition of wetting agents	No inclusion of wetting agents

cost. Thinning should be conducted within 3 weeks of full bloom and no later, otherwise there is no reduction in alternate bearing and fruit size will not be substantially increased.

As an alternative to hand-thinning, appropriate pruning can thin the shoots, reduce cropping and increase olive fruit size. Furthermore, this promotes shoot growth for the following year's crop.

Mechanical thinning

Mechanical thinning can be performed by the following methods:

1. Use of a hand-operated spray, with high-pressure water (Westwood, 1978).
2. Use of a mechanical shaker to shake the tree trunk. This method needs skill to avoid over-thinning. Furthermore, with this method the larger olives are removed since these fruits obtain a greater directional momentum than small ones. Also, fruits from the stronger branches are removed more easily. Therefore, with mechanical thinning it is difficult to achieve the degree of thinning the grower needs.

Chemical thinning

In recent decades, NAA, l-naphthaleneacetamide (NAAm) and ethephon-releasing ethylene have been tested for thinning table olives (Weis *et al.*, 1988, 1991). Chemical thinning has certain advantages over hand- or mechanical thinning:

- Low thinning cost.
- Greater size of olives, earlier maturation and better quality.
- Decrease in biennial bearing.

The disadvantages of this method include the following. Sometimes the results are variable and depend on various factors that may cause over-thinning. These factors include: (i) age of the tree; (ii) very low-vigour trees; (iii) light pruning; (iv) heavy bloom; (v) poor pollination; (vi) high humidity or high temperature; and (vii) high concentration of thinning agents. The lack of stable response is a common problem due to thinning (Westwood, 1978).

However, thinning of table olives produces certain benefits such as:

1. Large fruit size. This is a critical factor for the price per kg paid by the processor to growers. Trees overloaded give small fruits and enter into biennial bearing. Thinning increases the ratio of leaves per fruit, i.e. greater assimilate amounts are supplied to each fruit and the fruit obtains a greater size.
2. Annual bearing and precocious fruit ripening. A heavy crop of olives (on year)

is followed by a light crop (off year) or no crop at all. However, an average crop achieved with fruit thinning gives a satisfactory crop the following year. Therefore, fruit thinning results in bearing every year. Furthermore, after thinning, the crop is moderate, has more available carbohydrates per fruit and matures earlier.

3. Flesh:pit ratio and olive fruit quality. Thinned trees give fruits with a higher flesh:pit ratio and higher oil content. The harvesting cost per kg is significantly less from trees with moderate yield and large fruits than from trees with a big fruit load.

4. Flower bud differentiation. Flower bud differentiation increases as leaf assimilates and other accumulated reserve materials increase.

For the chemical thinning of olive, NAA is used. The time of application is very important in the effectiveness of thinning (Lavee and Spiegel-Roy, 1967). The best time for thinning is 12–18 days following full bloom. If NAA is sprayed earlier, this treatment over-thins the olive tree, while late application is not effective. The trees should be fully covered with the spray solution, the required quantity being 10–15 l/tree. The suggested NAA concentration is 10 ppm/day after full bloom, i.e. 100 ppm for spraying after 10 days or 150 ppm for spraying 15 days after full bloom.

The diameter of table olives should be 3–5 mm at the time of application. Furthermore, the greater the applied concentration of NAA the greater the thinning activity (Sibbett and Martin, 1981; Maranto and Krueger, 1994).

13

SYSTEMS OF PLANTING AND CANOPY TRAINING

SELECTION OF THE ORCHARD SITE

Since the olive tree grows and is more productive in heavy and fertile soils and less so in poor/dry ones, for the expansion of new olive plantings, fertile soils are preferred (Sibbett and Osgood, 1994). This does not mean that the modern olive culture should be expanded only to richer soils, since other crops in such soils may be more efficient than olives. Concerning the climatic conditions of the area for olive cultivation, these should be favourable for the growth and productivity of the trees. Furthermore, it is important that the selected area be irrigated and mechanically cultivated.

SOIL PREPARATION

Under favourable conditions the root system of olive trees, planted at distances of 6 × 6 m, covers within a period of 3–4 years the entire area of the orchard. Soil preparation includes levelling, deep ploughing, drainage, etc. and should be performed before planting (Sibbett and Osgood, 1994). Large stones should be removed to avoid damage to agricultural machinery.

SELECTION OF CULTIVAR AND ROOTSTOCK

Local experience plays a significant role in the selection of the best suited cultivars within a particular area. Furthermore, the new cultivation techniques of intense planting incorporate certain modifications of the traditional methods, and this affects the choice of cultivar. For example, large-fruited cultivars that are suitable for table olives and olive oil production are preferred, due to low cost and easiness of harvesting.

    The planting material is either budded plants or plants obtained asexually from leafy cuttings. In the traditional olive groves budded plants on *Olea sativa* (wild olive) are preferred, while in the dense planting systems, plants from leafy cuttings are the rule. However, until now little progress has been conducted in the direction of olive rootstocks, since the rootstocks used are seedlings from wild olives or cultivated varieties, giving plants with various characteristics and properties. The genera and species tested as olive rootstocks are the following: *Phyllirea, Ligustrum, Syringa, Chionanthus, Forsythia, Fraxinus forestiera, Olea verrucosa, O. chrysophylla* and *O. oblonga*. These rootstocks do not have good affinity with *O. europaea*. Furthermore, rooted cuttings from low-vigour cultivars could be used as dwarf rootstocks. The root system of rooted cuttings differs from that of grafted plants. Rooted cuttings have all their roots at the same level (base of cuttings) and tend to grow horizontally and close to the soil surface. In grafted plants, by contrast, the roots appear along an axis of 40 cm and grow at a greater depth. These differences may disappear after a few years, but they play an important role during the first few years of growth in the plant's tolerance to drought and frost. The planting material should be healthy and have adequate branching along the central axis.

SYSTEMS AND DISTANCES FOR PLANTING

The planting distances and the number of trees/1000 m² of olive grove are presented in Table 13.1. Before establishing an olive orchard it is very important to study the socio-economic conditions of the area. These conditions change with time and their trend should be determined. Local conditions include the examination of soil, climatic factors and cultivar characteristics. If the area under cultivation is covered with bushes, these should be removed and the area cultivated with annual crops for 2–3 years before planting. This decreases the danger from *Armillaria* and other soil fungi growing on roots. Levelling of the area follows and the irrigation system is established.

A planting, to be laid out in a regular pattern (square, rectangle or triangle), is started by establishing a straight base line, which is usually adjacent to a fence or roadway. Then, lines at right angles to the baseline are established at both ends and at one or two points in the middle of the plot. Right angles are usually established by using three ropes whose lengths are in the proportion 3:4:5 (e.g. 30, 40, 50 m). The 40 m rope is set along the baseline, then the 30 m one at 90° and finally lay the 50 m rope to close the triangle. In this way we can establish the positions of all trees. Small stakes can be used to mark the locations for tree holes.

When the slope of the area is greater than 3%, a contour layout is used. The first row is at the highest elevation and is staked out with all the points at the same elevation. Next, the steepest slope along this first row is found and the distance which was selected as the minimum distance between rows is measured down the slope. Continuing in the same way, the rows are established.

Table 13.1. Planting and spacing distances in the orchard.

Planting distance (m)	Area/tree (m²)	Trees/1000 m²	Spacing at full production (m)
3 × 3	9	111	6 × 6
3 × 4	12	83	6 × 4
3 × 5	15	67	6 × 5
3 × 6	18	55.5	6 × 6
4 × 4	16	62.5	4 × 8
4 × 5	20	50	8 × 5
4 × 6	24	41.5	
4 × 7	28	35.5	
4 × 8	32	31	
5 × 5	25	40	
5 × 6	30	33	
5 × 7	35	28.5	
5 × 8	40	25	
6 × 6	36	27.5	
6 × 7	42	24	For irrigated orchards
6 × 8	48	21	For irrigated orchards
7 × 7	49	20.5	For irrigated orchards
7 × 8	56	18	For irrigated orchards
7 × 9	63	16	For irrigated orchards
8 × 8	64	15.5	For irrigated orchards
8 × 9	72	14	For irrigated orchards
8 × 10	80	12.5	For irrigated orchards
9 × 9	81	12	For irrigated orchards
9 × 10	90	11	For irrigated orchards
10 × 10	100	10	For irrigated orchards

The planting of olive trees is conducted from November to February. In areas with mild winters the best period for planting is the autumn, while in those areas with cold winters planting is postponed until the end of winter.

PLANTING

For planting, trees in 3 l black plastic bags are used. Before planting, these bags are removed. The depth of planting is a little greater that in the nursery. The main object in planting is to protect the roots from drying or freezing and to get them into firm contact with moist soil. Tree holes should be large enough to accommodate the root system – a hole of depth 80–100 cm is adequate. Long roots can be cut back enough to balance the root system. Topsoil should be put around the roots and firmed in with the foot after the hole is about half-full of soil. The hole should then be filled the rest of the way up with loose soil. Weeds, manure and other organic materials should be kept out of the hole at planting time to avoid damage to the root system. In poor soils, a surface application of N fertilizer is recommended.

The most serious mistakes made during the establishment of an olive grove are improper irrigation and lack of weed control, which result in slow growth rate and tree death within the first few years of planting. During the early years of orchard life irrigation water should be applied many times, but the quantity of the water at each irrigation should be low to medium.

In traditional olive orchards planted on steep land, distances between trees should be long; short planting distances in muddy soils produce high trees with low leaf mass and very low productivity. As a general rule for the distances of planting, the distances between two trees should be double the radius of the top growth.

Planting distances are a function of the soil fertility, cultivar and planting system. The ratio of main cultivar:pollinizer may be 1:1, 2:1, 4:1, 8:1, 10:1 or 18.5:1 (see Fig. 13.1). The most common shape for the olive tree is the *open centre* tree. This shape allows light interception for the inner part of the top and increases tree productivity. The distance of planting with this system in areas of adequate precipitation is 10×10 m, in order to avoid tree crowding. However, when precipitation is not adequate and the soil is poor, greater distances of planting are preferable. Olive plants after planting are not pruned for 3–6 years, to speed up the onset of fruiting and to allow branches to develop at intervals along the trunk. Subsequently, three to five branches are selected and, if the central axis is not pruned, the top of the tree attains the form of a sphere (*spherical shape*).

PLANTATION SYSTEMS

Four major types of olive plantation systems are identified: traditional, semi-intensive, intensive and organic (Metzidakis and Koubouris, 2006). The classification of olive production systems on the basis of agro-ecological, technological and socio-economic criteria is analysed below.

1. Traditional. This is characterized by low-density plantations (< 100 trees/ha) of old trees (> 50 years) planted on moderately steep slopes. This system has low inputs, no mechanization and results in low yields not exceeding 1000–2000 kg/ha and low profitability.
2. Semi-intensive. The density in this system is variable, fluctuating between 100 and 150 trees/ha. These orchards have productive trees, yielding, for example, 2800–4000 kg/ha in Crete.
3. Intensive. Tree density is 250 trees/ha, yielding up to 5000 kg/ha and the system may be mechanized.

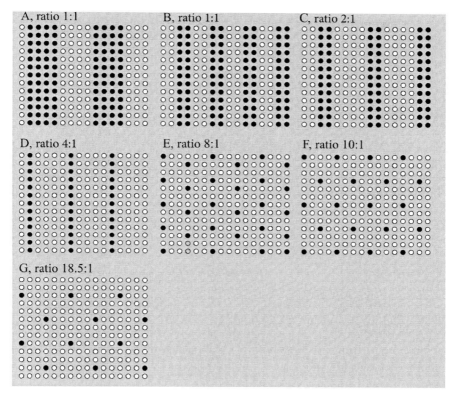

Fig. 13.1. Diagram of planting of two olive cultivars at ratios 1:1 (a, b), 2:1 (c), 4:1 (d), 8:1 (e), 10:1 (f) and 18.5:1 (g).

○, main cultivar; ● pollinizer.

4. Organic. These orchards do not have a mathematical form and can be established following mechanization of orchard management. This system is labour-intensive but it produces high-quality olive oil. Organic olive orchards in Greece represent ≥ 1% of total area planted.

MODERN TRAINING SYSTEMS

Vase system

This is the most common system in many areas of the world. Olive trees in the vase form have greater photosynthetic surface per tree in comparison with those trained in the globe system. The vase form includes the cone, inverted cone, the cylinder or multiple cones (Garcia-Ortiz *et al.*, 1997; Gucci and Cantini, 2004b).

Each tree has a single trunk and three to five branches forming an angle of 45°. The height of the trunk is 0.3–0.4 m for hand-harvested trees or 1.0–1.2 m for orchards adapted to mechanical harvesting. In mechanically harvested orchards the branches are more erect.

Polyconic system

This represents a modification of the vase form. Trees trained to a polyconic vase form have a single trunk (0.8–1.6 m) and three to six major branches, of varying orientation, in order to achieve optimum light and space utilization (Cain, 1972; Jackson, 1980; Connor, 2006). Each branch is trained to a cone. This system is widespread in Italy, although today its expansion is not recommended, since it requires high labour input for pruning and harvesting.

Central leader trees or monocone system

This is an alternative form of the polyconic vase and is recommended for high-density plantings, and is very common in Italy (see Fig. 13.2). The tree has a trunk up to 1.0–1.2 m in height; the primary branches have an elicoidal arrangement along the central axis. The length of lateral branches decreases

Fig. 13.2. Plantation of olive trees in double rows (6 × 3 m) in the monocone system.

from the base to the top of the tree and the tree becomes conical in shape. The main advantage of this system is that it is appropriate for mechanical harvesting with trunk shakers (Preziosi et al., 1994). However, it incurs a high pruning cost since the trees become tall. Trees of the 'Kalamon' cultivar trained in the monocone system and within a controlled irrigation regime produce good-quality olive oil (Pallioti et al., 1999; Patumi et al., 2002).

Vasebush system

The vasebush system has a vase form and the trunk and primary branches originate from the crown of the tree or from a very short trunk. Each plant has three to six branches. The height of the trees is relatively small, appropriate for hand-harvesting, and especially for table olives (Fig. 13.3). This system is not good for mechanical harvesting.

Planting of olive orchards in this form started in the 1950s/60s and the system continues to be expanded due to its following advantages:

- Rapid progress to fruit bearing (3–4 years).
- Triplicate yield/unit area in comparison with the free vase form planted at distances of 10×10 m.
- Reduction of harvesting cost by 70%.
- Ease of cultivation.

Fig. 13.3. Olive planting (cv. 'Chondrolia Chalkidikis') in high-density orchards and top pruning in dwarf bushes (vase form).

The planting distances are 5–7 m within rows and 5–7 m between rows. Smaller planting distances lead to shading after 10 years of growth. Greater planting distances entail inefficient use of the soil for the first 10 years.

After 10–15 years have elapsed since planting, rejuvenation of the treetops is necessary. In order to shape the tree into a bush vase, planting material is derived from leafy cuttings and the plants have many side-shoots. The trunk of the vasebush is short (< 50 cm), and from this grow three to four branches at spacings of 10–15 cm. For medium-vigour cultivars this small number of branches is adequate. Along the main branches, secondary branches develop at spacings of 50 cm, and their total number is six, eight, 12 or 16 depending on tree vigour.

The high-density system of olive production is efficient, and reliable annual yields of about 3–4 t/ha can be obtained in irrigated orchards with 20–40 trees/1000 m^2. These trees attain full productivity in about 8 years, and the orchard's yield is considerable. Harvesting is achieved manually, with trunk shakers or by over-the-row, comb-type harvesters, which are very efficient in fruit removal.

The failings of this system include the following:

- Long time (8 years) to achieve full production.
- Harvesting costs with trunk shakers are high and exceed 50% of the total production costs.
- Spray coverage and pest control are less efficient in comparison with those for smaller trees.
- A high-density orchard also requires very precise irrigation, good fertility management and sometimes more careful pest control and greater investment costs.

Palmette system

This system is an efficient method for other fruit trees such as pear and apple, and has been tested in olive culture. It incurs high costs for planting, pruning and maintenance.

Globe system

This system utilizes single-trunk trees and the shape of the canopy is round, ellipsoidal or hemispherical. The canopy is not open, unlike the vase form, and is occupied by secondary branches. Light penetration in this system is inadequate and yields are generally poor (Gucci and Cantini, 2004b). At planting the central axis is left alone and its terminal parts fill the centre of the canopy. To avoid excessive tree height the primary branches are shortened. The globe system is appropriate for traditional, non-irrigated orchards.

High-density system

High density is a recent olive-growing technique, and many workers have reported their experiences with it (Fontanazza and Cappelletti, 1993; Dag et al., 2006; Tous et al., 2006). The main features are listed below:

1. Increase in global food demands. The world population increase and the improvement in living standards has led to increased food consumption. Increase in total yield is achieved by both increasing the cultivated area and increasing the yield per unit of area, which is feasible with high-density systems.
2. Lack of labour. Technological, financial and social progress have, as a consequence, forced many rural manual workers to move to the cities for better conditions. The consequence of this is that the number of people working in olive groves is now significantly reduced.
3. Lifespan of the olive tree. Social and economical conditions change very rapidly, as do the preferences of consumers. For this reason the lifespan of olive trees today does not have the significance it previously had.
4. The time required for the olive trees to bear fruit. It is now more important to reduce the required time necessary for olive trees to start bearing than to achieve long-lived trees. In the traditional olive orchards full fruit bearing was achieved within 7–8 years of planting. This long period does not encourage olive planting in fertile soils where other crops might be more productive – densely planted orchards start producing within 3–4 years. In traditional olive orchards, parents plant olive trees for the benefit of their children, while in the modern, high-density systems orchards are planted for the benefit of the present generation and less so for the following.
5. Density of olive planting. The area of the orchard covered by the tree canopy depends on the tree age, shape of tree and the tree number/unit area. In traditional olive orchards many years are required for achievement of the maximum coverage of the area by the treetops. In high-density planting soil coverage is rapid, leading to optimum use of the soil.
6. Volume of the treetops. During orchard establishment the tree arrangement and the distance of planting are based on the maximum dimensions that the canopy allows. A smaller treetop volume secures denser planting. The productivity per unit area of the soil covered by the tops is greater in densely planted orchards. Hence, the correlation between the yield per unit soil area covered and the dimension of the tops is negative.

The establishment of dense plantings does not differ from the planting of classical olive orchards concerning the techniques used and soil preparation. In the case of dense planting intercropping is avoided. However, in the case of intercropping the following parameters should be taken into account:

- The differential requirements of olive and other crops concerning soil preparation.

- Fertilizers applied (type, quantity, time of application).
- Compatibility of insecticides and fungicides for olives and secondary crops.

APPROPRIATE VARIETIES

When selecting cultivars the best ones for a high-density system are the low-vigour varieties with compact growth habit, early bearing and fructification in clusters. Such cultivars are the clone 'Arbequina IRTA-1–18', 'Arbosana-6–43', 'Ioanenca', 'Canetera', 'FS-17', 'Urano', 'Askal' and 'Koroneiki'. Of those varieties the three most appropriate for super-high-density systems are 'Arbequina', 'Arbosana' and 'Koroneiki' (Tous *et al.*, 2004). These varieties grow upright (compact), have a central leader and very few lateral branches, are self-compatible, very precocious and produce olive oils of excellent quality. Due to heavy crop load and the annual bearing habit their size is medium. Among these three varieties 'Arbequina' is the most frequently planted variety for high- and super-high-density plantings in Spain. 'Arbequina' is considered frost hardy and well adapted to different soil types. It withstands temperatures down to −6°C (22°F) with no apparent problems. The fruit has an oil content of 20–22% and is quite aromatic and fruity, with very little pungency or bitterness and a short shelf life of about 1 year. The tree is relatively small, with low vigour and can be shaped and kept small by pruning. The inflorescences are long, with 22–27 flowers. The fruit is small (1.5–1.8 g) and round in shape. At maturation the fruit is uniformly dark black and the stone is large (18–20%).

The second recommended variety is 'Arbosana', which matures 3 weeks later and has 25% less vigour than 'Arbequina', is more sensitive to water stress and presents more vigourous regrowth after pruning. The oil of 'Arbosana' is more bitter to the taste than that of 'Arbequina', with a fruity and pleasant flavour. This cultivar is characterized by high productivity and resistance to low temperatures, leaf drop and olive knot (*Pseudomonas savastanoi*). The clone is propagated asexually and it produces 19–20% olive oil. Its leaves are small in size, 58–70 mm in length and 11–12 mm in width. The colour of the upper surface is green and that of the lower surface chrysolite green. The petiole is short, 5 mm. The fruit is early in maturation; it matures by the fourth week of October and is elliptical in shape. Composition of the oil is: palmitic acid (C 16:0), 1.3%; esteric acid (C 18:0), 2.0%; oleic acid (C 18:1), 74%; linoleic acid (C 18:2), 7.66%; polyphenols K225 (bitterness), 0.24%. The stability of this oil is 13.5 h at 120°C.

The third variety is the Greek cultivar 'Koroneiki', with a high percentage of fruit set and very small fruit. The oil from 'Koroneiki' is green in colour, very fruity, medium-bitter and pungent. Its shelf life is > 2 years.

With regard to the other varieties, 'Canetera' is more sensitive to *Verticillium dahliae* than is 'Koroneiki'. 'Arbosana' is sensitive to peacock spot, and 'FS-17' to *Colletotrichum gloeosporioides*.

SUPER-HIGH-DENSITY ORCHARDS

In the early 1990s, a new densely planted olive orchard appeared in Spain, which was later introduced to other countries, such as Portugal, Morocco, Tunisia, France, Chile, Argentina, California and Greece. Today, globally, there are about 30,000 ha planted to this system, with more than 60% being in Spain. This system allows mechanization of harvesting and significant cost reduction. This is accomplished by using continuous harvesting machines, thus reducing the harvesting cost by up to 50% in comparison with the system using trunk shakers. However, the drawbacks of this system are the high planting cost (many plants), the need for larger-scale orchards, the need for cultivars with medium to low vigour, the many problems of pests and diseases like *Cycloconium*, the need for irrigation and the capital required for the harvesting machines. To avoid problems with *Cycloconium*, drier soils are suggested, with wider tree spacings and application of copper fungicide sprays.

Tree spacing is about 1.3–2.0 m and about 3.0–4.3 m between rows. Modern plantations use greater tree spacing, i.e. 1.5–2.0 m × 4.0–4.5m (190–220 trees/1000 m^2) and rows are oriented east–west. After 3–4 years a hedgerow is formed and the system requires a central leader (monocone), with one stake per tree. The super-high-density system requires a trellis system or at least a stake for each tree (Iannotta and Perri, 2006; Vossen, 2006); most orchards use a single wire for support. Young trees should be tied to the stake about every 30–40 cm and remain unpruned the first 2 years. At the end of the second year, the lower (1 m) branches and any suckers should be removed to avoid interference with the central leader. After the first harvest (third year) the large lateral branches (> 2 cm diameter) are cut out; this allows adequate light interception and penetration into the treetops. Also, in the summer of the fourth year the trees should be topped to a height of 3.0–3.3 m; this reduces the shading of the lower branches and adjacent rows. From the fourth and subsequent years the trees are pruned by hand to maintain the lower part (0.6–1.0 m) clean (no branches), in order to achieve good harvester frame closure around the trunk. In order to reduce vigour the application of nitrogen is reduced after the fourth year. At full productivity (7–8 years old) the recommended amount of N applied is 5 kg/1000 m^2. With regard to P and K, standard rates are applied. The amount of water applied by drip irrigation is about 200 m^3/1000 m^2. Controlled-deficit irrigation is utilized after the fourth year to reduce vigour and improve olive oil quality.

YIELD PER UNIT AREA

The average production is 1.9 t/ha in the third year, 2.5 t/ha in the fourth year, 2.7 t/ha in the fifth year and 3.3 t/ha in the sixth year after planting. In orchards 5–8 years old the yield ranged from 1.2 to 7–8 t/ha depending on light penetration. The likely yield overall is around 4 t/ha, which is considered adequate if we take into account that the orchard does not have alternate bearing and the cost of harvesting is very low.

MECHANICAL HARVESTING BY OVER-THE-ROW HARVESTER

For mechanical harvesting, the over-the-row modified grape harvester is now used (Tombesi, 2006). This machine requires two people, one operator driving the harvester and another transferring the fruit in the field to the mill for processing. The capacity of a straddle harvester is 2 ha/h. The cost of mechanical harvesting comprises only 20% of the total production cost. Various types of harvesters are used, the most common being the Gregoire brand, which has beating bars that strike the tree from both sides, and collects the olives onto a moving conveyor belt that transports them to bins emptied periodically. Damage to olive trees by this machine is minimal. Other harvesting machines include the Korvan, Agright, Braud, New Holland and Pellenc.

14

PROPAGATION OF OLIVE TREES

INTRODUCTION

The olive tree, until the second half of the 19th century, had been propagated only asexually by using large cuttings, ovules or rooted suckers. Later, towards the end of that century, indirect methods of multiplication – such as grafting-on seedlings – were preferred in order to face the increasing demand for specimens for new plantings. If we consider the published papers over recent decades, we find that the most studied technique by far has been propagation by cuttings. Furthermore, interest in grafting and propagation by seed continues. Therefore, it is worthwhile studying all aspects of olive propagation, from traditional systems to *in vitro* propagation and other more advanced biotechnologies (Fabbri, 2006).

PROPAGATION BY SEED

Olives can be propagated very easily. The morphology of olive stones of nine olive cultivars is given in Fig. 14.1. There are a number of ways of propagating olive plants (Hartmann *et al.*, 2001; Fabbri *et al.*, 2004). Olive plants may be grown from seed, but a true cultivar will not be obtained from seed propagation. Olive seedlings are sometimes used as rootstocks to which some known cultivars are grafted. Seeds are cracked or treated with sulphuric acid as an aid to germination because the pits are very hard. Seed propagation aims at producing seedlings necessary for rootstocks. It is very simple and requires neither specific equipment nor experience. However, propagation by seed produces seedlings and grafted plants variable in vigour and size. Seedlings start fruiting after 12–15 years and they have a deep root system, while olive trees from asexual propagation have a shallow root system and start fruiting after only 3–5 years.

126   Chapter 14

'Gordal'   'Moresca'   'Grossa di Spagna'

'Moraiolo'   'Negral'   'Nocellara del Belice'

'Ghiacciolo'   'Giarraffa'   'Grignan'

Fig. 14.1. Olive stones from nine cultivars after separation from the fruit pulp and cleaning.

Seed collection, seed storage and breaking dormancy

For seed collection, the best sources are trees of certain cultivars, selected for their good characteristics and regular and high productivity. Selected trees should originate from a monocultivar orchard, in order to avoid cross-pollination that might lead to variability in seedling quality. These trees should be marked as being suitable for seed collection for several years. Cultivars used for seed collection include 'Arbequina' in Spain, 'Frantoio' in Australia, 'Oblonga' and 'Allegra' in the USA, 'Canino', 'Mignolo', 'Frangivento' and

'Maurino' in Italy and *Olea sylvestris*, 'Chondrolia Chalkidikis' and 'Koroneiki' in Greece. Other olive species such as *O. ferruginea*, *O. verrucosa* and *O. chrysophylla* have been used in the past as rootstock. Furthermore, some other species have been tested as rootstock, such as those of *Fraxinus* and *Syringa vulgaris*. The stones for seed propagation derive from fruits harvested when they are fully ripe, i.e. the period from October to January (northern hemisphere) (Adakalic *et al.*, 2004). Harvesting time affects the germination of olive seeds (Rinaldi, 2000). Full ripening results in accumulation of storage materials necessary for good germination. After fruit collection the separation of flesh proceeds, which can be achieved by two main methods: (i) hand separation; and (ii) water soaking of olive fruits for 10 days in order to render the fruit pulp soft, and afterwards flesh removal by equipment that separates the kernel from the flesh.

Following de-stoning, the kernels are cleaned to remove the remaining parts of the flesh and dried in the air or in a forced-draught oven at low temperature. Subsequently, the kernels are stored either under ventilation, refrigeration at 10°C or stratified with a sand:perlite (1:1) medium. Seeds maintain their germinability for 1 year or, in some cases, for up to 3 years, but with a small decrease in their germination percentage.

The seeds of olives require up to 4 years for complete removal of dormancy (Lagarda and Martin, 1983; Voyatzis, 1995). However, during seed storage a slow decrease in dormancy is observed. Ethylene *in vitro* promotes olive seed germination (Lambardi *et al.*, 1994).

The dormancy of olive kernels can be classified as below:

1. Mechanical or seed coat dormancy. Seed coats inhibit penetration of water and air and do not permit embryo growth and expansion during germination (Instanbouli and Neville, 1979).
2. Chemical dormancy (or embryo dormancy). This may be ascribed to substances accumulating in the embryo (coverings, seed coats, endosperm) which inhibit germination (Mitrakos and Diamantoglou, 1984). Removal of the seed coat overcomes dormancy.

In order to remove the problem of dormancy the following techniques may be used:

1. Mechanical scarification with a scarifier reduces the thickness of the stone and allows the entrance of water, and hence triggers seed germination.
2. Chemical treatment of kernels with concentrated sulphuric acid (H_2SO_4) for 1 h, which decreases the tolerance of the endocarp and aids germination. Kernels treated with H_2SO_4 germinate within 5–6 weeks. After this treatment, a period of cold stratification is unnecessary for the germination of olive s eeds. If kernels have been treated with sulphuric acid, the tip of the stone is broken with forceps until the seed appears. Consequently, the kernels are stratified in February and sowing follows at distances of 50–60 cm between

rows and 50–60 cm within rows. The seedlings are grafted into these sites after 2 years. Also, for chemical treatment, 5% w/v of sodium hydroxide (NaOH) could be used. Stones are immersed in 5% NaOH for 24 h, under continuous stirring.
3. Water soaking of kernels for a period of 20 days. The water should be replaced every 3 days.
4. Seed stratification in a sand:perlite (1:1) mixture in a cold room or outdoors during winter.

Seed sowing and seedling transplanting

Seed sowing may be performed in several ways:

1. Use of *cold frames* or plastic containers.
2. *Sowing outdoors* in a well-drained soil with adequate moisture and aeration. It is not necessary to provide darkness for good germination of olive kernels; however, shading during summer is beneficial. Seed sowing is conducted in summer (kernels from the previous year) and germination begins after 2 months.
3. Use of *seedbeds* with a bottom-heat system, in order to increase the germination percentage to 50–70%.
4. Following seed coat removal the seeds are *germinated in vitro* in Petri dishes or culture tubes. Seed disinfestation, with 2% sodium hypochlorite (NaOCl) solution for 20 min and several washes with sterile distilled water, then follows.
5. Use of *growth regulators*. To overcome chemical dormancy seeds may be treated with ethephon (200 mg/l), producing ethylene.

Whichever of the above methods is used, seedlings are transplanted to the soil after a light trimming. Seedlings of uniform quality are selected for transplanting in the nursery or in plastic bags. When the seedling base reaches the thickness of a pencil it is ready for budding.

PROPAGATION BY CUTTINGS

When propagating by cuttings the formation of adventitious roots is very important; these are roots arising from any plant part other than the root and its branches. Adventitious roots are distinguished by two types: *preformed roots* and *wound roots*. Preformed roots develop on a stem attached to the mother plant, while wound roots appear after cuttings are prepared. The following changes take place: (i) initially, the wound surface is covered with suberin, which protects the wound from desiccation; (ii) subsequently, some living cells divide and produce *callus*, while cells close to the vascular cambium and phloem start to produce adventitious roots (Ozkaya *et al.*, 1997). Callus is an irregular mass of parenchymal cells, which are produced from young cells at the base of the cutting. The first roots frequently appear through the callus. Callus formation is a prerequisite of adventitious root formation. However, in many species callus formation is independent of rooting. In the case of the olive, where layers of secondary xylem and phloem are present, adventitious roots arise from the living parenchymal cells and also from other tissues such as cambium, phloem and vascular rays. In some cultivars that are difficult to root, the presence of a sclerenchymatous ring between the phloem and cortex, outside the area of origin of adventitious roots, produces an anatomical barrier to rooting.

In rooting of stem cuttings it is important to place the cuttings in the rooting medium in the correct *polarity*. Stem cuttings form shoots at the *distal end* (close to the shoot tip) and roots at the *proximal end* (close to the crown of the plant). When we invert the position of cuttings (upside down) this does not alter this tendency.

Factors affecting olive propagation by cuttings

Significant differences exist between olive cultivars, in terms of propagation, as listed below:

1. Selection of material for cuttings.

- Physiological age (juvenility, maturation).
- Time of year during which the cutting is prepared.
- Type of wood.
- Physiological condition of the mother plant.

2. Environmental conditions during the rooting period.

- Water stress.
- Light (intensity, quality and duration).
- Rooting substrate.
- Temperature.

3. Treatment of cuttings with various chemicals (growth regulators, fertilizers, pesticides).

Juvenility

In the olive the use of juvenile leafy cuttings taken from the crown of the trunk root much more readily than those taken from the top of the tree (Porlingis and Therios, 1976). Experiments with leafy cuttings (cv. 'Chondrolia Chalkidikis') indicate that the ability of cuttings to root decreases with increasing height above the crown of the tree, i.e. by increasing the transformation from juvenile to mature characteristics. Juvenile olive shoots are characterized by their short leaf length and greater width, dark green leaves, change in phyllotaxy (two to three leaves per node instead of two). Some procedures, such as severe pruning, provide material with significant rooting potential, since this prevents the transformation from the juvenile to the adult form. The greater ability of juvenile olive cuttings to root may be ascribed to increased phenol concentration in comparison with that in mature plant material. Another possible reason is the greater concentration of rooting inhibitors that accompanies ageing.

For better results it is necessary to maintain the mother plantings. These plantings are maintained in the juvenile phase by spraying with growth regulators such as GA_3, cytokinin, TIBA or Alar. In olives, juvenile shoots are obtained from *sphaeroblasts*, which are overgrowths on those stems with meristematic tissues. Another method is to use mount-layering (stool beds) to root those cultivars considered difficult to root.

Choice of plant material for cuttings

The physical condition and any previous treatment of the plant material may exert a significant effect on the rooting of cuttings. Such factors include mineral nutrition, carbohydrate concentration and water stress. With regard to mineral nutrition, an average concentration of N in the leaves allows maximum rooting. High N results in excessive vigour and reduced rooting. Such tissues are succulent and their carbohydrate accumulation is low. The average N level determines the carbohydrate:nitrogen ratio; low N increases this ratio and results in better rooting. The level of carbohydrates is important for rooting. In order to obtain a high carbohydrate level the leaves of olive trees should be healthy and well illuminated. Factors such as infection from *Cycloconium oleaginum* resulting in leaf abscission, or very dense planting, lead to shading and decrease the level of carbohydrates. However, adequate levels of Zn increase the production of tryptophan necessary for IAA production and therefore may improve rooting. On the contrary a high Mn level activates IAA oxidase, leading to the destruction of IAA – important for rooting – and decreases rooting.

Water content of the cutting material should be adequate. Water-stressed cuttings encourage ABA production, leaf abscission and decreased rooting. Regarding the type of cutting and its siting depends on the season: in early summer basal (less succulent) cuttings are preferred while later (September), those from the middle and upper parts of juvenile shoots are preferred. When mature shoots are used for cutting production variations in rooting are recorded, with the best rooting in cuttings taken from the basal part of the shoot. When hardwood cuttings are used, cuttings with a small piece of older wood at the base (heel) give better results.

The selection of season when preparing leafy cuttings is very important for optimum rooting. Juvenile cuttings give optimum rooting irrespective of the time of collection, while mature cuttings rooting extremely well under mist can be obtained during late spring and summer. During winter, rooting is significantly poorer.

Treatment of cuttings and stock plants

Adventitious rooting may be induced by the presence of various chemicals, such as auxins (IAA, IBA, NAA, 2,4-D) and ethylene gas. The acid form of these compounds is relatively insoluble in water and therefore they previously should be dissolved in a few drops of ethyl alcohol before addition of water. The salts of the above-mentioned auxins are preferable, due to their greater solubility in water. For commercial-scale propagation it is more preferable to use IBA or NAA. The auxin IAA is more sensitive to light and is destroyed easily, while IBA and NAA are almost totally light stable. These latter compounds are also more tolerant to bacterial destruction and retain viability for a long period of time. Use of high auxin concentrations in olive cuttings, especially those from mature material, inhibits bud growth and no sprouting occurs under mist propagation. This does not apply to juvenile material, where the percentage of sprouting under mist is always very high.

Treatment of cuttings with fungicides results in better survival rates and improved rooting. Suitable fungicides include Captan and certain systemic fungicides such as Neotopsin, etc.

Wounding of the base of cuttings improves rooting, especially for those derived from mature stock material. *Wounding* promotes *cell division* due to auxin and carbohydrate accumulation in the wounded area, and exogenously applied auxin is absorbed more readily than that applied endogenously.

Environmental conditions

High air temperatures during rooting increase water loss from the leaves. The ideal air temperatures for rooting are 25–27°C during daytime and 15°C at night. Light exerts a strong influence on rooting (duration, intensity, quality). Many reports indicate that a relatively low light intensity given to both stock plants and cuttings increases the percentage of rooting. This is in accordance with the *etiolation* of stock plants or the base of shoots (Mencuccini, 2003), which has a remarkable action in adventitious root formation, as occurs in the method of propagation by *mound layering*. Furthermore, the *photoperiod* applied to mother plants has a significant influence on rooting. A long photoperiod increases carbohydrate accumulation and therefore rooting. Concerning light quality, the light colour affects rooting. However, the active spectrum depends on the species; in some it is orange–red, in others blue.

The *rooting substrate* supports the cuttings and provides moisture and oxygen at the base of the cuttings. The rooting substrate affects rooting percentage and root characteristics due to its porosity, pH and the level of the nutrients it contains. Rooting media with great porosity, such as sand or perlite, give an unbranched root system, while very acid or very alkaline media significantly reduce rooting. Furthermore, the nutrients that the substrate contains influence rooting.

Types of cuttings

In olives propagation by cuttings is divided into the following three categories: *hardwood, semi-hardwood leafy* and *softwood* (Khabou and Trigui, 1999). Of these, the most important in olive propagation is the semi-hardwood leafy, and less so hardwood.

Hardwood
These derive from mature hardwood, a method cheap and easy for propagation. Today, in various parts of Greece these cuttings are also termed *grotharia*. The necessary stock material is collected during pruning and derives from healthy, vigourous and typical specimens of the cultivar trees. The cuttings are prepared at the end of January–February from shoots 3–4 years old and 2.5–5.0 cm in diameter. The shoots are divided into segments, each of 30 cm, and the leaves are removed. The cuttings are planted with their physiological base downwards (true polarity). The base of the cutting is soaked for 24 h in a solution containing 13 ppm IBA. Subsequently, the cuttings are covered with moist material for 30 days at 13–21°C. During this period root initials develop and, after 30 days, the cuttings are planted in the nursery, with only 3–4 cm of their structure above the soil.

Leafy
Various factors affect rooting of leafy cuttings, such as: (i) the cultivar (Fouad *et al.*, 1990); (ii) the carbohydrate level (Rio *et al.*, 1991; Ozkaya and Celik, 1999); (iii) hydrogen peroxide (H_2O_2) (Sebastiani *et al.*, 2002a; Sebastiani and Tognetti, 2004); (iv) endogenous auxins (Gaspar *et al.*, 1994); (v) polyamines (Rugini *et al.*, 1992, 1997); (vi) free amino acids (Sarmiento *et al.*, 1990); (vii) light quality (Morini *et al.*, 1990); (viii) hormones (Fernandes Serrano *et al.*, 2002); and (ix) other factors (Gaspar *et al.*, 1997).

For propagation by leafy cuttings, two systems contribute significantly, the *mist* and the *fog* systems (Fouad *et al.*, 1990). In the case of mist propagation, cuttings are sprayed with water by an automatic system to maintain a film of water on the leaves, which reduces transpiration and lowers air and leaf temperature. The resultant temperature decrease is by between 5.5 and 8.5°C, in comparison with leaves not sprayed under mist. With mist propagation, *transpiration* is low, while light intensity is high. This increases the *photosynthetic rate* and rooting. To avoid problems of high moisture, the mist propagation system is operated only during the hours of daylight. The periods of spraying are controlled by a timer or by an *electronic leaf*, hence spraying is intermittent, water being applied at short and frequent intervals. The leafy cuttings are usually planted during June, July, August and September and the stock material is annual shoots, with a girth of 5–6 mm (see Table 14.2). The apical softwood part of the shoot is removed. The length of leafy cuttings is 14 cm and their leaf number is four to six. The basal 1.5 cm of each cutting is immersed in a 4000 mg/l solution of IBA for 5 s. Cuttings are planted in small containers in a 1:1 mixture of perlite:peat, and roots are initiated under mist propagation. Dipping of the cuttings into a fungicide helps to prevent disease attacks. The time required for rooting and root hardening is 3 months.

Subsequently, rooted cuttings (see Figs 14.2–14.4) are planted into plastic bags or small plastic containers and transferred to a shaded site (see Figs 14.5 and 14.6). Cuttings originate from annual shoots with either mature or juvenile characteristics. In comparison with cuttings exhibiting mature characteristics, juvenile cuttings initiate rooting faster and achieve a greater rooting percentage, more roots per rooted cutting and greater fresh weight of the root system per rooted cutting. The percentage of rooting follows a seasonal variation, with a maximum in summer and minimum during autumn and winter. The rooting percentage of juvenile cuttings is high and constant during the whole year (Porlingis and Therios, 1976). The percentage rooting of cuttings originated from the tree crown is greater in comparison with cuttings from shoots close to the treetop. The rooting media used and their properties are detailed below:

Fig. 14.2. Young adventitious roots a few days after emergence.

Fig. 14.3. Rooting of an olive micro-cutting under mist propagation.

1. Perlite is widely used as a rooting medium for leafy cuttings under mist, due to its high porosity. It is used mostly in combination, at various proportions, with peat moss.
2. Peat moss is combined with perlite in order to increase the water-holding capacity of the mixture. The proportions of perlite:peat are either 1:1, 2:1 or 1:3.
3. Vermiculite is used as a mixture of equal parts of vermiculite and perlite. This combination gives good results, achieving high porosity and adequate moisture-holding capacity and cation exchange capacity.

The properties of perlite and vermiculite are given in Table 14.1.
The advantages of perlite include the following:

- Improves aeration and drainage.
- Is an inorganic material and does not disintegrate with time.
- pH varies between 6.5 and 7.5.
- Reduces the temperature variation of the soil.
- Is free from diseases and weeds.
- Is clean, light and easy to use.
- Improves soil moisture and nutrient levels.

Fig. 14.4. Abundant root growth of leafy cuttings planted in plastic holder.

Table 14.1. The properties of perlite and vermiculite (from Therios, 2005a).

Characteristic	Perlite	Vermiculite
Inorganic material	–	Some samples contain B and F
pH	6.5–7.5	6.0–9.5
Sterilized material	When fresh it is sterilized for re-use	Sterilized when it is fresh
Purity, weight and safety	Light; safe for staff	Light; safe for staff
Scent	None	None
Colour	White	Honey-coloured
Aeration and drainage	Excellent	Excellent when fresh; disintegrates with time
Moisture and nutrient elements	Easily available nutrients	Easily available nutrients

Fig. 14.5. Olives planted in plastic pots and maintained outdoors.

Fig. 14.6. Olives planted in 2 l plastic pots.

Table 14.2. Rooting percentages of various cultivars under mist propagation (from Therios, 2005b).

Variety	Rooting success (%)	Date of planting	IBA concentration (ppm)
Greek varieties			
'Kalamon'	5–20	26 May	4000/5 s
'Chondrolia Chalkidikis'			
mature	60.6	26 May	4000/5 s
juvenile	95	26 May	4000/5 s
'Amphissis'	68.1	26 May	4000/5 s
'Ladolia Patron'	68.1	26 May	4000/5 s
'Kothreiki'	93.7	26 May	4000/5 s
Other varieties			
'Ascolana'	99	15 July	4000/5 s
'Leccino'	93	26 July	5000/5 s
'Coratina'	93	26 July	2500/5 s
'Frantoio'	96	26 July	5000/5 s
'Sevillana'	85	15 May	4000/5 s
'Manzanillo'	80	15 July	4000/5 s
'Pendolino'	67	10 February	100/24 h
'Mission'	66	15 July	4000/5 s

PROPAGATION BY SPHAEROBLASTS

Sphaeroblasts are ovular overgrowths of parenchymatic tissue located in the tree crown or in old wood in the upper part of the tree. They contain *latent buds* and are separated, cleaned from the wood and are divided into sections. They are cut from the tree during the autumn and are subsequently stratified, in a cool and dark room, into moist sand until March, and afterwards planted in the nursery. The planting distances in the nursery are 30–90 cm from each other within rows. The sphaeroblasts are planted with their spherical part upwards and are covered with soil. After 1.5–3 months the sphaeroblasts are sprouting and remain in the nursery for 1 year.

PROPAGATION BY SUCKERS

These develop at the *crown* of the tree and their origin is from sphaeroblasts. Suckers are taken from the tree when they have acquired an adequate root system. In order to speed the rooting process the base of the tree is covered with a thin layer of soil, encouraging root initiation. The rooted suckers are removed in the spring from the mother plant, with some old wood, and planted in the nursery. This method is slow and it cannot be used as a commercial method of olive propagation. The use of suckers is a common means of propagation of trees that have died after a severe frost. Furthermore, olive cultivars (e.g. 'Kalamon') that are hard to root could be propagated by soft layer or mound layer (Petridou and Voyiatzis, 1993, 2002).

PROPAGATION BY GRAFTING AND BUDDING

Grafting is the term used to describe the placing in contact of two portions or organs from two different trees, in order for them to heal and grow as a composite plant. The upper part is the scion and the lower one the rootstock. Various types of grafting are used in olives and are known under various names: chip and tongue grafting, side veneer grafting, saddle grafting, wedge grafting, crown grafting and bark grafting. Also, in olives certain types of budding are used, such as patch budding, ring budding and chip budding.

Many times we face the question of what is preferable – a grafted plant or a self-rooted olive plant? Substantial differences exist concerning the morphology and development of the root system. Olive plants originated from rooted cuttings develop a root system that is superficial but with a wide radius. With regard to seedlings, these develop their roots from an axis of 20 cm, have a narrower angle and grow to greater length in the soil, which is very important for sandy soils and dry areas. However, no significant differences were reported concerning productivity between grafted and self-rooted plants.

Grafting is recommended in the following cases:

1. For propagation of olive cultivars that cannot be propagated by cuttings. One example is the Greek table olive 'Kalamon'.
2. In using problematic soils (high calcium carbonate ($CaCO_3$), high moisture, salt stress and diseased) by using the appropriate rootstock.
3. To heal trunks damaged by rodents, diseases or agricultural machinery (bridge grafting).
4. To change cultivar, especially in the case of a lack of pollinizers or when the pollinizers are unsuitable.
5. To produce composite plants for plant physiology studies.

The graft union

Very few studies have been conducted concerning the anatomy of the graft union of olives. When a scion is grafted to a rootstock the two parts should be compatible. In fact, incompatibility is not common in *O. europaea* cultivars and between olive cultivars and wild olive (*O. sativa* or *oleaster*). However, incompatibility may occur when the olive is grafted on to other genera of Oleaceae, such as *Ligustrum, Forsythia, Syringa*, etc., and plants will die after a short period.

The graft union is completed by cells that develop after grafting. These cells in contact differentiate into a vascular cambium producing xylem and phloem elements and a complete vascular continuity by day 60.

A number of studies have been made on the healing of graft unions. The usual sequence of events in the healing of a graft union is as follows:

1. Freshly cut scion tissue capable of meristematic activity is brought into contact with cut stock in such a way that the cambial regions of both are in close proximity. The temperature and relative humidity should be such as to promote growth activity.
2. The outer layers of the cambial zone of both scion and rootstock produce parenchymal cells called callus tissue.
3. Some cells of the callus differentiate into new cambium cells.
4. The new cambium cells produce xylem towards the inside and phloem towards the outside, and establish a vascular connection between the scion and stock, a prerequisite of efficient graft union. It should be stated that, in the graft union, there is no intermingling of cell contents.

Temperature, moisture and other conditions during grafting

Temperature has a significant influence on callus tissue production. Outdoor grafting operations performed late in the spring, when high temperatures may occur, result in failure.

Moisture also has a pronounced influence, since the callus tissue cells are thin and not resistant to desiccation. This is common on healing olive grafts. Air moisture levels below the saturation point inhibit callus formation.

Furthermore, oxygen is necessary at the graft union for the production of callus tissue, since the rapid cell division is accompanied by high respiration levels, requiring O_2.

Growth activity of the scion and rootstock

Grafting depends upon the bark 'slipping'. This means that the cambium cells are actively dividing, producing thin-walled cells on each side of the cambium. These newly formed cells separate easily, i.e. the bark 'slips'. Initiation of cambium activity in the spring results from the onset of bud activity after the bud starts growing.

Polarity in grafting

It is essential that there is proper polarity in the graft union. In grafting, when two stem pieces are put together the morphologically proximal end of the scion should be inserted into the morphologically distal end of the rootstock. However, in grafting a stem piece (scion) on a piece of root, as in the case of root grafting, the proximal end of the scion should be inserted into the proximal end of the root piece.

Selection of scion wood and grafting time

One-year-old vigourous shoots, cut from mature and healthy plants, are the source of olive scion wood. The stock plants should be productive and free of diseases and viruses. Furthermore, the stock plants are marked and scions are cut from the same trees in a succession of years. These trees should be controlled and maintained under good sanitary conditions. The scion wood should be preserved for some days before it is used for scion preparations. The storage of scion wood for long periods should be conducted at 3–4°C in a refrigerator. Preservation for short periods is achieved in a room with a lower temperature than outdoors. In order to avoid dehydration, the scion stems are placed with their bases in a bucket containing 2–5 cm water.

The best time for grafting depends on the cambial activity, i.e. easy separation of the bark from the wood. This time coincides with the period April–May. Grafting is unsuccessful in hot and dry weather. For a successful grafting the following conditions are necessary:

1. Compatibility of rootstock and scion. This depends on the congeniality of the two parts.
2. Intimate contact of the cambial regions of stock and scion. The two surfaces should be held together tightly by wrapping. Rapid healing of the graft union is necessary.
3. Grafting or budding should be performed when growth activity is observed in both stock and scion.
4. Protection of the grafting area from drying out.

Methods of grafting

Cleft grafting
This method is adapted to topworking trees either in the trunk of a small tree or in the scaffold branches of a larger tree (Troncoso *et al.*, 1999). In topworking of olive trees, this should be limited to branches about 2.5–10.0 cm in diameter. Cleft grafting is successful if it is carried out in early spring. The scions are made

from 1-year-old wood, which should be used immediately or should be collected in advance and maintained under refrigeration.

The cutting of the branch for grafting should be made at 90° to the central axis. A large knife is used to make a vertical incision for a distance of 5–8 cm down the centre of the branch to be grafted. The knife is inserted with the help of a hammer. This incision should be in a tangential rather than in a radial direction in relation to the centre of the tree. In each split two scions are inserted, one on each side, in close contact with the cambium of the stock. The length of each scion is 8–10 cm, with a diameter of 10–12 mm and containing two pairs of leaves. The basal part of each scion should be cut into a sloping wedge, about 5 cm long. The scion is inserted into the split and, after removal of the grafting tool, the pressure of the split stock holds the scion in position. The bark of the stock is thicker than that of scion; therefore, the outer surface of the scion is placed in an appropriate position to match the cambium layers of the stock. Thorough waxing of the graft is necessary, and the top surface of the stock should be entirely covered. The tops of scions should also be waxed.

Bark grafting

This grafting technique is very simple and rapid, and is performed on stock from 2.5 up to 30 cm or more in diameter. For olive trees, freshly cut scion wood can be used, and several scions are inserted into each stub. In each scion a vertical incision of 2.5–5.0 cm is made. The scions are inserted into the stub and are held in place by wrapping string or adhesive tape. The wrapping material used after a successful budding must be removed in order to avoid constriction of the grafted area. All the cut surfaces – and also the end of the scions – should be waxed. A team of three people is required, one preparing the plants, one for tying and one for waxing. The output of such a team is in the order of 1500–2000 grafts per day.

With this method, bark grafting of wild olives is carried out in early spring and the wild olives are converted to productive trees of selected cultivars. Wild olives do not have a good taproot system and therefore do not possess tolerance to water stress.

Whip grafting

This method is extensively used to graft small seedlings of *O. sativa* and is also used in making root grafts. With this method, cambial activity is very efficient and the union is very strong. It is recommended that both scion and stock are equal in diameter. A long, sloping incision (5–6 cm long) is made at the top of the rootstock. An identical cut is made at the base of the scion. Another cut is made one-third of the length from the top to the base of the first cut in the rootstock and scion. The two parts (scion and rootstock) are inserted together in such a way that the cuts (tongues) interlock. Afterwards, the graft union is tied and covered with wax.

Other methods of grafting in olives, but of less significance, include the following:

Splice grafting
This is the same as whip grafting, with the exception that the cut is not made in the stock and scion.

Wedge grafting
This can be performed in late winter or early spring. With a sharp knife a V-wedge is made in the stock, 5 cm long. With a screwdriver the V-shaped chip is removed and in the space created the scion is inserted. The scion is 10–13 cm long and 10–12 mm thick, with two pairs of buds and four leaves. After the scions are put in place in such a way that the cambiums are in contact, the cut surfaces are waxed.

Bridge grafting
This is used when the trunk has been injured by agricultural machinery, winter damage or disease; in the case of bark damage, if the damage is extensive the tree will die. This type of grafting is conducted early in spring, i.e. the period when the bark is slipping. The damaged bark is removed down to the level of healthy tissue. Cuts are made in the bark at both top and bottom of the wound and scions are inserted into each slot. The scions are fixed in place by nails, and the unions at top and bottom are covered with grafting wax.

T-budding (shield budding)
T-budding is the most common method of budding in nurseries. The rootstock has a diameter of 6–25 mm and should be actively growing in order for the bark to separate easily from the wood. In the rootstock, horizontal and vertical cuts of 2.5–3.0 cm are made. The scion consists of a shield of bark containing one bud. The shield of bark is inserted by pushing it downward under the two flaps of bark, and the bud union is tightly tied with wrapping material or plastic tape. Two methods of preparing the shield are used; the first one contains wood and the second one does not.

Patch budding
In this method, a rectangular patch of bark is removed from the rootstock and in its place is placed a patch of the scion bark of the same dimensions (Krueger et al., 2004). This method is technically more difficult than T-budding and is used in trees with thick bark, like olives. Patch budding requires less scion wood, it is fast and can be conducted over a longer period. Patch budding requires that both scion and rootstock have good cambial activity, and therefore bark that slips. The time for patch budding extends from April to September, and this period is longer for wood of smaller diameter. The scion wood consists of vigourous 1–2-year-old shoots with the diameter of a pencil or greater. The leaves from the scion are removed and the scions are used

immediately after collection. In the rootstock, a strip of bark to prevent dying of the rootstock is maintained. For the cutting of patches special knives, double-bladed, have been devised to remove the bark segment from the stock and the bud stick. Using the double-bladed knife, one makes two horizontal cuts in the rootstock, then a vertical one at one end of the horizontal cuts and the bud is slipped into place. Afterwards, the inserted bud should be wrapped. In the case of patch budding of thick scaffold branches, the stock bark will be thicker than the bark of the inserted bud patch. This makes wrapping difficult when aiming to hold the bud tightly. Under such circumstances, it is necessary that the bark of the stock be trimmed around the bud patch in order that both components be the same thickness; thereby, the wrapping material will hold the bud patch tightly. For wrapping, adhesive tape may be used to prevent the entrance of air under the patch, which could lead to the death of the patch. The bud, however, is not covered during wrapping.

Another method is to use bark from the rootstock to hold the bud in place. The required time for healing is 1 month at a minimum, and it is often best to wait longer before forcing the buds. If budding is successful, the rootstock limb is cut above the inserted buds in order to force the buds to grow. Yet another way is to girdle the trunk above the inserted bud. The buds may remain dormant during the year of budding and are forced the following spring. On the same tree the scaffold branches are not all cut at the same time, but one to two scaffold branches are left on the southerly side of the tree as nurse shoots. These nurse scaffolds are removed after 2 years, when the buds provide shoots of adequate length.

IN VITRO CULTURE OF OLIVE TREES

This technique involves propagation under controlled conditions (temperature, light quality, intensity and photoperiod) (Amiri, 2004) of olive cultivars, to produce genetically homogenous, pathogen-free plants (Rugini, 1984; Cañas and Benbadis, 1988; Leifert *et al.*, 1992; Mitrakos *et al.*, 1992; Mencuccini and Rugini, 1993; Leva *et al.*, 1994, 2002; Ozkaya *et al.*, 1997; Chaari-Rkhis *et al.*, 1999; Briccoli Bati *et al.*, 2000; Garcia-Fèrriz *et al.*, 2000; Pritsa and Voyiatzis, 2002; Zuccherelli and Zuccherelli, 2002; Lambardi and Rugini, 2003; Giorgio *et al.*, 2006).

The main aims of micropropagation are the following:

- Rapid propagation of cultivars difficult to propagate by cuttings.
- Production of disease-free plants.
- Genetic improvement.
- Cryopreservation of variable germplasm (Martinez *et al.*, 1999; Lambardi *et al.*, 2002).
- Organogenesis.

- Somatic embryogenesis (Rugini and Tarini, 1986; Orinos and Mitrakos, 1991; Leva *et al.*, 1995, 2006; Rugini, 1995; Lambardi *et al.*, 1999).
- Micro-grafting.
- Synthetic seed production.

Organogenesis is the process by which undifferentiated cells differentiate and produce various organs such as roots, buds, flowers, etc. Differentiation originates from the cells of a tissue or from the cells of a callus that was formed (*indirect organogenesis*). Factors affecting regeneration include the cultivar, the substrate used and leaf position on the shoot.

Stages of micropropagation

The micropropagation process includes the following stages:

1. Excision and collection of explants from donor plants.
2. Disinfection of explants and culture initiation.
3. Shoot proliferation.
4. Shoot elongation and rooting of explants (Antonopoulou *et al.*, 2006).
5. Acclimatization of rooted explants and transfer to the field.

Disinfection of explants and culture initiation

The explants originate from stock plants of good growth and sanitary condition and with characteristics representing a certain cultivar and having genetic stability. The mother plants are maintained in the orchard or in a greenhouse. It is not suggested that plant material is used directly from olive groves, since this material is highly contaminated and contains a lot of tannins, leading to tissue oxidation when this material is cultivated *in vitro*. To reduce these problems, the mother plants are planted in containers and are maintained in a greenhouse. The stock plants are pruned to achieve rejuvenation, with explants characterized by juvenile characteristics and minimal oxidation. The stock plants are sprayed with systemic fungicides, insecticides and antibiotics (streptomycin sulphate) every week to control their diseases. For better results avoid moistening the leaves by overhead sprinklers. From the treated stock plants are cut vigourous, tender, apical twigs, which are divided into uninodal segments (1–2 cm, or longer, with leaves).

The disinfestation of the olive plant material is achieved by treating with sodium hypochlorite (NaOCl), at various concentrations and exposure times in order to select the most effective combination. Previously, the stock material was rinsed in running tap water, before sterilization. To avoid tissue browning due to phenolic oxidation, the culture medium is supplemented with antioxidants such as ascorbic acid (10–20 mg/l), or the tissue is pretreated in sterile water containing antioxidants, for 30 min. Another technique to avoid oxidation is frequent subculturing of explants very often until the substrate does not darken. After sterilization, the explants are transferred aseptically to test tubes, one per tube.

Shoot proliferation

The new shoots (*proliferation*) arise from axillary and adventitious buds due to exogenous application of cytokinin-like substances (Mencuccini and Rugini, 1993). Various substrates for proliferation may be used, such as MS (Murashige and Skoog medium), WPM (Woody Plant Medium) or OM (Olive Medium) (see Table 14.3).

Table 14.3. Composition of Olive Medium (OM).

Compound	Concentration (mg/l)
Macronutrients	
KNO_3	1100
NH_4NO_3	412
$Ca(NO_3)_2 \cdot 4H_2O$	600
$CaCl_2 \cdot 2H_2O$	440
KCl	500
$MgSO_4 \cdot 7H_2O$	1500
KH_2PO_4	340
Micronutrients	
$FeSO_4 \cdot 7H_2O$	27.8
Na_2EDTA	37.5
$MnSO_4 \cdot 4H_2O$	22.3
H_3BO_3	12.4
$ZnSO_4 \cdot 7H_2O$	14.3
$Na_2MoO_4 \cdot 2H_2O$	0.25
$CuSO_4 \cdot 5H_2O$	0.25
$CoCl_2 \cdot 6H_2O$	0.025
KI	0.83
Vitamins	
Myo-inositol	100
Thiamine-HCl	0.5
Pyridoxine-HCl	0.5
Nicotinic acid	5.0
Biotin	0.05
Folic acid	0.5
Amino acids	
Glycine	2
Glutamine	2194

For satisfactory growth, olives require both NO_3^- and NH_4^+, inorganic nitrogen sources and glutamine. The Ca:N ratio is very important; a ratio of about 1:11 is used in the OM. The OM contains macronutrients, micronutrients, vitamins and amino acids. As a carbon source, sucrose or mannitol is used. Mannitol has been proved more effective for olive shoot proliferation at a concentration of 30 g/l; agar concentration is 0.5–0.8%. The pH of the medium is adjusted to 5.7–5.8 before autoclaving at 121°C for 20 min. The OM also contains growth regulators. The best cytokinin for olive shoot proliferation is zeatin, at concentrations from 0.5 to 4.0 mg/l, depending on the cultivar. Other cytokinins such as 6-(γ,γ-dimethylallylamino)-purine (2iP) or BA (6-benzyladenine) are not very effective for olive shoot proliferation, since very short and thin shoots tend to be produced. Other cytokinin-like substances – such as thidiazuron (TDZ) – may be included in the substrate. For olive shoot proliferation glass jars of 500 cm^3 are used. These jars are closed with glass lids and wrapped in plastic membrane. In each jar, 100 ml of agar medium is added and ten to 20 explants. The jars are maintained in the growth room at 23–25°C, 16 h of photoperiod and 50 $\mu mol/m^2/s$ of PAR supplied by cool white or day light fluorescent lamps. Every 3–4 weeks the explants are transferred to a fresh substrate.

Shoot elongation
The shoots produced by proliferation are very short and not yet ready for the rooting stage. Elongation of short shoots is promoted by transferring the explants to another medium, similar to the previous but containing GA_3 (20–40 mg/l). When the plantlets obtain an appropriate length (four to five nodes) they are ready for the rooting stage. The media used for proliferation and rooting depend on the cultivar, and Table 14.4 gives the appropriate media for each cultivar as reported in the literature.

Rooting of explants
Rooting is achieved on the same shoot proliferation medium by decreasing the macro-element concentration to one-third or one-quarter of the initial. Sucrose was added at 2% concentration, plus 1 mg/l IBA or NAA. Rooting can be improved by maintaining the base of the shoots in darkness or by adding putrescine (1 mM) plus auxin to the rooting substrate. Auxin could also be applied by dipping the basal part of micro-cuttings in an IBA solution (100–200 ppm, for 10 s) and afterwards planting in auxin-free media. Rooting could also be achieved with the *Agrobacterium rhizogenes* strain 'NCP BB 1855' by wounding the shoot base with a blade contaminated by the bacteria.

Acclimatization of plantlets
Plantlets are removed from the flasks under the laminar air flow and their root system is washed to remove the gel, then dipped in a fungicide solution.

Table 14.4. Micropropagation of olive cultivars.

Cultivar	Proliferation medium	Rooting medium	Stage
'Arbequina'	OM (zeatin)	OM (IBA or NAA)	Rooting
'Dolce Agogia'	50% MS (zeatin + IBA + GA$_3$)	50% Knop/Heller (NAA)	Rooting
'Frantoio'	OM (zeatin + 2iP)	BN (NAA or IBA)	Rooting
'Carolea'	50% OM (zeatin)	50% OM (NAA/IBA + Putr)	Rooting
'FS-17'	mOM (zeatin + 2iP)	50% MS (NAA)	Rooting
'Kalamon'	WPM (BAP + IBA + GA$_3$)	WPM (IBA)	Rooting
'Moraiolo'	OM (zeatin or 2iP)	(NAA or IBA)	Rooting
'Maurino'	mMS (zeatin or TIBA)	mMS (IBA-dip method)	Rooting
'Nocellara Etnea'	mOM (zeatin)	mOM (NAA)	Rooting

Putr, the polyamine putrescine; mOM, modified Olive Medium; OM, Olive Medium formulation; WPM, Woody Plant Medium

Afterwards, the plantlets are planted in pots containing an easily draining substrate, consisting of perlite, vermiculite or peat, ideally 1:1:1. To avoid contamination the plantlets, plus the substrate, should be sprayed with the fungicide every week. Another method is to use inorganic substrate alone – such as perlite or vermiculite – and to moisten it with water containing a very low concentration of X-L-60 fertilizer. The pots with the plantlets are transferred to a bench and covered with transparent plastic film for acclimatization. To avoid overheating of this space the temperature should be controlled or this space should be shaded. Gradually, the relative humidity of this space is reduced by opening the plastic for longer periods. After acclimatization, the plantlets are grown for 1 year in the nursery to attain an adequate size for planting in the olive grove.

Somatic embryogenesis

Plant regeneration via organogenesis and somatic embryogenesis in the olive is a useful tool for breeding and somaclonal selection (Rugini and Tarini, 1986; Orinos and Mitrakos, 1991; Leva *et al.*, 1995, 2006; Rugini, 1995; Lambardi *et al.*, 1999). Regeneration via somatic embryogenesis has been reported using cotyledonary tissues (Orinos and Mitrakos, 1991; Leva *et al.*, 1995), immature zygotic embryos (Leva *et al.*, 1995; Garcia *et al.*, 2002) and immature embryo callus (Orinos and Mitrakos, 1991) in different olive cultivars.

Of the various explant sources (leaf blades, leaf petioles, hypocotyls of germinated seeds and roots of germinated seeds) roots gave the highest callus induction. Somatic embryogenesis was induced from root callus on embryogenesis induction medium containing 5.0 μM 2,4-D, 0.5 μM kinetin and 5.0 μM NAA, in darkness. Embryo regeneration was achieved by transferring the callus to a medium containing varying concentrations (0, 5.0,

10.0 and 15.0 μM) of 2iP, BA, thidiazuron (TDZ) and kinetin. Sucrose evoked higher embryogenesis than either fructose or glucose. Sorbitol and mannitol reduced embryogenesis. Somatic embryos were rooted by transferring them to a hormone-free medium.

Immature embryos possess a high embryogenetic capability, as shown in 'Frantoio' and other cultivars (Rugini, 1988). Furthermore, 75-day-old embryos were able to exhibit embryogenesis.

The study of the origin of somatic embryos (SEs) is of great importance both for the possible appearance of somaclonal variations and for their utilization in genetic engineering techniques.

Somatic embryogenesis occurs either indirectly from callus or directly from cells of organized tissues. Morphogenetic masses develop from both mature tissues (petioles) and originated primary embryos. By subculturing primary embryos, highly efficient continuous cycles of direct secondary embryogenesis were obtained. Histological observations showed early evidence of embryonic primordia as densely stained groups of five or more meristematic cells, located mainly in the epidermal layer of the primary embryos. These primordia evolved into globular embryos that grew to the cotyledonary stage, through the heart-shaped and the torpedo-shaped stages.

Other micropropagation techniques for the olive are described below.

Synthetic seed production

Somatic embryos or apical buds can be encapsulated in a substrate (sodium alginate) for protection and survival. These seeds can be used for micropropagation or for preservation of the genotype. Lambardi and Rugini (2003) reported a protocol for encapsulating SEs of the cv. 'Canino'. At least 49% of synthetic seeds remained viable after storage at 4°C. Synthetic seeds can be handled, stored or germinated in a way similar to that for a normal seed-derived seedling.

Germination of synthetic seeds

Unlike a zygotic embryo inside a true seed, the somatic embryo is heterophylic and may lack nutritional reserves. Therefore, the embryo culture medium has to contain carbohydrates which, under non-sterile conditions, will favour development of microorganisms. In order to monitor the conversion of SEs to plants we have to follow two steps: (i) coating of the SE, which represents the first substrate for the culture of the young plantlet; and (ii) sanitary protection of both the embryo and the developing plantlet. These may be achieved by several means:

1. The use of *micro-capsules* that progressively release sucrose into the alginate beds.

2. The use of *self-breaking beds*. Alginate beds obtained after hardening are rinsed thoroughly with tap water and then immersed in a monovalent cation solution followed by another rinsing with running tap water. After sowing in moist condition, such gel beds gradually swell, become brittle and finally split spontaneously, thus helping avoid oxygen deficiency.

3. Pharmaceutical-type capsules appear to be an efficient coating system for the production of synthetic seeds, allowing germination and subsequent development of plantlets. The capsule is filled with 500 µl of germination medium containing gelrite, cotton or vermiculite. Finally, a torpedo-shaped SE is placed on the internal medium without special sterilization. The capsule is closed with a cap. Pharmaceutical-type capsules provide the nutritional requirements of germinating somatic embryos by limiting sugar loss during planting and germination.

Protoplast culture

The protoplasts isolated from olive organs or tissues can regenerate cell walls and divide and form colonies and cells, followed by plant regeneration under proper culture conditions (Adiri, 1975; Mencuccini, 1991; Perri *et al.*, 1994a,b).

The regeneration process from protoplast to plant can be divided into several phases: (i) cell wall regeneration; (ii) initiation of sustained divisions; (iii) cell colony and callus formation; and (iv) plant regeneration from these tissues. Viable protoplasts from seedlings (hypocotyls, cotyledons or cultivar-micropropagated leaves) were obtained by 1.5% Driselase after incubation with 0.6 mM mannitol and 500 mg/l each of calcium chloride ($CaCl_2$)·$2H_2O$ and potassium chloride (KCl). Some divisions were recorded in a BN (a nutrient medium), which was supplemented with NAA, 2,4-D and zeatin riboside (Rugini, 1986). Furthermore, Cañas *et al.* (1987) separated protoplasts from cotyledons by addition of 0.03 mM ornithine.

Growth regulator is the most critical medium component for protoplast culture. In order to induce sustained divisions, an auxin-type growth regulator and a cytokinin are generally required. The type and concentration of the growth regulators vary with variety. In addition, plant hormones have some effect on protoplast culture. In many cases glucose seems to be a better carbon source and osmoticum (a controller of water movement, or osmosis) than other sugars. The most frequently used techniques are: liquid culture, liquid over agar (or agarose) culture and protoplast-embedded method.

Cryopreservation

Cryogenic methods have been applied to olives for long-term conservation of germplasm (Lambardi *et al.*, 2002). A 30% survival (Martinez *et al.*, 1999) was observed for 'Arbequina' shoot tips after their desiccation to 30% moisture content and direct immersion in liquid nitrogen (–196°C), followed by a period at room temperature. The shoot tips that survived remained green and started to grow after 4 weeks. Furthermore, 'Canino' cultures, with SEs at various stages of development, can be cryopreserved by vitrification (Lambardi *et al.*, 2002), with a high percentage of the cryopreserved cultures surviving. These cultures had a good proliferation rate and morphogenetic potential.

IRRIGATION OF THE OLIVE

INTRODUCTION

Water is fundamental for the physiological functions of plant cells and organs, and it also represents a significant proportion of the protoplast. Water represents 85–90% of the weight of all living tissues. When the water content of cells is reduced, physiological activity is reduced and plants begin to wither. Subsequently, and following resumption of water supplies, the plant returns to its initial physiological condition. The water content of the various parts of olive trees is given in Table 15.1.

The water content of olive plants varies, due either to water absorption via the roots or from water loss by transpiration. The water required for the physiological functioning of plants is less than 5% of the total quantity of water absorbed: most of the absorbed water is lost in the form of transpiration. The role of transpiration is in maintaining the thermal balance of plants, since transpiration absorbs heat and cools the plant. The transpiration quotient is the amount of absorbed water (l) to produce 1 kg of dry matter. Factors affecting water consumption are the cultivar, the relative humidity, wind, temperature and sunshine. Soil factors, such as fertilization, affect the transpiration rate. Trees grown in fertile soils require less water. The water needs of olive trees in areas with a precipitation of 450–650 mm/year are completely fulfilled. When precipitation is > 650 mm/year, olives can be replaced by other, more efficient, crops. In contrast, when precipitation is < 450 mm/year, irrigation is necessary.

Table 15.1. The percentages of water, organic matter and ash of olive wood, leaf and fruit.

Component	Wood	Leaf	Fruit
Water	32	54	53
Organic matter	66	43	45
Ash	2	3	2

© CAB International 2009. *Olives* (I. Therios)

In solutions with a high concentration of nutrient elements, the needs of olive plants are covered by the absorption of a smaller quantity of H_2O. Ignoring transpiration, to calculate the water requirements of olive trees, water loss from the soil surface is included. The sum of transpiration plus water loss from the soil surface by evaporation is termed evapotranspiration.

Water availability is fundamental for starch hydrolysis and contributes to cell turgor, which controls the stomatal mechanism. If the leaf is not turgid then the stomatal apertures are closed, and this reduces both CO_2 supply and the rate of photosynthesis. Therefore, water availability, stomatal opening, Pn and final yield are closely linked. Furthermore, water is a significant solvent for photosynthesis, and in combination with CO_2 produces carbohydrates according to the equation:

$$nCO_2 + nH_2O \rightarrow nO_2 + C_n(H_2O)_n$$

Another role of water is as a solvent of nutrient salts and affecting their transport to cells and tissues. The soil solution and nutrient elements are easily absorbed from the plant providing that the osmotic pressure of the solution within the plant counteracts the force of soil nutrients. When soil water is retained by the soil with a force greater than the force with which it is absorbed, the plant withers through lack of water.

WATER QUALITY

Rainfall is traditionally the main water supply in many areas of the world in arid and semi-arid regions. In these areas, irrigation from underground sources contains excess salts, affecting plant growth (Klein *et al.*, 1994). Furthermore, the quality of surface water is sometimes compromised by its mixing with poor-quality drainage water. A serious problem nowadays is the over-pumping of groundwater, which lowers the water table and leads to the entrance into aquifers of sea water. The salt content of irrigation water is usually expressed in terms of conductivity (Ecw); the units are dS/m at 25°C and measurements are conducted by using conductivity meters. Furthermore, the water quality for irrigation and the evaporation from the soil surface lead to salt accumulation in the root zone. The consequence of this is that the Ec of the soil solution is significantly higher than that of the irrigation water. The reduced yields in olives due to salinity can be predicted when the Ec of a soil solution reaches approximately 3–6 dS/m. The suitability of irrigation water depends on the total amount and the type of salts present. Sodium, chloride and boron ions at high concentrations are toxic. The suitability of irrigation water also depends on the concentrations of Na, Ca and Mg (me/l), which are used to determine the specific absorption rate with sodium adsorption ratio (SAR), according to the equation:

$$SAR = (Na)/\sqrt{(Ca + Mg)/2}$$

SAR values greater than 3 lead to breakdown of soil aggregates and dispersion of clay minerals; such soils are known as sodic soils. The guidelines for determining irrigation water quality are given in Table 15.2.

WATER REQUIREMENTS

Most of the global olive grove hectareage is not irrigated. Furthermore, in areas where oil or table olive cultivars are under irrigation the water supply is scarce. However, the new super-high-density plantings require irrigation, and efficient water use is very important. The water needs of olives are a function of

Table 15.2. Characteristics of water quality for irrigation (from Ayers and Westcot, 1985).

		Degree of restriction on use		
Irrigation problem	Measurement	None	Slight to moderate	Severe
Salinity	Ecw (dS/m)	< 0.7	0.7–3.0	> 3.0
	TDs (mg/l)	< 450	450–2000	> 2000
Infiltration	Ecw (dS/m)	< 0.5	< 0.5	< 0.2
	SAR (0–3)	> 0.7	0.7–0.2	< 0.2
	SAR (3–6)	> 1.2	1.2–0.3	< 0.3
	SAR (6–12)	> 1.9	1.9–0.5	< 0.5
	SAR (12–20)	> 2.9	2.9–1.3	< 1.3
	SAR (20–40)	> 5.0	5.0–2.9	< 2.9
Specific ion toxicity				
Sodium (Na)				
Surface irrigation	SAR	< 3	3–9	> 9
Sprinkler irrigation	meq/l	< 3	> 3	–
Chloride (Cl)				
Surface irrigation	meq/l	< 4	4–10	> 10
Sprinkler irrigation	meq/l	< 3	> 3	–
Boron (B)	mg/l	< 0.7	0.7–3.0	> 3.0
Miscellaneous ions				
Nitrate (NO_3^-N)	mg/l	< 5	5–30	> 30
Bicarbonate (HCO_3)	mg/l	< 1.5	1.5–8.5	> 8.5
pH	Normal range 6.5–8.4			

evaporation (E) of water from the orchard soil and the canopy transpiration (T). Hence, the required water is the sum of these two and is named evapotranspiration (ET) (Testi et al., 2006). Factors affecting water use are: solar radiation, temperature, humidity and wind speed. From the total precipitation only 50–60% is stored in the soil, while the rest is subject to evaporation. The intensity and duration of precipitation determines the percentage of water stored in the soil, and the relationship between them is negative.

The rate of water evaporation from the soil surface is high when the soil is irrigated and decreases as the soil dries out. Therefore, the greater the orchard area that is wetted the greater the ET is. The modern trend in irrigation is localized, such as drip irrigation. Planting cover crops increases ET, and their use is avoided in areas with scarce irrigation water due to the high cost.

Transpiration is affected by various factors, such as total leaf area, solar radiation intensity and duration, wind speed, etc. The total leaf area is affected by the size of the tree canopy, the density of planting and the size and health of leaves. The percentage of the orchard soil shaded by the olive trees correlates well with the ET.

For *ET estimation* two methods are used (Testi et al., 2006). The first of these measures water loss from a round metal pan 1.2 m in diameter and 25 cm in depth. The pan is placed in an irrigated area of cut grass. Every day the water loss from the pan is measured, the values being known as *Epan*. These values are measured at meteorological stations. In the second method, the *modified Penman equation* is used, which is based on measurements of solar radiation, temperature, relative humidity, wind speed and direction. The values calculated give an estimation of water use for a close-cut grass referred as ET_o. This method gives a better estimation of water use than Epan. However, these values are only an estimation, and research is necessary in order to determine the actual water requirements of various crops (ET_c). The relationship between ET_o and ET_c is constant, and the ratio $ET_c:ET_o$ is used to develop crop coefficients (K_c) (Michelakis et al., 1994; Orgaz et al., 2006). For olives in California, USA, a K_c of < 0.65 leads to water stress of mature olive trees; a value of ⩾ 0.75 produces the best results and verifies efficient water use (see Table 15.3).

A new crop coefficient, $K_c = Et_c/ET_o$, can be calculated as the sum of three components: tree transpiration (K_p), evaporation from the soil (K_{s1}) and evaporation from the areas wetted by the emitters (K_{s2}) (Orgaz et al., 2006; Testi et al., 2006). With the method of Orgaz et al. (2006) the calculation of the needed irrigation amount (IA) is improved since various factors, such as soil, plant and weather, are taken into account (Fernández, 2006). The K coefficients, as calculated by Orgaz et al. (2006), are presented in Table 15.4.

The ET_o ranges between 1000 and 1400 mm/year for an average rainfall in the Mediterranean region, equal to 500 mm/year, and for mature olive trees under localized irrigation the ET_c is 600–700 m³/1000 m², so 300–400

Table 15.3. Average water requirements for clean-cultivated olive orchards in the Sacramento area, California (from Beede and Goldhamer, 1994).

Month	ET_o (mm/month)	K_c	ET_c (mm/month)
January	30.4	0.75	22.86
February	40.6	0.75	30.48
March	73.7	0.75	53.34
April	116.8	0.75	86.36
May	154.9	0.75	116.84
June	185.4	0.75	139.70
July	216.0	0.75	162.56
August	185.4	0.75	139.70
September	137.1	0.75	121.14
October	91.4	0.75	68.58
November	40.6	0.75	25.40
December	25.4	0.75	29.32
Total	1297.1		

Table 15.4. Water requirements of an olive orchard with mature 'Picual' trees planted at 10 m × 10 m in southern Spain (from Orgaz and Pastor, 2005).

Month	K_p	K_{S1}	K_{S2}	K_c	ET_o (mm/month)	ET_c (mm/month)
January	0.18	0.67	0.00	0.85	33.2	28.2
February	0.19	0.65	0.00	0.84	45.9	38.6
March	0.20	0.40	0.00	0.60	87.1	52.4
April	0.23	0.25	0.04	0.51	110.4	56.8
May	0.27	0.13	0.03	0.43	154.1	66.8
June	0.32	0.05	0.03	0.40	169.5	67.3
July	0.32	0.04	0.03	0.39	210.8	81.3
August	0.31	0.05	0.03	0.38	182.3	70.1
September	0.28	0.18	0.03	0.49	122.1	59.9
October	0.31	0.38	0.04	0.73	80.6	58.8
November	0.28	0.68	0.00	0.96	42.6	40.9
December	0.18	0.72	0.00	0.90	29.6	26.9

K_p, coefficient related to tree transpiration; K_{S1}, coefficient related to soil evaporation; K_{S2}, coefficient related to evaporation from the soil surface wetted by the emitters; K_c, crop coefficient; ET_o, potential evapotranspiration; ET_c, actual water requirement.

m³/1000 m² must be applied by irrigation. These values are not applied for super-high-density (SHD) plantings or for young olive trees. According to Grattan *et al.* (2006), with regard to SHD plantings of 'Arbequina I-18' (1.5 m × 3.9 m), the ET_c was 600 m³/1000 m², while the required irrigation was 75% of ET_c, i.e. 450 m³/1000 m².

THE NEED FOR IRRIGATION

Olive leaves are covered on the lower surface with peltate trichomes which protect them and reduce transpiration. This permits olive cultivation in very dry areas. However, for good productivity the application of water is necessary. Irrigation is essential in the following circumstances:

- The rainfall is inadequate.
- Precipitation distribution is not appropriate. Therefore, water shortage is recorded in the critical periods during spring and summer.
- In light-textured soils, with a low water-retaining capacity.

Irrigation is necessary in both table and olive oil varieties. However, irrigation is more important in table olives to achieve large fruits in dense and high-density plantings. Table olives should be irrigated during the third stage (cell expansion) of fruit growth to increase their size. In contrast, over-irrigation increases shoot growth, produces waterspouts and increases the sensitivity of vegetative growth to winter frosts.

The methods applied for olive orchard irrigation include flood, furrow, sprinkler, surface, sub-surface and drip irrigation (Fereres and Castel, 1981; Bonachela et al., 2001). For drip irrigation one or two driplines are used per row.

THE ROOT SYSTEM OF AN OLIVE TREE

Olive trees do not have a dominant taproot system, and they develop many lateral roots. The root system is shallow and is confined to the top 0.7–1.0 m of soil; the lateral root system exceeds the projection area of the canopy. The root pattern of the olive indicates that this system is not efficient in an avoidance response to drought, since it is shallow and cannot reach water from greater depths during dry periods. Root length increases under water stress and is 2.5 times longer than on irrigated trees. When olive trees are water-stressed for a long period it is important to supply water to the entire root system (Bongi and Palliotti, 1995). Experiments with split root systems and irrigation at two different values of soil water potential indicated that olive plants grew less even when a small portion of the root system was exposed to low Ψ, in comparison with plants where both portions were irrigated with a high soil water potential. This can be ascribed to the production of ABA by the stressed root system and its transport to leaves (Zhang and Davies, 1989).

IRRIGATION AND DAILY VARIATION IN TRUNK GIRTH

During periods of water stress, olive trees experience a reduction in transpiration, stomatal conductance and net photosynthesis (Angelopoulos et *al.*, 1996). However, environmental factors affect H_2O and CO_2 exchange to varying degrees, causing variations in water use efficiency. Water stress limits stomatal conductance, photosynthesis, non-stomatal factors and assimilation (see Figs 15.1–15.4). Sap flow patterns showed a steep diurnal morning increase and a pronounced seasonal reduction throughout summer. The daily alterations in stomatal conductance in olive leaves (cm/s) are presented in Fig. 15.1. Reduction in soil water from 100% of evapotranspiration to 66%, and afterwards to 33%, significantly reduced stomatal conductance. The maximum stomatal opening was recorded at 09.00 h and the minimum at 15.00. With regard to water potential, this was reduced (more negative values) at 12.00 h, but later was increased due to stomatal closing (see Fig. 15.2).

Measurement of Pn in the cultivars 'Koroneiki' and 'Mastoides' indicated that the maximum photosynthetic rate (Pn) was 12–14 $\mu mol/m^2/s$ (see Fig. 15.3). Furthermore, when a decrease in water potential value (Ψ) was observed, the pressure potential was reduced (see Fig. 15.4) in those cultivars.

The changes in Ψ values in eight olive cultivars indicate varying osmoregulatory capacity. Cultivars whose Ψ value decreases rapidly under salt stress conditions – such as 'Chondrolia Chalkidikis', 'Manzanillo' and 'Amygdalolia' – are able to absorb water under adverse environmental conditions and to adjust osmotically fairly rapidly; therefore, these cultivars suffer less. In contrast, cultivars such as 'Megaritiki' and 'Amphissis', with a constant Ψ leaf value as the Ψ of the soil solution is reduced, suffer more under conditions of increased salinity (see Table 15.5). The olive tree is characterized by its ability to adapt to water-poor environments (Connor, 2005).

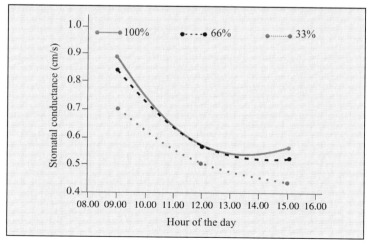

Fig. 15.1. Diurnal changes in stomatal conductance (22 September) in the leaves of the cultivar 'Koroneiki' as a function of the irrigation regime (100, 66 and 33% of evapotranspiration).

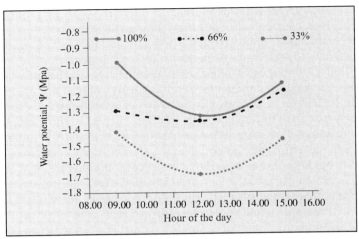

Fig. 15.2. Diurnal changes in water potential in the leaves of the cultivar 'Koroneiki' as a function of the irrigation regime (100, 66 and 33% of evapotranspiration).

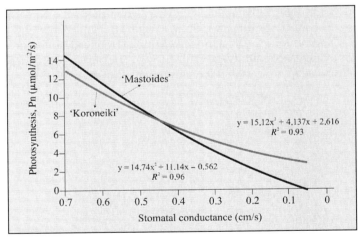

Fig. 15.3. The relationship between photosynthetic rate (Pn) and stomatal conductance in the cultivars 'Mastoides' and 'Koroneiki' (data from an experiment with various levels of water application).

The hydraulic resistance of the whole tree, estimated from sap flow and leaf water potential, revealed changes in the liquid/water path from soil to leaves. Stem and root hydraulic conductivity in the olive can be reduced under drought stress due to cavitation. The maximum value of stomatal conductance was recorded early in the morning and decreased throughout the day, with the lowest values in

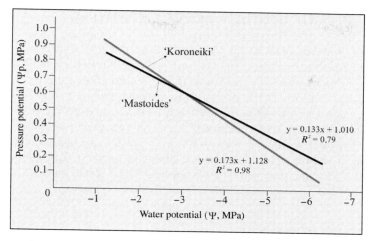

Fig. 15.4. The relationship between pressure potential (Ψp) and water potential (Ψ) in the cultivars, 'Mastoides' and 'Koroneiki' (data from an experiment with various levels of water application).

Table 15.5. The osmoregulatory capacity of eight selected Greek olive cultivars (from Therios, unpublished data).

Cultivar	Osmoregulatory capacity (−MPa)
'Chondrolia Chalkidikis'	0.80
'Manzanillo'	1.30
'Amygdalolio'	0.95
'Frantoio'	0.30
'Adramyttini'	0.50
'Megaritiki'	0.20
'Amphissis'	0.15
'Koroneiki'	0.40

the non-irrigated trees. Positive correlation between stomatal conductance and leaf water potential indicates that olive trees are partially able to restrict water loss by closing stomata. Furthermore, a close correlation between photosynthesis and stomatal conductance was recorded (Moriana and Fereres, 2002).

During irrigation of olive orchards it is important to wet the entire area covered by the root system; patches of non-wetted soil may restrict performance due to ABA signalling. More productive olive plantations could be achieved on soils with increased hydraulic conductivity and through appropriate orchard management (pruning, planting density, etc.).

PRINCIPLES OF IRRIGATION PROGRAMMING

There are various methods for the scheduling of water application in olive orchards (Fereres and Castel, 1981; Fereres et al., 1981; Beede and Goldhamer, 1994). The most common method is to rely on previous experience and to monitor soil moisture levels by hand, or with instruments, and to add the necessary amount of water. The other way is to measure crop water use. Increasing efforts are being invested in the development of new techniques for a more precise irrigation of olive orchards. Among those, leaf or stem water potential, trunk diameter changes, sap flow and infrared thermography are included. Therefore, the tree is used as a sensor of soil water status.

Measuring soil water

Many instruments measure Ψ and soil matric potential. Such measurements are useful in monitoring soil water status and in programming irrigation. These sensors are very cheap and may be automated.

Measuring leaf water potential (Ψ_{leaf} and Ψ_{stem})

Leaf water potential has been used extensively to monitor tree water status and to determine the optimal time for irrigation (Sorrentino et al., 1999). In recent years stem water potential – which is less sensitive to variations in atmospheric conditions – has been used extensively. According to Fernández et al. (1997), −0.46 MPa is a constant average pre-dawn value of Ψ_{leaf}, and a Ψ_{leaf} value of −0.05 MPa can be considered as a threshold for satisfactory water recovery overnight. Based on measurements in central Spain, midday Ψ_{stem} values of −1.2 MPa for an 'off' year and −1.4 MPa for an 'on' year are recommended as thresholds for irrigating mature olive orchards. Measurements are labour intensive, and the threshold values may have to be adjusted, depending on the orchard.

Measuring variations in trunk diameter

This technique can be used to schedule irrigation when deficit irrigation is used (Goldhamer and Fereres, 2001; Michelakis and Barbopoulou, 2002; Moreno et al., 2006). The fundamentals of this technique are the shrinkage of the trunk during stress and expansion during rehydration. This method is suitable for scheduling irrigation.

Measuring sap flow

Transpiration and water uptake by the roots could be estimated from sap flow measurements (Fernández et al., 1999, 2001; Giorio and d'Andria, 2002; Tognetti et al., 2004).

Infrared thermography

With this method we can determine the appropriate time for irrigation, but not the required amount of water. This technique is expensive, requires image analysis with the appropriate software but is suitable for use in hot and dry conditions. Furthermore, we can determine variability within the orchard due to both soil and tree conditions.

THE LOW-VOLUME IRRIGATION TECHNIQUE

In recent years low-volume irrigation (LVI) systems have been introduced into olive orchards, with favourable results. These involve drip irrigation and mini-sprinklers, which offer the possibility of applying the irrigation water directly to the root zone of the olive trees, without wetting the soil between the rows. Furthermore, the water can be applied frequently at low application rates and the system allows fertilizer to be incorporated into the irrigation water. Therefore, the cost of fertilizer application is low and the fertilizer is applied directly to the active root zone of the trees. The frequent and low-dose irrigation creates a relatively shallow and fibrous root system, which is very efficient with regard to water and nutrient absorption. Therefore, the trees exhibit a rapid growth rate and produce high yields. On heavy and shallow soils it is preferable to irrigate with the drip system. However, the irrigation method depends on soil depth, texture and water quality. In light-textured soils or when planting at large spacings, trees should be irrigated with jet-stream sprayers or mini-sprinklers. Spray irrigation is advantageous on sloping areas. In areas with limited water resources during the vegetative period, extra irrigation at the critical stages (flower bud differentiation, etc.) makes a significant contribution to the annual yield (see Table 15.6).

Irrigation of young olive trees by the low-volume irrigation technique

In orchards irrigated by the LVI system water is applied every 3–7 days, depending on the soil type and environmental conditions affecting the

Table 15.6. The critical periods of water stress for olives, and their effects.

Growth stage	Effect of soil drought
Flower bud development	Reduction of flower bud formation
Flowering	Pistil abortion
Fruit set	Reduction in fruit set
	Alternate bearing
Shoot growth	Reduced shoot growth
First stage of fruit growth	Reduced cell division
Cell division	Small fruit size
	Shrivelling of fruit
	Poor shoot growth
Cell enlargement	Small fruit size
(3rd stage of fruit growth)	Reduced cell expansion
	Shrivelling of fruit
	Poor shoot growth

evapotranspiration rate. When drip irrigation is applied, two emitters per tree should be used and, in the case of mini-sprayers, these are positioned upside down to reduce the wetted radius. Nitrogen application through the irrigation system increases the vegetative growth of young trees.

As trees grow, their root system grows too, and a wider area of the olive grove has to be irrigated and the water quantity increased. This can be accomplished by placing the mini-sprayers in an upright position and, in the case of drip irrigation, by adding another drip laterally.

IRRIGATION OF MATURE OLIVE TREES

The quantities of irrigation water required by olives are calculated based on the daily evaporation of the Class A pan. A pan means the evaporation of free water from a pan that has been used for vegetation by the introduction of empirical coefficients. The potential ET from an area with a green crop well supplied with water is assumed to be 0.6–0.8 of the free water evaporation depending on season, area, length of daylight, etc. These values serve as a basis for determining olive grove ET by multiplying the evaporation from the Class A pan by the appropriate crop coefficient. This coefficient depends on the stage of growth and whether the fruit is intended for olive oil or table olives.

THE PARTIAL ROOT ZONE-DRYING IRRIGATION TECHNIQUE

General

The root system of plants in contact with drying soil is known to produce chemical signals that travel via the xylem and effect a restriction in stomatal aperture and leaf expansion rate. Therefore, modern irrigation systems exploit this chemical signalling capability in reducing transpirational water loss and crop water requirement (Goldhamer et al., 1994; Centritto et al., 2005). In the early 1990s the use of a 'split-root system' was established. In this system the roots of experimental plants were divided between two pots; one part of the root system was irrigated in order to deliver sufficient water to the shoots and to sustain photosynthesis and water relations, while the rest of the root system was maintained in dry soil. The principle of the split-root technique is to minimize the negative effects of deficit irrigation, by sustaining shoot water equilibrium. The split-root technique was expanded into a field-based deficit irrigation system called 'partial root zone drying' (PRD); this reflects the alternate drying and re-wetting of the two halves of the root system in order to sustain water relations and yet to generate root signals that limit transpirational water loss. This is accomplished by allowing part of the root system periodically to dry the soil in which it is rooted.

With this method, transpiration is restricted; leaf area is also restricted while no negative effects on yield were recorded. Furthermore, decrease of the vegetative vigour improved the olive quality. Also, water use efficiency (WUE) was improved, meaning a decrease in the required amount of water, which today is scarce and very expensive.

Effects of partial root zone drying on olive oil quality

The oil content and acidity of olive fruit showed no significant differences between partial root zone drying (PRD) treatment and full irrigation. The PRD irrigation strategy, which restricts water irrigation supply, reduces the quantity of the salt incorporated within the root zone and permits the sustainability of the agricultural system.

Very few studies have been conducted on the effect of saline water and PRD strategy on oil quality (Klein et al., 1994). PRD treatment leads to an increase in oleic acid content, in the same way as rainfed conditions.

Deficit irrigation

Due to irrigation water scarcity, growers must apply less water, i.e. follow a deficit irrigation approach that provides less than the olive tree needs, in such a way as to achieve close to maximal production (Goldhamer et al., 1994). Three systems are described:

1. Sustained deficit irrigation (SDI). The grower applies a reduced percentage of ET_c during the whole irrigation period.
2. Low-frequency deficit irrigation (LFDI). The soil is left to become dry, since all the available water has been absorbed. Consequently, the soil is irrigated to field capacity and again is left to dry.
3. Regulated deficit irrigation (RDI). This method supplies up to 100% of ET_c during the stage that the crop is sensitive to water stress and reduced water (by 30–40%) during the remaining growth period. Hence, the required water for irrigation is about 50% of ET_c (Fernández et al., 2006). This method has, as a prerequisite, a thorough knowledge of the physiology of the crop.

THE EFFECT OF IRRIGATION ON THE QUALITY OF TABLE OLIVES AND OLIVE OIL

Irrigation induces greater shoot growth and increased total leaf surface area, photosynthesis and transpiration. Irrigation does not influence fruit shape but increases fruit weight, volume and pulp:pit ratio. The larger fruit size is primarily the result of both a larger number of cells and the positive effect of water availability on cell division rather than cell expansion. Irrigation has no substantial delaying effect on ripening. With irrigation, pulp water content increases and firmness decreases slightly. Irrigation results in less sugar content, which can be unfavourable to fermentation. Therefore, it could be useful to add sugars to the brine to produce good lactic fermentation, which is important for good fruit storage.

Olive oils produced without extra irrigation and based completely on annual precipitation have higher polyphenol contents (Motilva et al., 2002), bitter index and stability. Furthermore, those from non-irrigated orchards have higher levels of oleic and linoleic acids (Castro et al., 2006).

The quantity of the water applied to trees exerts a profound influence on olive oil quality (Patumi et al., 1999; D'Andria et al., 2002; Faci et al., 2002; Motilva et al., 2002; Berenguer et al., 2004, 2006; Grattan et al., 2006; Herenguer et al., 2006). Producers who wish to optimize oil production – which is a function of both yield and percentage oil content – need to take into account the benefits of moderate irrigation, since excessive irrigation drastically reduces oil yield. Moderate water stress gives a better quality of olive oil, i.e. greater polyphenol content, greater stability and better flavour; irrigation increases productivity but has a negative effect on oil quality. The total polyphenol content (K_{270}, K_{232}) and oxidative stability (K_{225}) are dramatically affected by the level of irrigation. The more water applied the more likely it is that the oil will have a significant reduction in both polyphenol content and stability; since polyphenols are water soluble the water content of the fruit can influence the amount of polyphenols remaining in the oil after processing. Increasing the degree of irrigation may affect oil flavour, becoming bland. Minimally to moderately irrigated trees produce oils that are fruitier, with a balanced ratio of bitterness and pungency (see Table 15.7).

If a producer wants to increase the intensity of bitterness and pungency they should limit water application. An irrigation system supplying 40% of ET_c gives good oil extractability and maintains excellent oil chemical parameters.

IRRIGATION BY WASTE WATER

The scarcity of water and the expected decrease of precipitation in coming years will create problems for olive grove culture. In this respect, many countries are devising ways of using waste water for irrigation, since waste water is an additional alternative and valuable source for coping with the scarcity of irrigation water. However, re-use of waste water for irrigation may involve risks for both environmental and human health. Therefore, waste

Table 15.7. Chemical analysis of olive oil (2003) from the cv. 'Picual' under four different irrigation treatments (Irrigation Maximum, I_{max}; Progressive Deficit Irrigation, D_1; Linear Deficit Irrigation, D_2; and Rainfed, R (from Castro et al., 2006).

Chemical	I_{max}	D_1	D_2	R	Significance
α-Tocopherol (mg/kg)	220.6	230.2	225.0	210.0	
β-Tocopherol (mg/kg)	2.4	1.4	2.2	2.4	
γ-Tocopherol (mg/kg)	11.4	10.6	11.2	8.6	
Total tocopherols (mg/kg)	234.4	242.2	238.4	221.0	
Polyphenols (mg/kg)	359.6	447.0	542.6	1183.8	***
Bitter index (K225)	0.32	0.29	0.37	0.57	***
Stability (h)	156.70	149.90	169.98	228.49	***
Carotenoids (mg/kg)	4.28	3.74	4.40	5.24	
Chlorophylls (mg/kg)	5.26	7.96	4.44	5.64	
C16 (%)	13.80	13.11	12.65	11.36	**
C16:1 (%)	1.12	1.16	1.04	1.03	
C17 (%)	0.07	0.05	0.07	0.06	
C17:1 (%)	0.13	0.11	0.11	0.09	*
C18 (%)	3.40	3.25	3.31	3.48	
C18:1 (%)	77.0	77.9	78.7	79.0	**
C18:2 (%)	2.99	2.98	2.74	3.52	**
C18:3 (%)	0.61	0.64	0.59	0.67	**
C20 (%)	0.39	0.43	0.40	0.45	*
C20:1 (%)	0.30	0.27	0.28	0.23	
C20:2 (%)	0.17	0.09	0.13	0.10	*

$^*P \leq 0.05$; $^{**}P \leq 0.01$; $^{***}P \leq 0.001$.

water should be appropriately treated before use, for both health protection and avoidance of crop damage. Only in the USA, Australia and South Africa have any specific regulations been established for re-use of waste water. In several EU countries regulations and codes of practices concerning the use of treated waste water for irrigation are under preparation or revision (Angelakis *et al.*, 2002). Irrigation with waste water is now a common practice in many countries.

The nutrient content of waste water is beneficial for olives. However, high nutrient concentrations may constitute a source of pollution or may be toxic to plants. This may cause underground or surface water pollution, excessive vegetative growth and reduction in quality.

The nutrients in waste water essential for plants are nitrogen, phosphorus, potassium, zinc, boron and sulphur. The N content of municipal waste water after secondary treatment ranges from 20 to 60 mg/l. Therefore, the N in treated waste water may be in excess of crop needs. Phosphorus levels in secondary treated waste water vary from 6 to 15 mg/l, and those of K range from 10 to 30 mg/l; with regard to other nutrients, waste water contains adequate amounts of S, Zn, Cu and B. Treated waste water contains enough B for plant nutrition and, in some cases, B is often found in excess, causing toxicity. The most toxic ions in treated waste water are Na, Cl and B. Olive is tolerant to waste water containing B at a level of up to 1 mg/l. The greatest concern with regard to the re-use of waste water, however, is the presence of pathogens.

16

WATER USE EFFICIENCY BY THE OLIVE

INTRODUCTION

The olive tree is very efficient in its use of water. As an illustration, it requires 312 g water to produce 1 g dry matter, whereas other evergreen plants such as *Citrus* and *Prunus* species require 400 and 500 g, respectively. Water consumption by the olive is 30 and 40% of that of *Citrus* and *Prunus* species, respectively. The mechanisms conferring high water use efficiency (WUE) include the ability of the olive leaves to tolerate very low Ψ values (Connor, 2005) and osmotic adjustment via the synthesis of osmoregulatory solutes (Morgan, 1984). Furthermore, the use of growth retardants increases the drought resistance of olive plants (Frakulli and Voyiatzis, 1999).

WHAT IS WATER USE EFFICIENCY?

The ratio between the amount of water transpired by plants and dry matter accumulation is known by various terms such as transpiration coefficient, transpiration ratio or water requirement of plants, while the reciprocal is known as the efficiency of transpiration. Nowadays the term WUE is used most frequently (Bacon, 2004).

Terrestrial plants have evolved mechanisms that enable control of water loss, without any decrease in photosynthesis. This is very important, since the available water is the determining limiting factor for agricultural productivity. These mechanisms maintain survival of plants and productivity under limited water supply and include the following:

1. Mechanisms of avoidance of water deficits:
- Short growth cycle; the plant completes its life cycle before the appearance of any water deficit.
- Limited leaf area (small leaves), closure of stomata and thick cuticle.

2. Mechanisms of tolerance to water stress:

- Osmoregulation, elastic modulus of elasticity and turgor maintenance.
- Synthesis of organic solutes to maintain turgor (sucrose, mannitol, proline, etc.) and water stress-tolerant proteins and enzymes.

3. Efficient use of irrigation water:

- Genotypes with high WUE.
- Appropriate irrigation systems and methods of achieving efficient water use (see Table 16.1) (Vermeiren and Jobling, 1980; Motilva *et al.*, 2000; Fernández *et al.*, 2003, 2006).

Various definitions of WUE are to be found in the literature (Centrito *et al.*, 2002):

1. WUE is the ratio of the produced biomass to plant transpiration.
2. WUE_t; this is known as the transpiration efficiency. The WUE can also be expressed as the ratio of total biomass, shoot biomass or yield harvested to total ET or plant transpiration (Ep). Furthermore, WUE could be expressed in terms of dry mass, fresh mass or glucose equivalent of those masses.
3. WUE_e is the ratio of net CO_2 assimilation rate to transpiration. The values of WUE, which are based on gas exchange, can be converted to WUE based on dry matter, and the factor 0.61–0.68 kg dry matter/kg CO_2 fixed. The value of WUE at a given site is stable for a range of cultivars or similar species. The WUE for a wide range of C4 plants ranges from $2.41 \cdot 10^{-3}$ to $3.88 \cdot 10^{-3}$. These values are significantly greater than the values for C3 plants ($0.88 \cdot 10^{-3}$–$2.65 \cdot 10^{-3}$). Furthermore, Crassulacean acid metabolism (CAM) plants exhibit up to ten times the values recorded for C4 plants ($20 \cdot 10^{-3}$–$35 \cdot 10^{-3}$).

The equations describing net assimilation rate, transpiration and WUE are presented (Jones, 2004):

$$\text{Net assimilation rate (A)} = \frac{p_a - p_i}{P_a (r'_a + r'_s + r'_m)}$$

Table 16.1. Efficiency of various irrigation systems.

Irrigation system	Efficiency (%)
Drip-microsprinkler	90–95
New sprinkler	75–85
Older sprinkler	65–80
Basin	75–80
Contour flood	60–65
Furrow	40–60

where p_a = atmospheric partial pressure; p_i = internal partial pressure of CO_2 in leaves; P_a = atmospheric pressure; r'_a, r'_s and r'_m = the resistances (boundary layer, stomatal resistance, mesophyll resistance).

The equation for transpiration is $E_e = \dfrac{e_l - e_a}{P_a (r_a + r_s)}$

where e_l, e_a = saturated water vapour pressure of leaves and air, respectively.

After combining these two equations the following ratio is obtained (Jones, 2004):

$$WUE = \frac{A}{E} \frac{(p_a - p_l)}{(e_l - e_a)} \times \frac{(r_a + r_s)}{(r'_a + r'_s + r'_m)}$$

This ratio indicates that WUE is expected to increase as stomata are closing.

The measurement of WUE is based on measurements with the carbon isotope ^{13}C. The principle of this measurement is based on the discrimination by photosynthesis between ^{12}C and ^{13}C. Therefore, less ^{13}C is expected to be found in the dry matter than in the air. This discrimination is usually expressed as ‰ and is described by the symbol Δ.

$$\Delta = \frac{(^{13}C/^{12}C)\ air}{(^{13}C/^{12}C)\ dry\ matter} - 1$$

The Δ value ranges between 13 and 28‰ for C3 plants and between 1 and 7‰ for C4 plants. Another way of expressing Δ is:

$$\Delta = 0.0044 + 0.0256\ p_i/p_a$$

where p_i = internal space partial pressure of CO_2; p_a = atmospheric partial pressure of CO_2.

PLANT NUTRITION AND WATER USE EFFICIENCY

The response to less than optimum nutrient supplies leads to growth rates below maximum and reduces WUE. There is also a parallel between WUE and relative growth rate, with increased WUE at higher growth rates. With regard to sources of N, lowest water use efficiency is for N_2 in symbioses relative to $NO_3 + NH_4$ N sources. Of all N sources ammonium gives lower water use efficiencies than does NO_3. Nutrients can influence the development of stomata, altering stomatal density and hence potential maximum leaf conductance.

In k-deficient plants stomatal density is lower, leading to transpiration decrease. Calcium is another ion that functions in the regulation of stomatal aperture, due to its concentration in the apoplast of the leaf epidermis. In the case of calcifuges, high levels of Ca^{2+} in the rhizosphere could reduce the coupling of stomatal conductance to assimilation, and this reduces water use efficiency by increasing apoplasmic Ca^{2+} in the leaf. Nitrogen sources could alter stomatal conductance by activating nitrate entry into guard cells. Another inorganic anion involved in stomatal opening is Cl. Therefore, the stomatal conductance of NH_4^+-fed plants is 20–40% lower than that of those fed on NO_3^-.

Deficiency of certain elements is accompanied by organic carbon efflux, as soluble, low-relative molecular mass compounds (Raven et al., 2004). This is especially true for deficiency of phosphorus and Fe. This mechanism is significant.

Water use efficiency can be increased by following more intensive cropping systems in semi-arid environments and by increasing plant population density in more humid environments. Important factors determining water and nutrient use by crops include the physical and chemical properties of soil (water retention, capacity, available water, microbial biomass, soil organic matter).

Certain practices increase nutrient use by plants and include the form and quality of added fertilizer and the timing of its application. In order to maximize the use of applied fertilizer, precision farming tools such as GPS digital maps of soils and crops, and also the use of optical sensors and growth models, have been applied.

METHODS OF IMPROVING WATER USE EFFICIENCY

Introduction

More than 75% of fresh and good-quality water used throughout the world is destined for agriculture. Today more than 3100 km^3 of water is used in agriculture for the irrigation of approximately 240 million ha of crops. The sources of irrigation water are either surface or groundwater. Due to overuse of irrigation water a significant reduction in groundwater reserves has been recorded in many agricultural areas. Therefore, it is necessary to improve water use efficiency (Fernández and Moreno, 1999), by which we mean the amount of water absorbed necessary to produce 1 unit of biomass or harvested crop.

Regulated Deficit Irrigation (RDI) and Partial Rootzone Drying (PRD) irrigation

Crop performance is affected by many factors such as cultivar genotype, humidity, precipitation and wind, pruning, nutrition and irrigation. Of these factors irrigation is the most important, and growers are now exploring new ways of manipulating the water use efficiency of crops by using new irrigation techniques.

One such technique is Regulated Deficit Irrigation (RDI) (Loveys *et al.*, 2004; Tognetti *et al.*, 2005). This is an irrigation technique based on the science of phenology regarding vegetative and reproductive development. Plant vigour is very important for fruiting, and excessive vigour results in shading and decreased fruiting. Reduced vigour can be achieved through pruning, application of growth inhibitors (Frakulli and Voyiatzis, 1999) or appropriate fertilization and irrigation. By judicious use of these methods, manipulation of vigour through irrigation management offers more flexibility, with minimal labour input (Dichio *et al.*, 1994; Centritto *et al.*, 2005). The main principle of RDI is that the quantity of applied water is reduced during the period of high vegetative growth and low fruit growth rate. Later, when fruit growth rate is high, a normal irrigation regime is resumed. Due to water deficit the vegetative growth is reduced while fruit growth and development is least affected. Due to water deficit the canopy size and leaf stomatal conductance are reduced and intermittent drought affects leaf anatomy (Chartzoulakis *et al.*, 1999; Dichio *et al.*, 2003). RDI is a method developed for high-density orchards. Although the root system remained dry, the parameters of photosynthesis, stomatal conductance and growth regained pre-stress values within a few weeks, although one part of the root system still remained dry (Dry and Loveys, 2000).

Stomatal closure is due to the production of ABA in the dry part of the root system (Stoll *et al.*, 2000). These responses were not evident in plants where the same half of the root system was irrigated throughout the growing season and the remainder of the roots were dry. This suggests that the root-derived signals during the stomatal responses were not sustained, in the same way that had been observed in various plants. Furthermore, PRD did not result in great changes in leaf ABA in comparison with plants subjected to an overall water deficit (Stoll *et al.*, 2000). We may assume that stomatal control is achieved with small alterations in leaf ABA, when ABA synthesis is initiated by changes in water availability to only part of the root system. Measurement of ABA content after PRD indicates that its content is almost equal after extraction from roots and petioles, and twice as great from leaf blades.

The pH of the sap from PRD-treated plants was higher in comparison with controls. This increase in the pH of xylem sap reduces partitioning of ABA into the symplasm, i.e. away from the sites of action of guard cells. The ABA concentration of xylem sap varies, depending on the cultivar and species. Furthermore, the pH of xylem sap varies depending on plant species (Wilkinson, 2004). When the pH of the xylem sap is close to the pK_a value of ABA (4.8), very small changes in sap pH will create large changes in the proportion of ABA present as an ion, while greater pH values have less effect on ABA ionization. Under high VPD stomatal conductance was low in both control and PRD trees on the western side of the trees. Furthermore, the difference in stomatal conductance between control and PRD trees was evident in the morning, on the side of the tree receiving less direct sunlight.

In response to water stress roots synthesize the ethylene precursor 1-aminocyclopropane-1-carboxylic acid (ACC) (Wilkinson, 2004). This is transported to shoots, where ethylene, released from ACC, can induce ABA synthesis. PRD results in stomatal closure, and exogenous application of the cytokinin benzyladenine can overcome the inhibitory effect. Furthermore, PRD reduces endogenous concentration of zeatin and zeatin riboside in both roots and shoots.

Partial root drying is an effective technique for improving WUE, by changing the relationship between stomatal conductance and ambient evaporative condition. Furthermore, PRD increases root growth in deeper soil layers. This influences the WUE, since more water is available from greater soil depths. PRD is effective in crops which, after stress, produce ABA to elicit a stomatal response. In this case another mechanism is active, which is the accumulation of osmotically active solutes maintaining stomatal conductance, even at low leaf water potentials. This is achieved through a decrease in osmotic potential and the maintenance of turgor at low water potentials.

PRD as a system for irrigation requires that the irrigation system is installed in such a way that 50% of the root system is irrigated and 50% is dry, in any given period. Above ground drip-pipes are used.

WATER USE EFFICIENCY AND CHEMICAL SIGNALLING

Plants require CO_2 via open stomatal pores on the leaf blade and, at the same time, they also risk dehydration via water loss through transpiration. Therefore, plants lose water in order to fix CO_2. To enable that, plants have been adapted to allow control of water loss and this is achieved by controlling stomatal aperture, through pairs of guard cells. The effect of stomatal closure in improving WUE occurs during the first stages of water stress. However, during severe stress photosynthesis is severely inhibited and WUE decreases again. At leaf level the plant achieves water balance by leaf rolling, leaf epinasty, denser leaf trichomes and a thicker cuticle, reduced leaf area and shedding of older leaves. This reduces the surface area from which water can be lost and redistributes nutrients to the stem, roots, fruits and leaves. Another response is the increase in the root:shoot ratio. Under water stress shoot growth is reduced while root growth is increased. This allows the plant to scavenge moisture from a larger volume of soil, while minimization of water loss from the aerial part of stomatal closure and leaf growth inhibition is among the earliest responses to drought, and can occur after very slight soil drying.

Many plants will regulate stomatal aperture independently of hydraulic signals but in response to chemical signals. These chemical signals are generated by the interaction between the root and the drying soil, controlling WUE as water deficit develops in the soil. Chemical signals involve changes in the transport of growth regulators or other substances via the xylem, as described below.

ABA is synthesized in the root, as soil drying progresses, and is subsequently transported to the leaf, via the xylem vessels (Wilkinson, 2004). In leaves, ABA accumulates in the guard cells and induces stomata to close. ABA accumulation induces a signal transduction inside the guard cell, which involves an increase in cytoplasmic Ca. This Ca increase, and also an increase in cytoplasmic pH, depolarize the guard cell plasma membrane causing outward K^+ and Cl^- channels to open, followed by efflux of these ions. This leads to decreased water potential of the cell, reduced turgor and stomatal closure. The synthesis and transport of ABA take place before a significant decrease in soil or leaf water, i.e. before the appearance of hydraulic signals. Other chemical signalling molecules include cytokinins, ethylene, NO_3 and pH. Soil drying reduces transport of growth-promoting cytokinin from the root to the shoot.

The basic ABA signal can also be influenced by soil drying via synthesis of ABA conjugates such as ABA-glucose ester (ABA-GE) and their transport via the xylem to leaves. ABA-GE is lipophilic and unable to cross lipid cell membranes, and is transported to the shoot with no loss to stem parenchyma. Therefore ABA-GE is a more efficient form of transporting ABA from the root to the shoot. When ABA-GE reaches the leaf apoplast, enzymes such as esterases, hydrolases and/or glucosidases cleave the conjugate and release free ABA to the cells.

Furthermore, soil drying and flooding increase ethylene levels by promoting the synthesis of the ethylene precursor ACC and its transport from roots to leaves. Water stress, soil flooding and nutrient deficiencies can induce changes in the xylem sap pH. The first change measured is an increase in xylem pH after 24 h of exposure to soil drying and prior to the reduction in stomatal aperture. Ethylene generation by leaves occurred after 3–4 days. Furthermore, xylem sap pH becomes more alkaline after soil flooding. Such changes in pH can induce stomatal closure by affecting ABA compartmentation. Increases of apoplastic sap pH from 5.5 to approximately 7.0 in water-stressed plants have been reported. The same trend regarding pH was exhibited by nitrate-deficient soil. Changes in xylem and symplastic pH may be linked to effects of cellular dehydration or the activity of H^+-ATPases associated with xylem parenchyma. Furthermore, soil drying-induced changes in ionic composition of xylem sap consequently affect its pH. Reduction of NO_3 in the xylem sap is a sensitive chemical change that precedes increases in xylem sap pH and stomatal closure. Soil drying induces xylem alkalinization, which may result in a decrease in nitrate reductase (NR) activity from roots to leaves.

This can be promoted by a reduced supply of nitrate from the root to the leaf. Malate and citrate are loaded into the xylem instead of NO_3^-, which increases pH. This happens in species which transport NO_3 to the leaves and assimilate it there. Species with NR located in the root transport organic products of root NO_3 assimilation via the xylem, and sap pH will remain stable during soil drying. This is common with deciduous woody species.

Another chemical signal involves cytokinins. Root apical meristems are a major site of synthesis of free cytokinins. Cytokinins are transported through the xylem to the shoot as zeatin riboside and are released as free zeatin in the leaves. Cytokinin is synthesized and transported to the shoot at a rapid rate in non-water-stressed plants. Upon drying of the soil the cytokinin content of xylem sap is reduced due to reduced synthesis. This reduction sends a negative message to the shoot to reduce growth and stomatal opening. High cytokinin concentrations induce increased k concentrations within guard cells and increase their turgor.

17

STRESS-INDUCED ACCUMULATION OF PROLINE AND MANNITOL

INTRODUCTION

Accumulation of proline has been proposed as a protector of membranes, polyribosomes and enzymes during water or salt stress. Proline is an osmoregulant and contributes to 45% of total osmotic adjustment, and can also detoxify free radicals. Most proline is accumulated in the vacuoles, but some is in the cytoplasm.

Proline is accumulated after stress by both increased synthesis and reduced degradation (Van Rensburg *et al.*, 1993; Hare and Cress, 1997; Yoshida *et al.*, 1997). Proline may be synthesized from both glutamate and ornithine. The intermediates in proline synthesis are glutamic γ-semialdehyde (GSA) and Δ'-pyrroline-5-carboxylate (PCP). The enzymes involved in synthesis are P5C reductase and proline dehydrogenase.

COMPATIBLE SOLUTES

Olives are grown mostly in arid and semi-arid regions, where they are subjected to high temperatures and scarcity of water. Osmotic homeostasis requires an increase of osmotic pressure in cells by either uptake of soil salts or the synthesis of metabolically compatible compounds. These solutes are commonly carbohydrates (sugar or sugar alcohols, such as mannitol), amino acids and quaternary ammonium compounds (Hare and Cress, 1997; Hare *et al.*, 1999). Besides their role in osmotic adjustment (Hare *et al.*, 1998), compatible solutes also have osmoprotective functions – protecting membranes – thus preserving the biological functions of these, and seem to play a role in hydroxyl radical scavenging, hence being active in protecting plants against oxidative damage (Smirnoff and Cumbes, 1989).

Sucrose is the most abundant carbohydrate and is an important factor in micromolecule and membrane stabilization. Sugar alcohols too, such as mannitol (Bieleski, 1982), constitute an important group of compatible solutes.

Mannitol is an osmoprotectant and also serves as a storage compound and a redox agent. Mannitol degradation produces, in its initial stage, NADH and its oxidation in mitochondria yields up to three molecules of ATP (Stoop et al., 1996). Fructose represents those reducing sugars frequently involved in osmotic adjustment in stressed cells.

Proline is one of the most common compatible solutes accumulated in plants under stress conditions (Yoshida et al., 1997). It contributes to osmotic adjustment, preserves enzymes and other important cellular structures and is a reserve for carbon and nitrogen following stress relief. Increasing osmotic stress stimulates the accumulation of mannitol as a compatible protective compound in olive.

The contribution of the ornithine pathway to salt-induced proline synthesis is less important. The oxidation of proline to glutamate is restricted to the mitochondria. Proline accumulation is an excellent means of storing energy, since the oxidation of one molecule of proline can yield 30 ATP equivalents.

Six enzymes participate in proline biosynthesis and degradation (Kavi Kishor et al., 1995; Peng et al., 1996; Savouré et al., 1997; Hare et al., 1999):

- PCP synthetase.
- PCP reductase.
- Proline dehydrogenase.
- PCP dehydrogenase.
- Ornithine δ-aminotransferase.
- Arginase.

The reactions involving these enzymes are presented in Table 17.1.

Table 17.1. Enzymatic reactions of proline biosynthesis and degradation in plants.

Enzyme	Reaction(s) catalysed
Δ^1-pyrroline-5-carboxylate synthetase	(i) Glutamate + ATP → Glutamyl-γ-phosphate + ADP (ii) Glutamyl-γ-phosphate + NADPH+H$^+$ → glutamic-γ-semialdehyde
Δ^1-pyrroline-5-carboxylate reductase	Δ^1-pyrroline-5-carboxylate + NAD(P)H+H$^+$ → proline + NAD(P)$^+$
Proline dehydrogenase	Proline + ½O$_2$ + FAD → Δ^1-pyrroline-5-carboxylate + H$_2$O + FADH$_2$
Δ^1-pyrroline-5-carboxylate dehydrogenase	Δ^1-pyrroline-5-carboxylate + NAD(P)$^+$ → glutamate + NAD(P)H + H$^+$
Ornithine δ-aminotransferase	Ornithine + 2-oxoglutarate → glutamic-γ-semialdehyde Δ^1-pyrroline-5-carboxylate + glutamate
Arginase	Arginine + H$_2$O → ornithine + urea

MANNITOL CONTENT OF OLIVES

Mannitol is a widespread polyol throughout the plant kingdom, occurring in more than 100 species of vascular plants. Mannitol may participate in many physiological processes in plants, such as: (i) carbon storage and translocation of products (Flora and Madore, 1993; Conde *et al.*, 2007); (ii) regulation of reductive power by production of NADPH; (iii) osmoregulation and regulation of H_2O status (Peltier *et al.*, 1997); and (iv) a hydroxyl radical scavenger and a cryoprotectant. In the family Oleaceae mannitol represents a significant proportion of the carbohydrate pool. Mannitol content increases in response to conditions of increased drought or salinity (Peltier *et al.*, 1997).

The mannitol content of the leaves of olive plants under salinity stress increased from 160 to about 220 µmol/g d.w. (Tattini *et al.*, 1996), and mannitol represents 70% of the soluble carbohydrates in olive leaves. The mannitol level in olives is more or less constant during the year (Oddo *et al.*, 2002). The poor correlation (Oddo *et al.*, 2002) between mannitol content in *Olea* and rainfall or temperature suggests that mannitol does not play a fundamental role in the response to drought or high temperature.

18

MINERAL NUTRITION OF THE OLIVE

CHAPTER OVERVIEW

Olive trees in many areas of the world are among the least fertilized trees. Therefore, biennial bearing is very common. Although olive trees represent a large global hectareage, our knowledge of their mineral nutrition and salt tolerance is very limited. Hence, this chapter represents an attempt to provide some information in this area.

The topics discussed in this chapter are the following:

- The needs of olives for N, P, K and their concentration in various plant parts.
- The annual needs for N, P, K and Ca for various fruits and the annual variations in N, P, K, Ca and Mg concentrations in leaves. Nitrogen distribution in fruiting olive shoots.
- The concentrations of various nutrients considered as being adequate.
- Nutrient stress and growth.
- Nutrient deficiency or excess and photosynthesis.
- Nitrogen excess and quality of olive oil and table olives.
- Roles of N and its assimilation, kinetics of nitrate (NO_3) absorption and effect of N form on growth.
- Olive culture and the nitrate accumulation problem.
- Nitrates leaching from olive orchards.
- Roles of K, P, Ca and Mg and their absorption.
- Roles of B and Mn.
- Response of olive trees to salinity.
- Olive fertilization and slow-release fertilizers.

INTRODUCTION

Olive trees are not big feeders. They are hardy plants that will tolerate poor growing conditions – especially low fertility – better than almost any other fruit tree. They also tolerate a very wide range of soil pH and tend to fruit better under conditions of average vigour and nutrition. The olive is a tree that, in general, receives minimal fertilization. This unfavourable treatment of olives concerning the application of organic or inorganic fertilizers derives from the opinion that they do not require adequate fertilization, because of their extensive root system. This opinion is misdirected, since the volume of soil around the root system is limited. As a consequence of such inadequate fertilization, biennial bearing is quite frequent. There is a lack of information in international literature concerning the behaviour of olive cultivars with regard to mineral nutrients (Marschner, 1997). Therefore this chapter aims at giving information pertinent to olives in respect of nutrition. A significant portion of these data are derived from research projects of the author or PhD theses conducted under his guidance.

THE MINERAL REQUIREMENTS OF OLIVE TREES

The needs of olive trees with respect to N, P, K are comparable to the needs of other fruit trees such as deciduous trees (Gavalas, 1978). The quantities of nutrient absorbed from orchard soils are 0.50 kg N/100 kg of fruit and 1.0 and 0.76 kg N/100 kg of leaves and wood, respectively. For K the required quantities are 0.95, 0.56 and 0.39 kg, respectively. The large fruit load, the significant amount of leaves and wood removed every year by pruning and the leaf drop due to diseases and water stress entail, as a consequence, the loss of a large amount of nutrients from trees. Therefore, it is of vital importance that the lost nutrients are replaced through appropriate fertilization. For an orchard of 1000 m^2 the nutrients absorbed by olives per annum amount to 1.5–3.5 kg N, 0.8 kg P, 1–5 kg K and 2–5 kg Ca. The nutrients (N, P, K and Ca) required for 100 kg of fruits, 50 kg of leaves and 50 kg of wood are given in Table 18.1.

Concentrations within the various plant parts of N, P and K are presented in Table 18.2. Nitrogen concentration decreases in young leaves and stems in spring and summer and increases during the autumn in both 'off' and 'on' years. Nitrogen concentration in old leaves and stems remains almost constant during the off years and is mobilized during the on years to support growth.

Table 18.1. The N, P, K and Ca (kg) required every year by an olive orchard for the production of fruit, leaves and wood.

Nutrient	Fruit (100 kg)	Leaves (50 kg)	Wood (50 kg)	Total (kg)
N	0.500	0.500	0.380	1.380
P	0.120	0.120	0.150	0.390
K	0.950	0.280	0.195	1.425
Ca	0.960	0.500	0.300	1.760

Table 18.2. Concentration of N, P and K within the various plant components of olive trees.

Component	N (%)	P (%)	K (%)	Ratio (N:P:K)
Secondary roots	0.33	0.113	0.402	2.9:1:3.5
Primary roots	0.37	0.123	0.477	3.0:1:3.8
Trunk	0.26	0.070	0.219	3.7:1:3.1
Main branches	0.29	0.090	0.354	3.2:1:3.9
Secondary branches	0.23	0.099	0.191	1.2:1:1.9
Shoots	0.64	0.179	1.000	3.5:1:5.5
Leaves, 1 year old	1.63	0.271	0.994	6.0:1:3.6
Leaves, 2 years old	1.24	0.201	0.678	6.1:1:3.4
Unripe fruits	0.90	0.333	2.760	2.7:1:8.3
Ripe fruits	0.97	0.397	3.220	2.4:1:8.1

Leaves have a larger concentration of N than stems, and fruits produced during the on years are the main sink for N. Furthermore, N also may be mobilized from organs other than leaves to support fruit growth.

METHODS OF SOLVING NUTRITIONAL PROBLEMS

For determination of the various nutritional problems, three methods of application are available: leaf analysis, soil analysis and visual symptoms.

Leaf analysis

The nutrient concentrations within olive leaves depend on the cultivar (Jordao et al., 1999), the level of fertilization and area. In the literature are to be found leaf analysis data from various regions of the world (Fernández-Escobar et al., 1994; Dimassi et al., 1999). The season of leaf sampling is that when the nutrients attained have a more or less constant concentration. Starting early in spring, the concentrations of N, P, K and Mg (in leaves) are more or less constant until a drop in August. Subsequently the concentration increases again and stabilizes from October to the start of the next vegetative cycle (see Fig. 18.1). The same trend is observed in leaves of 1, 2 or 3 years old. The winter period is characterized by stable concentrations of N, P, K, Ca and Mg, and this period is considered as the optimum for leaf analysis in the olives. However, other workers prefer to collect leaf samples in July from the current year's growth. The required number of leaves per sample is 80–100, and these leaves are harvested from shoots with no fruits. The nutrient concentrations determining deficiency or sufficiency of nutrients are presented in Table 18.3. The chemical composition of olive leaves varies depending on the cultivar and time of sampling (Table 18.4).

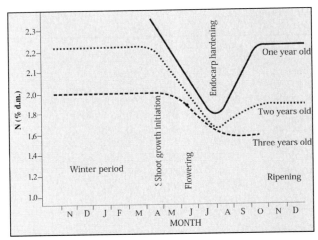

Fig. 18.1. Seasonal variation in N concentration of olive leaves 1, 2 and 3 years old (from Bouat, 1961).

Table 18.3. The nutrient concentrations in leaves determining deficiency, sufficiency or excess for olives (from Therios, 2005b).

Nutrient	Deficiency	Relative lack	Sufficiency	Excess
N (%)	< 1.20	1.20–1.60	1.60–1.80	> 2.20
P (%)	< 0.07	0.07–0.09	0.09–0.11	> 0.14
K (%)	< 0.50	0.50–0.70	0.70–0.90	> 1.10
Mg (%)	< 0.07	0.07–0.10	0.10–0.30	> 0.30
Ca (%)	< 0.50	0.50–1.00	1.00–2.50	–
S (%)	< 0.05	0.05–0.10	0.10–0.25	–
Cl (%)	–	–	0.10–0.40	> 0.80
Fe (ppm)	–	20–50	50–150	–
Mn (ppm)	–	5–20	50–150	–
B (ppm)	–	15–20	20–50	–
Zn (ppm)	–	5–10	10–30	–
Cu (ppm)	–	< 5	5–20	–
Mo (ppm)	–	< 0.03	–	–

NUTRIENT ELEMENTS

Nitrogen

Roles of nitrogen

Nitrogen is necessary for plants since it is a primary component of nucleic acids, proteins, amino acids, purines, pyrimidines and chlorophyll. Nitrogen exerts a significant effect on plant growth, as it reduces biennial bearing and

Table 18.4. Variation in chemical composition of cvs 'Chondrolia Chalkidikis' and 'Amphissis' leaves 30–120 days after full bloom (from Chatzissavvidis et al., 2004).

Nutrient	Days after full bloom	'Chondrolia Chalkidikis'	'Amphissis'
N (% d.m.)	30	1.22	0.93
	60	0.73	0.78
	90	0.83	0.76
	120	0.90	0.65
P (% d.m.)	30	0.15	0.11
	60	0.11	0.08
	90	0.10	0.11
	120	0.11	0.06
K (% d.m.)	30	0.94	0.65
	60	0.89	0.73
	90	1.71	1.18
	120	1.63	1.32
Mg (% d.m.)	30	0.09	0.06
	60	0.06	0.06
	90	0.07	0.05
	120	0.06	0.05
B (ppm)	30	47	52
	60	47	61
	90	96	84
	120	122	44
Fe (ppm)	30	56	58
	60	84	97
	90	18	27
	120	17	29

increases the percentage of perfect flowers. The olive is a fruit tree requiring enough N and irrigation in order to avoid biennial bearing. Therios and Sakellariadis (1988) and Fernández-Escobar et al. (2004a) found that the increase of N supply increased olive growth. Lack of N leads to decreased growth, shorter length of annual shoots (< 10 cm), fewer leaves, reduced flowering and decreased yield. Nitrogen concentrations in the leaf of < 1% lead to the formation of imperfect flowers. However, excess N application creates problems, while the application of slowly available fertilizers has beneficial effects (Fernández-Escobar and Marín, 1999; Garcia et al., 1999). The time of N fertilization in the non-irrigated orchard is December–February, in the form of ammonia or urea. In irrigated orchards N should be supplied in three doses, i.e. December–February – as ammonia – and the other two doses later in NO_3 form. The amount of precipitation is the determining factor for the quantity of N applied: if the annual precipitation is < 400 mm we supply 100 g N/tree/100 mm rainfall, while if precipitation is > 700 mm we supply

150 g N/tree/100 mm rainfall. In the latter case the total amount of N applied would be 1.5 kg N/tree or 15 units N/1000 m².

Nitrogen should not be applied every year and an optimum level of N fertilization is required (Marín and Fernández-Escobar, 1997). In sandy soils N tends to leach out, causing groundwater and runoff pollution. Therefore, lighter and more frequent applications are preferable. Conventional N fertilizers such as urea, ammonium nitrate, ammonium phosphate, ammonium sulphate, potassium nitrate or calcium nitrate are all suitable. Conventional N fertilizers have traditionally been used because they are cheap, less bulky and easy to apply. When combining N fertilization with organic materials the total cost is considerable. Furthermore, most of the organic materials release their N within the first year or two. However, organic materials have the benefit of being slowly released and are less likely to leach into ground or surface waters.

Forms of soil nitrogen
The N found in the soil can generally be classified as either inorganic or organic (Tisdale et al., 1993). By far the greater total amount of N occurs as a part of the soil organic matter complex. The inorganic forms of soil N include NH_4^+, NO_2^-, N_2O, NO and elemental N, which is inert. From the standpoint of soil fertility the NH_4^+, NO_2^- and NO_3^- forms are of greatest importance.

Nitrogen transformation in soils
Olives absorb most of their N in the NH_4^+ and NO_3^- forms. The quantity of these two ions presented to the roots depends largely on both the amounts supplied as commercial N fertilizers and released from reserves of the organically bound soil N. The amount released from the organic reserves or fertilizers depends on the balance that exists between the processes of N mineralization, immobilization and soil losses. The mineralization of organic N compounds takes place essentially in three reactions: aminization, ammonification and nitrification.

Aminization: protein → RNH_2 + CO_2 + energy + other products
Ammonification: RNH_2 + HOH → NH_3 + R-OH + energy
Nitrification: $2NH_4 + 3O_2 \rightarrow 2NO_2^- + 2H_2 + 4H^+$ (*Nitrosomonas*)
$2NO_2^- + O_2 \rightarrow 2NO_3$ (*Nitrobacter*)

The ammonia released is subjected to conversion to nitrite and nitrates, for direct absorption by olive trees; it may be utilized by heterotrophic organisms and it can be fixed in the lattice of expanding-type clay minerals. The NH_4^+ permits its absorption and retention by soil colloidal materials in the negative charge of clays and is not subject to leaching by water, as is the NH_3^- form. The more unfavourable the conditions for nitrification and the greater the exchange capacity of minerals, the longer is the retention time. After nitrification ammonia is subject to leaching. Of the three forms of N, NO_3^- and urea are completely mobile forms in soils and move largely with the soil water. Also, during dry weather capillary and upward movement of water containing nitrates is possible, leading to NO_3^- accumulation in the upper soil zone. Furthermore, it is possible for NH_4^+ to be fixed by the 1:2 clay minerals having an expanding lattice. The fixed ammonia can be replaced by cations that expand the lattice (Ca^{2+}, Mg^{2+}, Na^+, H^+). Nitrogen losses in irrigated olive orchards may occur under the following conditions:

- Denitrification, which is a biochemical reduction of nitrates under anaerobic conditions; the N losses occur in the form of N_2 or N_2O.
- After surface application of NH_4^+ or urea in soil with alkaline pH, N is lost in the form of NH_3.

Nutrient stress and growth

When mineral nutrient levels are insufficient for vegetative and reproductive growth in olive trees, a series of events commence. Hence, N deficiency reduces vegetative growth and N excess promotes vegetative growth, susceptibility to various pathogens and disturbs the nutrient balance with other elements – especially P. Leaf concentrations of N, P and K are high before maximum yield. The high N:P ratio reduces lateral shoot growth. Boron concentrations < 15 ppm lead to leaf chlorosis, necrosis, leaf drop, absence of flowering and 'monkey-faced' fruits. Low and high Mn concentrations significantly affect the chemical composition of leaves in cultivars tested, i.e. 'FS-17', 'Manaki', 'Kalamon', 'Koroneiki' and 'Picual'. A genotypic difference was recorded between cultivars, with 'Picual' accumulating the minimum amount in comparison with other cultivars. Iron, Zn and B concentrations were not affected; furthermore, Ca was not affected by Mn levels, although the cvs 'Manaki', 'Kalamon' and 'Koroneiki' accumulated the minimum amount of Ca.

Nutrient deficiency/excess and photosynthesis

Plants require various nutrient elements in order to effect efficient photosynthesis, and their responses depend on a particular nutrient. The effect of a nutrient depends on the photosynthetic step at which it operates.

Nitrogen is a macroelement, and photosynthesis requires significant amounts of N. Nitrogen is required in photosynthesis, since it is necessary for 50% of the proteins in thylakoid membranes and in Rubisco, which represents 50% of the total soluble proteins in leaves. Nitrogen deficiency reduces protein and chlorophyll content per unit leaf area, resulting in decreased photosynthesis.

Phosphorus also affects photosynthesis. Phosphate deficiency decreases the number of leaves, leaf size and the concentrations of proteins and pigments. However, the photosynthetic component is more or less constant. The ATP content of tissues is reduced and this leads to a reduced RuBP biosynthesis. Consequently, the activity of RuBisCO decreases. Manganese concentration affects the rate (Pn) of photosynthesis. However, a significant genotypic difference is obvious. In 'Koroneiki' 640 mg/l Mn increased the Pn. In the other four cultivars mentioned above, high Mn increased both transpiration rate and stomatal conductance (Chatzistathis *et al.*, 2006).

Nitrogen excess and quality of olive oil and table olives
Fertilization and irrigation practices can modify the quality of both table olives and olive oil. Excess N fertilization in olive fruit results in a significant decrease in the content of polyphenols, which are the main antioxidants. Furthermore, a decrease in polyphenols induces a significant decrease in the oxidative stability of the oil and accentuates its bitter taste. In contrast, tocopherol content increases with increasing N concentration in fruit. An excess of N fertilization does not affect pigment content – particularly carotenoid and chlorophyll pigments – nor fatty acid composition.

Excess N fertilization can negatively affect olive production and delay fruit ripening. A high degree of fertilization reduced polyphenol content and decreased levels of K_{225}, oxidative stability, the ratio of mono- to poly-fatty acids and reducing sugars. The reducing sugar content, expressed on a dry weight basis, was about 17% lower in the two higher fertilizer treatments (i.e. 400 and 600 g N/tree) in comparison with control. Texture, which is an important organoleptic characteristic in table olives, showed a linear decrease when the amount of fertilizer was increased. Furthermore, a negative correlation was found between texture and N content. The loss of texture by increasing the amount of fertilizer could be due to partial solubilization of cell wall polysaccharides, through increased water content. Polyphenol content (mg/kg) was reduced by 33% in the high N treatment, in comparison with control. The oxidative stability was also reduced, by 34%.

Olive culture and nitrate assimilation
Nitrate is a major form of N available to plants in many environments. Before NO_3^- is taken up by the plant it should be reduced to NO_2^- and, subsequently, to NH_4^+. The enzyme catalyzing the reduction of NO_3^- to NO_2^- is *nitrate* reductase (NR), while the reduction of NO_2^- to NH_4^+ is catalysed by the enzyme *nitrite* reductase. For NO_3^- assimilation into organic N two electrons are required, and six electrons for NO_2^- to NH_4^+ reduction. Furthermore, ATP is required for the assimilation of NH_4^+ into amino acids and proteins.

Species that reduce NO_3^- predominantly in their shoot may have the advantage of being able to use excess reductant released from photosynthesis (Andrews, 1986; Touraine *et al.*, 1994). In contrast, species reducing NO_3^- mainly in their roots must obtain reductant from glycolysis and the pentose phosphate pathway (Oaks and Hirel, 1986). Therefore, the carbon budget is affected by the plant's site of NO_3^- reduction. Furthermore, ammonium is incorporated into amino acids by the glutamine synthetase, glutamine-2-oxoglutarate transaminase (GS-GOGAT) enzyme system (Therios, 2005a).

Under stress conditions nitrite is also reduced to nitric oxide by NR. Nitric oxide constitutes a significant component.

A minor amount of nitrite reduction may also occur during darkness, at the expense of reducing equivalents generated by the oxidative pentose phosphate pathway and starch breakdown. The NADH required for NO_3^- reduction in the cytosol could be provided by photosynthetic electron transport via malate–oxaloacetate or from mitochondrial substrate oxidation.

Olive culture and the nitrate accumulation problem

Nitrate is essential to life, but a hazard when applied in the wrong place and at the wrong time. Nitrate is present in most natural waters – in rain, rivers, lakes, sea and in the water stored in porous rocks. Nitrate concentrations in natural water are steadily rising with the increasing use of nitrogenous fertilizer by olive growers. Furthermore, the European Union (EU) has established a limit of 50 mg/l for NO_3^- in drinking water, or 11.3 mg/l N.

Due to the common agriculture policy (CAP) of the EU, which ensured a fair price for agricultural products and the use of extra N, this has resulted in many cases in excess production; high rates of N fertilization have resulted in NO_3^- pollution. The solution is not simple, since a reduction in use of nitrogenous fertilizer results in lower agricultural income. Furthermore, lowered productivity raises food prices in the supermarket.

Nitrate itself is not toxic but becomes problematic after its conversion into nitrite, which is responsible for blue baby syndrome, or methaemoglobinaemia, and stomach cancer. The blue baby syndrome is common in children under 1 year old who consume a certain level of nitrates. Microorganisms convert NO_3^- to NO_2^- and NO_2^- reacts with haemoglobin, which transports oxygen through the body, and haemoglobin is transformed into methaemoglobin. The Fe in haemoglobin is in the ferrous form, but in the ferric form in methaemoglobin. Therefore, the O_2-carrying capacity of haemoglobin is reduced. Stomach cancer has also been linked to the NO_3^- concentration in drinking water. Nitrite produced from NO_3^- reacts with secondary amines, producing N-nitroso compounds that can modify DNA. The World Health Organization (WHO) has set two limits, one of 50 mg/l – the recommended one – and a maximum of 100 mg/l (22.6 mg/l of N).

When more N is applied, the increase in yield per kg unit of N reduces (diminishing returns). When the soil contains adequate N the curve attains the form of a straight line, which declines continuously as N increases. Therefore, additional N reduces yield. The third form of N curve is that in which additional N increases the yield steadily, up to a point above which it sharply declines.

Nitrate leaching from olive orchards

Nitrate-containing fertilizers, when applied to the soil surface and after their entrance into the soil by irrigation or rainfall – undergo a process known as 'dissociation', i.e. separation into anion and cation. The soil solution should contain equal numbers of positive and negative ions, due to the principle of electroneutrality.

The ammonium ion (NH_4^+) is the other ion containing N, and its relation to ammonia (NH_3) is given by the equation

$$NH_3 + H_2O \leftrightarrows NH_4OH \leftrightarrows NH_4^+ + OH^-$$

Therefore, when ammonia (NH_3) is dissolved in water this gives an alkaline solution. Reactions proceed in both directions – towards the left/right or clockwise/anti-clockwise, meaning that in alkaline soils NH_4^+ can be converted to NH_3 gas and therefore lost to the air. The clays in soil readily accumulate electrical charges, the overall charge being positive at very acidic pH and negative at pH in the range 5.5–8.0. This indicates that positively charged ions are attracted to the surface of the clay, but negatively charged ions such as NO_3^- are repelled and are more vulnerable to downward leaching.

Ammonium is strongly attracted to the negative charges of clay particles and does not leach out in soils with adequate clay content. Use of nitrification inhibitors reduces the speed of NH_4^+ transformation to NO_3 by soil bacteria and reduces NO_3^- leaching. Such inhibitors are based on pyridine and are circulated under the commercial names N-Serve or Dicyandiamide (DCD). Furthermore, such transformation in the soil is inhibited by the alkaline soil reaction. Leaching of NO_3^- requires liquid water (not gas or ice) and a soil saturated or partly saturated. Therefore, we can describe two flows of water in the soil – saturated and unsaturated. The main force of NO_3 leaching is gravity, or downward movement, of water and NO_3^-.

Water containing NO_3^- can leach downwards by pushing down on the water containing NO_3^- like a piston on an equivalency basis. For example, 20 mm of precipitation or irrigation displaces an equivalent soil water volume. If we hypothesize that soil porosity is filled with water and air 50/50, then the depth of NO_3-containing leaching water is 20/0.5 = 40 cm. Nitrate loss by leaching is greater in soils with low water-holding capacity and in soils with rapid water movement, i.e. in soils with high hydraulic conductivity.

Therefore, soil texture and its volumetric H_2O content are determining factors of NO_3^- leaching. Other factors affecting NO_3 leaching include: (i) the microorganism population; (ii) the organic matter content of those olive groves with a high degree of organic culture; and (iii) a higher root density of olive groves, which increases NO_3 absorption and reduces the risk of NO_3 leaching.

The simplest model is the 'piston flow', the analogy of which is a piston pushing out or displacing fluid from a cylinder. This model is expressed by the equation

$$Zp = \Omega/\Theta$$

where Zp = the depth to which the front of the displacing solution penetrates; Ω = the quantity of water or solution necessary for displacement;

Θ = the volumetric moisture content, i.e. the percentage of soil volume that can hold water.

This equation means that, the more water the soil can hold the smaller the downward movement that is caused by a given amount of applied water. This model works best in sandy soils, and it deals only with a single application of NO_3^-. The main source of NO_3 leakage is organic material in the cultivation system called organic farming.

Organic olive orchards

Organic farmers point out that the large inputs of chemicals to the soil–plant system upsets the natural ecosystem. Organic farmers use manures to fertilize olive orchards or they grow a legume crop, such as beans, in order to bring N into the soil. Plants must take up mineral N whether they are grown conventionally or organically. The high availability of N from chemical fertilizers encourages rapid growth and large yields. Organic farmers produce fewer yields but of higher quality and, for these products, people are prepared to pay more. To obtain yields similar to those of conventional farmers, organic farmers should supply with organic matter a similar amount of mineral N.

Organic olive farming has increased in Greece and represents 2.5% of the total area of olive cultivation and about 50% of all organic farms. The use of composts in the organic groves supplies the olive tree with all the essential nutrients for achievement of adequate nutritional status, i.e. normal productivity. The use of compost increases the soil organic matter content. The soil pH values of organic orchards are higher than those of the conventional, the lower pH values of the latter being the result of the application of mineral fertilizers and more frequent cultivation. Concerning the N level of leaves this is marginal in organically grown olives, while P and K concentrations in the leaves are greater in organic farming.

With regard to micronutrients, Fe, Mn, Zn and Cu can be bound to chelates in organic matter, and their concentration in leaves is lower in comparison with non-organic olive groves. Boron levels are higher in organic than in conventional orchards. Furthermore, irrigated conventional and organic farming orchards have higher concentrations of leaf P than the non-irrigated.

Nitrate leaching losses as affected by traditional and slow-release fertilizers

Olive trees require N in greater amounts than other mineral nutrients, but less than 20% of the N applied to the olive orchard is taken up by the fruit trees. The use of traditional fertilizers increases N leaching, reduces N utilization efficiency (NUE) and also increases the pollution of ground water. The use of slow-release fertilizers achieves a step-wise supply of nutrients for a long period. This increases NUE and reduces NO_3 leaching. However, although the use of slow-release fertilizers is advantageous, their use is very limited due to the economic cost in comparison with the low income from olives, especially of non-irrigated olive groves.

Some of the best-known slow-release N fertilizers are circulated under the names Floranid, Multicote, Basammon, etc. (Fernández-Escobar et al., 2004a). Floranid significantly increases N concentration in leaves in comparison with other fertilizers such as ammonium sulphate, ammonium nitrate, calcium nitrate and Basammon. Most of the total N (> 60%) of olive plants is found in the leaves. Nitrate is the most common form of N leaching. However, NO_2–N leaching can be observed when the orchards are fertilized with urea or ammonium sulphate. Another form of N leaching is that of NH_4 when the fertilizer Basammon is used. With respect to traditional fertilizers, urea and ammonium sulphate result in lower N losses.

KINETICS OF NITRATE ABSORPTION BY OLIVE PLANTS

Nitrate is the final N form in soils well aerated and drained, as occurs with olive orchards. The factors affecting NO_3 absorption by olive plants include: (i) induction period; (ii) NO_3 and Ca concentrations; (iii) pH of the nutrient solution; (iv) temperature in the root environment; (v) N deficiency; (vi) NH_4 concentration; and (vii) light intensity (Therios, 1981). These measurements were conducted in a hydroponic system (see Fig. 18.2).

Induction period

The rate of NO_3 absorption is significantly less in the first hours than later on. This reduced rate of absorption (lag phase) is the required time for induction of the enzyme NR. Nitrate absorption has two stages, the first due to active absorption and the second due to the induction of NR (reduction); the ratio between absorption and reduction is 3:1.

The effect of substrate nitrate concentration on absorption rate

When the external NO_3 concentration in solution changes from 0.06 to 0.60 mM, the rate of NO_3 absorption is a function of the external concentration (Therios, 1981). The maximum rate of absorption (V_{max}) is 0.7 μmol/g/h f.w. of roots and the K_m value is 0.23 mM. When we calculate $1/V$ versus $1/S$ a straight line is produced, indicating that 'Chondrolia Chalkidikis' follows Michaelis–Menten kinetics. Increase in the external concentration to 2 mM NO_3 increases the rate of NO_3 absorption above the V_{max} value. These data indicate the existence in olives of two mechanisms at work: (i) at lower concentrations than 0.6 mM; and (ii) at greater concentrations than 0.6 mM. The V_{max} of olives is 25–50% of the NO_3 absorption rate reported in *Prunus*

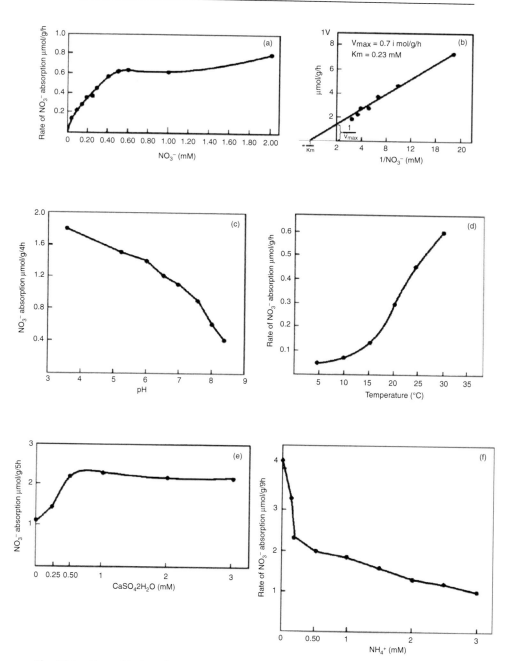

Fig. 18.2. Rate of NO_3^- absorption by olive plants. (a) Effect of NO_3^- concentration; (b) Lineweaver-Burk plot; (c) pH effect; (d) effect of temperature; (e) effect of calcium concentration; and (f) effect of NH_4^+ concentration (from Therios, 2006).

domestica (Therios *et al.*, 1979) and 10% of that reported for barley. The low V_{max} value and the high K_m value (0.23 mM) indicate that olive trees are not efficient in NO_3 absorption. Therefore, in olives NO_3 fertilization would be expected to lead to greater N losses due to leaching, in comparison with Marianna 2624, with a K_m value of 0.06 mM. This point is not very important in non-irrigated olive orchards, while in irrigated ones more NO_3 losses by leaching would be expected. Hence the fertilization programmes for olives should be carefully planned: N should be applied in many and small doses in order to avoid leaching losses. A good solution is to apply N in the form of NH_4 and to use inhibitors of nitrification, such as N-Serve (nitrapyrin) or other forms of slow-release N fertilizers.

pH and absorption of nitrate

Nitrate absorption is greater in acid nutrient solutions and decreases by increasing the pH values. No optimum pH value for N absorption has been recorded for olives. At pH 8.0 NO_3 absorption is 30% of that absorbed at pH 6.0. Nitrate absorption is affected more by the OH^- than by H^+ ions. With regard to the NH_4 form, its absorption is less at acidic pH (6.0), due to H^+ competition, and greater at alkaline pH. The root has the capacity to modify the pH in its microenvironment. Hence, when roots absorb more cations than anions the pH of the surrounding medium is reduced due to H^+ extrusion.

The effect of temperature on nitrate absorption

The rate of NO_3 absorption viz-à-viz soil solution temperature shows that the relationship tends to be linear at temperatures of 15–30°C. Temperatures <15°C significantly reduce the rate of absorption, which is almost zero at 5°C. The Q_{10} quotient for temperatures of 20–30°C is about 2.0. This indicates that NO_3 absorption in olives is a metabolic process. These data indicate that NO_3 fertilization of the olive is ineffective during low winter temperatures. Furthermore, during this period the danger of NO_3 leaching is increased.

The effect of calcium concentration on nitrate absorption

An increase in Ca concentration results in an increased rate of NO_3 absorption. The relation between V (rate of absorption) and Ca concentration is a hyperbola. The optimum concentration of Ca is 0.5 mM; greater Ca concentrations do not affect the rate of NO_3 absorption. When Ca is excluded from the nutrient solution the rate of NO_3 absorption is 50% less than the maximum value. Calcium exerts a significant effect on plasmalemma selectivity and semipermeability.

The effect of various anions and cations on nitrate absorption

The cations K^+, Ca^{2+}, Mg^{2+} and NH_4^+ affect NO_3 absorption. Greater NO_3 absorption is found when K is used as the carrier of NO_3. Nitrate absorption is 90% when Mg and Ca are the NO_3 carriers. The minimum NO_3 absorption occurs when NH_4^+ is the accompanying cation. The rates of NO_3 absorption are: 77% from $NaNO_3$, 90% from $Ca(NO_3)_2 \cdot 4H_2O$ or $Mg(NO_3)_2$ and 46% from NH_4NO_3. The anions Cl^- and SO_4^{2-} do not reduce NO_3 absorption (see Table 18.5).

The effect of nitrogen deficiency on nitrate absorption

Nitrate exclusion or its low concentration in soil solution for a number of days increases the rate of NO_3 absorption. The maximum rate of NO_3 absorption was recorded on the tenth day of deficiency.

The effect of ammonium on nitrate absorption

Ammonium decreases the NO_3 absorption rate. By increasing NH_4^+ concentration to 0.5 mM in the nutrient solution the rate of NO_3 absorption decreases by 46% (Therios, 1981). At greater NH_4^+ concentrations the NO_3 absorption rate is 25% of the control. The decrease in NO_3 absorption is related to either the reduced rate of utilization of NO_3 by plants or the effect of NH_4 on the NO_3 reduction system. The decrease of NO_3 absorption in the presence of NH_4 may be ascribed to the acidity created due to faster absorption (more cations) of NH_4 in comparison with NO_3. This results in a reduction of plasmalemmal permeability. Furthermore, NH_4 may displace Ca from plasmalemma, or the NO_3 efflux in the presence of NH_4 may be reduced. Moreover, NH_4 may inhibit non-cyclic phosphorylation and therefore NO_3 absorption.

Table 18.5. The effects of various cations and anions on NO_3 absorption by olive plants (from Therios, 1981).

Ion	NO_3 absorption (μmol/g/10 h)	Relative rate of NO_3 absorption (K = 100%)
K^+	5.2	100
Na^+	4.0	77
Ca^{2+}	4.7	90
Mg^{2+}	4.8	92
NH_4^+	2.4	46
SO_4^{2-}	5.3	102
Cl^-	5.1	98

The effect of light intensity on nitrate absorption

Light affects NO_3 absorption since it controls photosynthesis and the production of ATP (energy) required for active absorption. Light is also important for the induction of the enzyme nitrate reductase (NR).

Ammonium absorption

The classical enzyme kinetics is followed in NH_4^+ absorption, which resembles the absorption of other monovalent ions and especially that of potassium. This is the reason that a high NH_4^+ concentration competes with K^+, although the affinity for the common point of attachment is only 10% that of K^+. Plants growing under acidic soil pH are obliged to absorb NH_4^+ rather than NO_3^--N.

Furthermore, the absorption of NH_4^+-N by olive plants is greater than the NO_3 or urea N forms. Two systems of NO_3 transport have been proposed: the first of these has a high affinity for NO_3; it works at 5–100 μM NO_3 concentration and consists of the *NRT1* genes. The V_{max} ranges between 3 and 8 μmol/g/h. The second, the low-affinity system, uses the *NRT2* genes and works at concentrations > 250 μM NO_3. Both systems are activated by the presence of NO_3 in the external solution. The lack of NO_3 induces NO_3 absorption and increases the rate of its absorption.

THE EFFECT OF NITROGEN FORM ON THE GROWTH AND MINERAL COMPOSITION OF OLIVE PLANTS

Experiments with several plant species comparing the three N-forms (NO_3-N, NH_4-N and urea-N) have shown that the form of N supplied exerts a pronounced effect on both growth (Tattini *et al.*, 1990; Garcia *et al.*, 1999; Therios, 2006) and chemical composition, as also noted in olives (Therios and Sakellariadis, 1988; Tsambardoukas, 2006).

Recently, the rapid expansion of olive culture has demonstrated the need for integrated studies concerning their growth and nutrient requirements. One such study involved the growth response of olives to N forms. The mean fresh weight per plant was increased significantly by increasing the N level from 1 to 8 mM. This increase in N concentration induced significant increased growth for the NH_4-N and urea-N treatments, while with NO_3-N growth was reduced. Root morphology was affected to a lesser degree by the NH_4-N forms, producing at 16 mM a root system with shorter and thickened feeder roots in comparison with those treated with NO_3-N and urea. The growth reduction at 16 mM NO_3-N was probably due to limited P absorption by the olive plants. The presence of Ca in all the nutrient solutions in our study minimized the toxic effect of NH_4-N. Based on the fact that the greatest growth coincides with the greatest N concentration in leaves and also that most olive orchards are well supplied with Ca, we may conclude that the NH_4-N form is the most preferable source of N for olives. Other processes should also be considered, such as a pH decrease with increased $(NH_4)_2SO_4$ supply or the possibility of NH_3 loss at higher pH values and dry soil conditions. Addition of Ca to the NH_4-N form may increase ammonium absorption and accumulation. In the presence of an adequate supply of carbohydrates, NH_4-N assimilation should proceed more rapidly than NO_3-N assimilation. It is therefore not surprising that greater concentrations of N were found in the NH_4-N-fed plants.

The N form did not significantly affect K and Fe concentrations. However, Ca concentration was affected by the N form. In our experiments we tested the effect of NO_3, NH_4, urea or $NO_3 + NH_4$ on the mineral composition of leaves. Again, the greatest accumulation of N was recorded with 16 mM NH_4 or $NH_4 + NO_3$ (see Table 18.6). The same was recorded for K. However, the greatest accumulation of P was recorded with urea 1–16 mM or 1 mM $NH_4 + NO_3$. Also, the greatest Mg concentration was recorded with 1 and 8 mM $NH_4 + NO_3$. The greatest Mn concentration was recorded with 16 mM NH_4 in the top and basal leaves. Zinc concentration was greatest with 1 mM urea or 1 mM $NH_4 + NO_3$. Urea and $NH_4 + NO_3$ resulted in the greatest concentration of Fe. Our data also indicated an effect of the N form on the photosynthetic rate (Tsambardoukas, 2006).

Nitrogen assimilation

Ammonium absorbed by olives is a reduced form of N and could be used by the plants to produce amino acids. Nitrate absorbed should previously be reduced

Table 18.6. Effect of N form (NH_4^+, urea and $NH_4^+ + NO_3^-$) on mineral composition of olive leaves (cv. 'Kalamon') at three concentrations (from Tsambardoukas, 2006).

Mineral	NH_4^+			Urea			NH_4^+-NO_3^-		
	1 mM	8 mM	16 mM	1 mM	8 mM	16 mM	1 mM	8 mM	16 mM
N*	1.66^b	1.98^{cd}	2.21^e	1.61^b	1.96^{cd}	2.07^{de}	1.66^b	1.97^{cd}	2.19^e
P*	0.21^{ab}	0.23^{bc}	0.32^{de}	0.37^e	0.36^{de}	0.39^e	0.39^e	0.29^{cd}	0.14^a
K*	1.10^{abc}	1.21^{bcd}	1.48^e	1.31^{de}	1.25^{cd}	1.20^{bcd}	1.06^{bc}	1.26^{cd}	1.45^e
Mg*	0.57^{de}	0.50^{bcd}	0.40^a	0.46^{abc}	0.43^{ab}	0.54^{cde}	0.48^{abcd}	0.60^e	0.46^{abc}
Fe**	40^{ab}	62^{bc}	67^c	69^c	68^c	65^c	56^{bc}	62^c	65^c
Mn**	10^a	27^c	56^e	15^{abc}	17^{abc}	42^d	15^{ab}	22^{bc}	49^{de}
Zn**	8^a	17^c	17^c	24^e	17^c	18^d	24^{de}	14^{bc}	16^c

*%; **ppm. Superscript letters in the same column denote statistically significant differences for $P \leq 0.05$ (Duncan's multiple range test).

to NH_4 in order to be assimilated (Pilbeam and Kirkby, 1992). This NO_3 reduction to NH_4 is completed in two stages. The first stage is catalysed by the enzyme nitrate reductase (NiR) located in the cytoplasm, which converts NO_3^- ions to NO_2^- (Campbell, 1988). The second stage is catalysed by NiR, which converts NO_2 to NH_3 and is located in the chloroplast.

The NO_3 reduction is expressed by the equation:

$$NO_3^- + 2e^- (NADH\ or\ NADPH^+) \xrightarrow{NiR} NO_2^- + 6e^- (fd) + 7H^+ \xrightarrow{NiR} NH_3 + 2H_2O$$

Subsequently, NH_4^+ is incorporated into organic molecules with the help of four enzymes and reactions (Nathawat et al., 2005):

- Glutamic dehydrogenase (GDH).
- GS.
- GOGAT.
- Asparagine synthetase.

The following equations are then used:

α-ketoglutaric acid + NH_4^+ + NAD(P)H → L-glutamic + NAD(P)

Glutamic acid + NH_4^+ + ATP → Glutamine + ADP + H_3PO_4

Glutamine + α-ketoglutaric acid → 2Glutamic acid

Aspartic acid + NH_4^+ + ATP → Asparagine + ADP + P

Potassium

Roles of potassium

Potassium is important for its involvement in the following processes (Tisdale et al., 1993).

- Carbohydrate metabolism.
- Metabolism of N and protein synthesis.
- Enzyme activity.
- Regulation of the opening and closing of stomata.
- Improvement in fruit quality and disease tolerance.
- Alteration of photosynthesis and respiration and activation of the enzymes peptase, catalase, pyruvic kinase, etc.

Potassium plays an important role in olive nutrition. Potassium deficiency represents 62% of nutrient deficiencies in olives. The optimum K concentration of leaves is 0.7–0.9% (leaves collected in winter from the middle of the last flush of growth, 5–8 months old).

Potassium in the soil

Potassium is absorbed by olive plants in greater amounts than any other element, excepting N. Potassium is present in relatively large quantities in most soils; the K content of the earth's crust is 2.4%. Of the total amount of K in most soils, only a fraction can be immediately utilized by plants. Soil K exists in three forms: (i) relatively unavailable; (ii) slowly available; and (iii) readily available. These three forms are considered to be in equilibrium (Tisdale *et al.*, 1993).

Unavailable K occurs as a part of the crystal structure of unweathered primary and secondary micaceous and feldspathic minerals. Readily available K occurs in the soil solution and on the exchange complex and is readily absorbed by plants. Slowly available K becomes available to plants slowly and over a longer period of time: slowly available $K^+ \leftrightarrow$ exchangeable $K^+ \leftrightarrow$ water-soluble K^+.

Potassium can be fixed in soils high in 2:1 minerals, especially those such as illite. Ammonia is almost of the same ionic radius as the K^+ ions and is subject to fixation by 2:1 clays. Because NH_4^+ can be fixed by clays in a manner similar to the fixation of K, its presence will alter both the fixation of added K and the release of fixed K. An interaction of NH_4NO_3 and K nutrition has been reported (Hagin *et al.*, 1990).

Potassium needs and fertilization

When the K concentration of leaves drops to 0.3% and symptoms of K deficiency appear, the application of 6–20 kg K_2SO_4/tree is recommended; a greater amount for heavy soils. For a concentration of 0.3–0.5%, 4–10 kg K_2SO_4/tree is suggested; for a concentration of 0.7% K is applied at twice the rate of N; and for > 0.9% no K fertilizer is added. Potassium fertilizers are applied during autumn or winter and are incorporated into the soils. Some suggest the addition of 0.8–1.0 kg K/tree for non-irrigated olive orchards and 1.0–1.5 kg K for irrigated. Fertilization with K is necessary in shallow soils with high $CaCO_3$ and also in acid soils.

Phosphorus

Roles of phosphorus

Phosphorus is a component of high-energy substances such as ATP, ADP and AMP; it is also important for nucleic acids and phospholipids (Olsen and Sommers, 1982). Phosphorus affects root growth and maturation of plant tissues and participates in the metabolism of carbohydrates, lipids and proteins. Due to its extensive root system the olive tree absorbs adequate quantities of P. Therefore, some investigators have proposed that P fertilization is not necessary. The ideal P concentration of leaves is 0.09–0.11%. However, P fertilization is applied only when there are certain indications of a true response to P. When P concentration is raised above the critical value, K concentration is reduced. Phosphate fertilization could be tried under the following circumstances: (i) in shallow and poor soils; (ii) in orchards fertilized for long periods only with N fertilizers; (iii) in soils high in $CaCO_3$; and (iv) in soils of low pH. When P fertilization is necessary the ratio of P:N in the fertilizer should be 1:3.

Soil and phosphorus fertilizers

The total P content varies from soil to soil (Tisdale *et al.*, 1993). The available P level in fertilized and cultivated soils is high because little of this element is lost in percolating water from most soils and crops. Soil P can be classed generally as either organic and inorganic. The organic fraction is found in humus, while the inorganic occurs in numerous combinations with iron, aluminum, calcium and other elements. These compounds are usually only very slightly soluble in water. Phosphates also react with clays to form insoluble clay–phosphate complexes. Phosphorus is absorbed by plants largely in the form of the ions $H_2PO_4^-$ and HPO_4^{2-} that are present in the soil solution. The surface actively absorbing P on plant roots is near the tips.

Organic soil phosphorus

Organic P occurs in the form of phospholipids, nucleic acids and inositol phosphates. The mineralization of organic P has been studied in relation to the ratio C:N:P in the soil. A C:N:P ratio of 100:10:1 for organic matter has been suggested.

Inorganic soil phosphorus

Plants absorb P largely in the forms $H_2PO_4^-$ and HPO_4^{2-}. The concentration of these ions in the soil solution at any time is low, generally never more than a few parts per million. The crop removal of P is usually in the range 0.4–2.0 kg/1000 m^2 of orchard. Phosphorus in the soil solution must be continuously replaced or the crop would not have sufficient P to grow to maturity. The concentration of the various phosphate ions in the solution is intimately related to the pH of the medium. The $H_2PO_4^-$ ion is favoured in more acid media, whereas the HPO_4^{2-} is favoured at pH above 7.0. Phosphate retention occurs in acid soils after reaction with Fe and aluminum. In alkaline soils, which are very common in olive culture, precipitation of calcium phosphate is very common.

Magnesium

Magnesium plays an important role in chlorophyll synthesis and also as an enzyme activator. Healthy olive leaves contain 0.1–0.3% Mg, and below 0.1%

Mg deficiency occurs. The important factor is not the concentration of Mg but the Ca:Mg ratio (Demetriades et al., 1960b). Some cases of abnormal olive nutrition have been reported, even with 0.64% Mg, due to the antagonism of Mg to Ca (Therios and Sakellariadis, 1982). Magnesium deficiency problems are solved by application of 2.0 kg/tree of $MgSO_4 \cdot 7H_2O$ or by spraying with a 2% solution of $MgSO_4 \cdot 7H_2O$.

Calcium

Calcium is the element that participates in the integrity and semipermeability of the plasmalemma. Calcium also regulates soil acidity and improves soil texture. The Ca concentration of leaves is 1.0–4.5%. Deficiency is recorded when Ca is < 0.5%. In order to redress Ca deficiency we add 200–1000 kg/1000 m² of $CaCO_3$ dust during the autumn and incorporate this into the soil.

Micronutrients

Iron
Iron is one of the most common of the metallic elements in the earth's crust. Its content ranges from a low of 200 ppm to more than 10%. It occurs in soils as oxides, hydroxides and phosphates, as well as in the lattice structure of primary silicate and clay minerals. Total Fe content is of no value in diagnosing Fe deficiencies.

Iron deficiency is believed to be caused by: (i) an imbalance of metallic ions, such as Cu and Mn; (ii) an excessive amount of P in soils; (iii) high pH; (iv) high soil moisture; (v) cool temperatures; and (vi) high levels of HCO_3 in the rooting medium (Benitez et al., 2002). Iron deficiency chlorosis in olive could be corrected by using a low-pressure trunk injection method (Fernández-Escobar et al., 1993). High HCO_3 levels in the soil increase the solubility of P and result in a large uptake of this element, which interferes with Fe metabolism in the plant. Excessive phosphate fertilization will also induce Fe chlorosis.

Manganese
Soil Mn originates from the decomposition of ferromagnesian rocks. The quantities vary from a trace to several thousand kg/ha. Manganese deficiency has been reported and may be induced by an imbalance of other elements such as Ca, Mg and ferrous Fe. Manganese in the soil is generally considered to exist in three valency states: (i) divalent Mn^{2+} soil solution; (ii) trivalent Mn existing in the form of the oxide, Mn_2O_3; and (iii) tetravalent

Mn (Mn^{4+}), which exists as the very inert oxide MnO_2. The three forms exist in equilibrium.

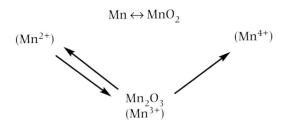

Manganese deficiency is very common at high pH values. Low and high Mn concentrations (10, 40, 160, 640 µM) significantly affect the chemical composition of the leaves in the cvs 'FS-17', 'Manaki', 'Kalamon', 'Koroneiki' and 'Picual' (Chatzistathis *et al.*, 2006). However, a genotypic variation has been recorded among the cultivars, and the cv. 'Picual' accumulated the least amount in comparison with the other tested cvs. Iron, Zn, B and Ca were not affected by Mn levels. Manganese concentration affected the Pn rate (Pn), and a significant genotypic variation was recorded, with 'Koroneiki' increasing its Pn with 640 µM Mn. The leaf fluorescence (F_v/F_m) was decreased by increasing the Mn concentration (see Table 18.7).

Boron
Boron occurs in most soils in extremely small quantities, ranging from 20 to 200 ppm, and B deficiency is common in olive trees (Demetriades *et al.*, 1960a). During flowering and fruit development mobilization of B has been recorded in olive trees (Delgado *et al.*, 1994). In most cases B does not normally occur in toxic quantities in most arable soils, unless it has been added in excessive amounts with commercial fertilizers. However, there are arid regions in which soils contain this element in toxic quantities, but these areas are few in number and of little agricultural value. Native B in most humid-region soils is in the form of tourmaline.

Most of the available soil B is held by the organic fraction and it is retained rather tightly. As the organic matter decomposes the B is released; part is taken up by plants and part is lost by leaching. Some B is held by clay or adsorbed by the inorganic fraction of soils. The principal sites of such adsorption are thought to be: (i) Fe and Al-hydroxy compounds present as coatings on or associated with clay minerals; (ii) Fe or Al oxides; and (iii) clay minerals, especially of the micaceous type. Coarse-textured, well-drained, sandy soils are low in B. The fact that clays retain B more effectively than sandy soils does not necessarily imply that plants will absorb this element from clays in greater quantities than from sands. On the contrary, plants will take up much larger quantities of B from sandy soils. The practical implication of these data is of

Table 18.7. Effects of four Mn concentrations on the mineral composition of the olive cultivars 'FS-17', 'Manaki', 'Kalamon', 'Koroneiki' and 'Picual' (from Th. Chatzistathis, personal communication).

Mn concentration (μM)	Mn*	Fe*	Zn*	Ca**	Mg**	K**	P**	B*
'FS-17'								
0 (control)	24^a	84^a	14^a	2.67^b	0.12^a	2.09^a	0.16^c	40^a
10	24^a	67^b	16^a	2.64^b	0.11^a	2.06^a	0.22^{bc}	31^b
40	46^a	68^b	19^a	2.70^{ab}	0.07^b	2.06^a	0.31^a	39^a
160	147^b	59^b	16^a	2.68^b	0.10^a	2.05^a	0.27^{ab}	38^a
640	591^c	67^b	21^a	2.81^a	0.12^a	1.74^b	0.31^a	38^a
'Manaki'								
0 (control)	20^a	85^{ab}	12^c	1.11^b	0.11^b	1.90^a	0.27^{ab}	44^{ab}
10	44^{ab}	88^{ab}	16^b	1.18^a	0.12^{ab}	1.91^a	0.25^{ab}	37^b
40	72^{ab}	98^a	15^{bc}	1.17^a	0.12^{ab}	1.88^a	0.28^{ab}	41^{ab}
160	125^b	80^{ab}	15^{bc}	1.08^{bc}	0.12^{ab}	1.90^a	0.24^b	43^{ab}
640	734^c	63^b	21^a	1.04^c	0.13^a	1.88^a	0.30^a	47^a
'Kalamon'								
0 (control)	15^a	81^a	13^a	0.92^a	0.12^a	1.59^{ab}	0.15^a	23^a
10	21^a	84^a	15^a	0.95^a	0.11^{ab}	1.57^{ab}	0.18^b	23^a
40	44^a	86^a	14^a	0.91^a	0.12^a	1.60^{ab}	1.15^a	25^{ab}
160	129^b	93^a	15^a	0.84^a	0.10^b	1.68^a	0.14^a	24^{ab}
640	625^c	62^b	13^a	0.89^a	0.11^{ab}	1.48^b	0.17^{ab}	27^b
'Koroneiki'								
0 (control)	13^a	63^a	11^b	1.40^a	0.16^{bc}	2.50^a	0.18^b	30^{ab}
10	20^a	53^a	15^a	1.43^a	0.20^a	2.25^a	0.15^b	35^a
40	25^a	58^a	12^b	1.37^a	0.17^b	2.17^a	0.26^a	35^a
160	78^b	60^a	9^b	1.32^{ab}	0.16^{bc}	2.59^a	0.23^a	31^a
640	708^c	56^a	12^b	1.19^b	0.14^c	2.41^a	0.24^a	25^b
'Picual'								
0 (control)	8^a	87^a	18^{ab}	1.68^{ab}	0.19^a	1.47^a	0.16^a	46^a
10	15^a	69^a	22^a	1.78^a	0.21^a	1.35^a	0.15^a	42^{ab}
40	24^a	78^a	16^{ab}	1.55^{ab}	0.18^a	1.76^b	0.13^a	33^c
160	40^a	80^a	19^{ab}	1.39^b	0.20^a	1.59^{ab}	0.13^a	32^c
640	189^b	74^a	15^b	1.56^{ab}	0.19^a	1.49^{ab}	0.16^a	35^{bc}

Superscript letters in the same column denote statistically significant differences for $P \leq 0.05$ (Duncan's multiple range test). *, microelements; **, macroelements.

course that rates of applied water-soluble B fertilizer should be lower on coarse-textured sandy soils than on fine-textured soils for the same degree of expected plant uptake.

Boron is one of the essential nutrients necessary for plant growth. In nature B toxicity is not as common as B deficiency. Boron toxicity that limits plant growth can be observed under certain conditions such as soils with high B content, overdosing with fertilizers and irrigation with water high in B

(Chatzissavvidis, 2002; Chatzissavvidis et al., 2005). Safe concentrations of B in irrigation water range from 0.3 mg/l for sensitive plants to 1.0–2.0 mg/l for semi-tolerant plants, like olives, and to 2.0–4.0 mg/l for tolerant plants. In closely related species, or in cultivars of the same species, genotypes susceptible to B toxicity generally give higher B concentration in plant tissues than do tolerant genotypes. Rootstock affects B concentration at the canopy of the olive tree (Chatzissavvidis and Therios, 2003).

An olive culture area with irrigation water high in B (3.6 m/l) exists in northern Greece (Chalkidiki). Leaf samples were collected over a period of 24 months and analyzed for N, P, K, Ca, Mg, Fe, Mn and Zn. The results indicated that olive is a species relatively tolerant to excess soil B, and no visual symptoms of B toxicity were observed on leaves. The two tested cultivars ('Chondrolia Chalkidikis' and 'Amphissis') followed a similar pattern of fluctuations in B concentration over the summer (see Figs 18.3 and 18.4) (Chatzissavvidis et al., 2007). The maximum B concentration in leaves was 175 mg/kg dry matter in 'Chondrolia Chalkidikis' and about 70 mg/kg dry matter in 'Amphissis'. Boron concentration in flowers was higher than that in the leaves. The seasonal variation in B concentration is also presented in Figs 18.3 and 18.4. The minimum B concentrations in the previous year's leaves of 'Chondrolia Chalkidikis' were found during flowering (see Fig. 18.5) (Chatzissavvidis et al., 2005). The number of florets per inflorescence declined from 13.4 to 10.8 for 'Chondrolia Chalkidikis' and from 21.5 to 19.7 for 'Amphissis' (Chatzissavidis et al., 2004). This decrease could be attributed partly to the relatively low leaf N (1.53%) during flowering. Fruit B concentration (see Table 18.8) increased by only 160% for 'Chondrolia Chalkidikis', with a maximum value of 122 mg/kg. The form of nitrogen (NO_3, NH_4, etc.) under a high B content of the medium modified the nutrient levels of olive plants (Chatzissavvidis et al., 2007).

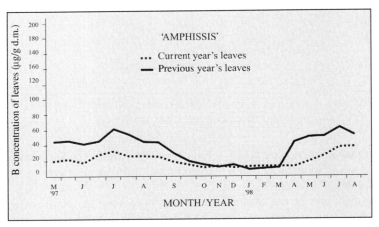

Fig. 18.3. Seasonal variation in B concentration in leaves of the current and previous years (cv. 'Amphissis') (from Chatzissavvidis et al., 2004).

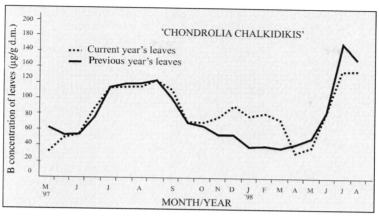

Fig. 18.4. Seasonal variation in B concentration in leaves of the current and previous years (cv. 'Chondrolia Chalkidikis') (from Chatzissavvidis et al., 2004).

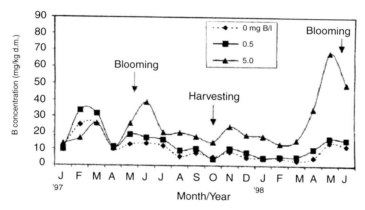

Fig. 18.5. Seasonal variation in B concentration in leaves of cv. 'Chondrolia Chalkidikis' as affected by B level in the nutrient solution (from Chatzissavvidis et al., 2005).

RESPONSE OF OLIVE TREES TO SALINITY

It is generally accepted that many stresses inducing salinity also induce an overproduction of reactive oxygen species (ROS) such as hydrogen peroxide, superoxide radical and hydroxyl radicals, and these compounds are thought to be responsible for the oxidative damage associated with plant stress.

Olive is moderately salt tolerant, although salinity may be a problem due to the high salt concentration in the irrigation water used in the Mediterranean area. Olives can be irrigated with water containing up to 3200 ppm of salt with

Table 18.8. Seasonal variation in concentrations of selected nutrient elements in olive fruits (cv. 'Chondrolia Chalkidikis') (from Chatzissavidis et al., 2004).

Nutrient element	Days after full bloom	Nutrient concentration
B (mg/kg d.m.)	30	47[a]
	60	47[a]
	90	96[ab]
	120	122[b]
N (% d.m.)	30	1.22[f]
	60	0.73[b]
	90	0.83[cd]
	120	0.90[de]
P (% d.m.)	30	0.15[d]
	60	0.11[c]
	90	0.10[bc]
	120	0.11[c]
K (% d.m.)	30	0.94[bc]
	60	0.89[ab]
	90	1.71[e]
	120	1.63[e]
Fe (mg/kg)	30	56[ab]
	60	84[b]
	90	18[a]
	120	17[a]
Zn (mg/kg)	30	18c
	60	16[bc]
	90	13[b]
	20	11[b]

Superscript letters in the same column denote statistically significant differences for $P \leq 0.05$ (Duncan's multiple range test).

a SAR of 26 or lower. It is estimated that approximately one-third of the world's irrigated lands and half the lands in semi-arid and coastal regions are affected by salinization; 10 million ha of irrigated lands are abandoned annually because of excessive salinity. Hence, an effective way must be found to use soil of high salinity by the cultivation of tolerant cultivars (Tabatabaei, 2006). Tolerance to salt appears to be cultivar-dependent (Therios and Misopolinos, 1988). For example, 80 mM NaCl may increase or decrease the percentage of perfect flowers in different cultivars. Salinity reduced the growth of all cultivars tested to varying degrees. A reduction in growth and total leaf area was recorded by increasing the concentration of NaCl in the nutrient solution (Vigo, 1999; Vigo et al., 2005). Furthermore, Pn was reduced and stomata were closed above a threshold salinity. The cultivar 'Kalamon' proved to be resistant to salinity, 'Megaritiki' and 'Kothreiki' salt-tolerant and 'Mastoidis', 'Amphissis' and 'Koroneiki' less tolerant (Chartzoulakis et al., 2002, 2004, Chartzoulakis, 2005). The degree of susceptibility to salinity stress varied between the tested cultivars. Tolerance to sea water follows the order: 'Kalamon' > 'Chondrolia Chalkidikis' > 'Manzanillo'. A decline in tolerance with age and time of exposure to salt was reported for the cv. 'Arbequina' (Aragüés et al., 2005). The effect of six different sea water dilutions (0, 4.3, 8.5, 12.8, 17.1 and 21.3% sea water) on total plant mineral content, leaf number, leaf area and leaf development, of the cvs 'Chondrolia Chalkidikis', 'Kalamon' and 'Manzanilla de Sevilla' was measured (see Tables 18.9 and 18.10).

Plants grown under NaCl salinity show an increase in Na and Cl concentrations and a decrease in plant growth and K, Ca and Mg concentrations (Vigo et al., 2002). The decrease in K, Ca and Mg concentrations is due to the antagonistic effect of Na with these three elements. A decrease in leaf B concentration is also usually observed (Vigo et al., 2002). These results show that sea water decreases plant growth, increases Na and Cl concentrations in all plant parts and decreases K, Ca and B concentration (see Tables 18.9 and 18.10).

Table 18.9. The effect of irrigation with six sea water dilutions on total plant content of P, K, Ca, Mg, Na and Cl (mg) and Fe (μg) in the olive cvs 'Chondrolia Chalkidikis', 'Manzanilla de Sevilla' and 'Kalamon' (from Vigo et al., 2005).

Sea water dilution (%)	P	K	Ca	Mg	Na	Cl	Fe
'Chondrolia Chalkidikis'							
0	29.3	200.0	94.8	32.8	28.6	60.3	5,721
4.3	32.7	207.2	72.8	35.6	73.2	95.1	6,308
8.5	30.2	169.1	65.4	31.9	79.4	84.6	3,879
12.8	21.3	118.8	51.9	26.5	67.6	67.6	3,262
17.1	20.3	96.7	47.1	23.4	68.9	75.8	2,013
21.3	19.3	96.5	49.2	26.6	86.2	123.5	3,898
LSD (5%)	7.59	48.11	20.86	NS	18.06	28.17	2,598.00
'Manzanilla de Sevilla'							
0	33.6	173.4	104.7	30.4	25.5	43.4	6,584
4.3	32.8	129.9	76.4	30.8	73.2	93.9	6,160
8.5	27.8	98.8	60.1	26.1	83.6	104.9	35,222
12.8	31.4	97.6	55.0	26.8	81.0	100.9	3,555
17.1	28.1	84.5	55.0	26.5	92.3	109.4	4,172
21.3	18.2	54.4	39.0	19.4	86.0	105.0	2,673
LSD (5%)	6.85	28.61	15.57	6.88	16.95	25.54	1,622.90
'Kalamon'							
0	11.5	78.9	50.2	12.2	9.8	19.2	1,317
4.3	10.3	56.8	36.3	13.1	26.7	27.2	1,761
8.5	11.2	70.6	37.1	15.8	39.3	37.0	989
12.8	8.7	51.8	31.3	13.0	32.4	29.0	745
17.1	8.3	52.6	28.7	12.6	32.1	30.1	704
21.3	5.9	42.5	23.2	10.4	30.0	31.4	547
LSD (5%)	NS	NS	13.18	NS	10.34	NS	600.50

LSD, least significant difference for $P \leq 0.05$; NS, not significant.

Table 18.10. The effect of irrigation with six sea water dilutions on total leaf number, final leaf area and number of days from initiation to final area of newly formed leaves[a] in the olive cvs 'Chondrolia Chalkidikis' ('Ch. Ch.') and 'Manzanilla de Sevilla' ('Mz. S.') (from Vigo et al., 2005).

Sea water dilution (%)	Total leaves (n)		Final leaf area (cm^2)		Time from leaf initiation to final leaf area (days)	
	'Ch. Ch.'	'Mz. S.'	'Ch. Ch.'	'Mz. S.'	'Ch. Ch.'	'Mz. S.'
0	23	22	4.9	4.1	20	20
4.3	19	20	5.3	3.8	20	20
8.5	17	19	5.2	3.4	21	20
12.8	16	16	4.9	3.6	22	21
17.1	13	15	4.7	3.1	22	21
21.3	15	14	4.7	3.0	22	22
LSD (5%)	3.6	3.0	NS	0.40	NS	NS

[a]Average of all newly developed leaves on the main shoot during the 157 days of the experiment.
LSD, least significant difference for $P \leq 0.05$; NS, not significant.

With regard to the distribution of Na and Cl, even at the highest sea water concentration most of the Na and Cl was found in base stems and roots, indicating that the plants of the tested cultivars excluded Na and Cl ions from their leaves. Thus it is clear that salt exclusion from the shoot is the mechanism of salinity tolerance in olives. Tolerance to sea water of the tested cultivars seems to be related to total plant K content, K concentration and the K:Na ratio of leaves. Also, the relatively high Ca concentration of the roots of 'Kalamon' plays an important role in improving tolerance to salinity. Calcium maintains the integrity and function of cellular membranes. In that way root selectivity for K instead of Na is maintained. The increase in leaf thickness as a response to saline treatment seems to increase the internal surface per unit of leaf area in which CO_2 and water vapour diffusion takes place, so reducing the internal resistance to CO_2 assimilation. The results indicate that, in coastal areas with light-textured soils, diluted sea water could be used for irrigation of olive plants. Finally, the correlation coefficients of mineral concentrations with (i) time (see Table 18.11); and (ii) B concentration (see Table 18.12) are presented.

FERTILIZATION OF OLIVE ORCHARDS

Spatial variability in soil fertility

The most common practice in olive orchards is to apply fertilizer without taking into consideration the variability in soil fertility (López-Granados et al.,

Table 18.11. Correlation coefficients of mineral concentrations with time, experimental results (from Chatzissavvidis et al., 2005).

Mineral	Boron concentration in nutrient solution (mg/l)		
	0	0.5	5.0
B	−0.574*	−0.498*	0.483*
N	0.600**	0.862**	0.684**
P	NS	NS	NS
K	0.648**	NS	NS
Ca	NS	NS	NS
Mg	NS	NS	NS
Fe	0.646**	0.730**	0.655**
Mn	NS	0.551*	0.530*
Zn	NS	NS	NS

*Significant for $P \leq 0.05$; **significant for $P \leq 0.01$; NS, not significant.

Table 18.12. Correlation coefficients between B concentration and concentration of the other elements (Chatzissavvidis et al., 2005).

Mineral	Boron concentration in nutrient solution (mg/l)		
	0	0.5	5.0
N	−0.593**	−0.567*	NS
P	NS	NS	NS
K	NS	NS	NS
Ca	−0.471	NS	−0.736**
Mg	NS	−0.457*	NS
Fe	−0.632**	−0.576*	NS
Mn	−0.632**	−0.644**	NS
Zn	NS	NS	NS

*Significant for $P \leq 0.05$; **significant for $P \leq 0.01$; NS, not significant.

2004). This process may lead to under-application or over-application of nutrients in some olive orchards. As a consequence under-fertilized orchards do not give the optimum yield, while in over-fertilized sites there is the danger of pollution with NO_3 and also the high fertilization cost (Bouma, 1997). Therefore, olive growers should be aware of the varying fertility levels of individual orchard sites and manage their fertilization appropriately.

Nitrogen fertilization should be conducted at the end of winter, supplying ammonium sulphate or urea to the soil or urea in the form of a foliar spray. For olive growers, therefore, in order to both protect the environment and increase their income it is necessary to adopt a strategy that adjusts fertilization according to the fertility requirements of each olive grove. Hence, N fertilizer is not supplied when leaf N is greater than a threshold value; such a limit, according to Fernández-Escobar et al. (2002), is ⩾ 1.5%. Nitrogen levels above this limit adversely affect the polyphenol level of the oil, negatively affecting the quality of both the oil and the groundwater. The quality of table olives is also affected by fertilization (Morales-Sillero et al., 2006). Furthermore, adequate K fertilization increases drought tolerance and adaptation (Restrepo et al., 2002).

Nutrient uptake by olive

Many olive growers believe that olive plants have low requirements for nutrient elements, and so under-fertilize olives. Others used to apply fertilizer quantities in excess of their actual needs. The total N absorption required to cover the annual needs of growth ranges from 9.2 kg/ha during the second year of growth to 185.9 kg/ha in the sixth. The respective quantities for K_2O are 8.6–163.1 kg/ha, and for P 1.3–30.7 kg/ha.

The N, P, and K requirements (percentage of the total quantity) during the stages of fruit set, pit hardening and maturation are given in Table 18.13.

Estimation of fertilizer requirements

The optimal rate of fertilizer application to olives is that which results in the maximum income. This rate is smaller than the fertilizer quantity required to produce the maximum yield. The most important way to increase olive production within a particular olive culture is to recognize the nutrient deficiencies and to apply the required quantities of selected fertilizers. Therefore, a relationship should be established between the rate of fertilizer and yield. In order to do this, experimentation should be planned to give the required information.

This relationship could be expressed by the equation $G=f(N)$, where G is the profit from the product and $f(N)$ is the rate of the application of one fertilizer (N=nitrogen, P=phosphorus, K=potassium). The optimal rate is calculated by solving the equation $dG/dN=0$, where N is the amount of N producing the

Table 18.13. N, P and K requirements of olive trees during three important stages in the fruiting process.

Nutrient	Fruit set (%)	Pit hardening (%)	Maturation (%)
N	39.5	28.5	32.0
P	24.6	38.9	36.5
K	33.5	31.4	35.1

maximum profit and dG is the differential equation indicating profit increase per unit nitrogen, dN. For concurrent application of two or more nutrients the equation becomes G=f(N, P, K), where N, P and K are the optimal rates of nitrogen, phosphorus and potassium. The optimal rates are calculated by solving the equation dG/dN=dG/dP=dG/dK.

Foliar application of urea to olive

With the exception of N, macronutrients cannot be applied to fruit trees foliarly. Urea contains a high percentage (46%) of N. Furthermore, uptake of urea is rapid after its application. When conditions are favourable 60–70% of the urea applied can be absorbed by the olive leaves. When foliar application is conducted during October the N content of the olive leaves increases by 47%, is subsequently reduced over the next 6 months and redistributed within the plants in order to effect various physiological processes. Therefore, accurate timing of foliar fertilization is important in order to effect certain processes.

The use of ^{15}N-labelled urea allows the measurement of redistribution of N within the plant (Weinbaum, 1984). Foliar uptake of urea is not influenced by leaf N status. Translocation of urea N from mature leaves to vegetative tissues in N-deficient plants, however, is lower than that in N-sufficient plants. Foliage-applied urea is translocated acropetally and basipetally. Translocation of foliage-applied urea N to roots of N-deficient plants is less affected than translocation to shoots.

19

GROWTH AND SALT TOLERANCE OF THE OLIVE

INTRODUCTION

Salinization of lands is increasing progressively throughout the world. It is estimated that approximately one-third of the irrigated lands on a worldwide basis and one-half of the irrigated lands in semi-arid regions are salinized. It is also estimated that 10 million ha of these irrigated lands are taken out of production due to high salinity (Epstein *et al.*, 1980). Of the total land area cultivated, about 5% (Munns, 1993) is affected by salt, and therefore, it is important to study salinity and to measure the salt tolerance of the various existing cultivars.

With respect to olives there is not enough information available. Therefore, this chapter presents the available knowledge concerning the behaviour of olives to salinity.

PLANT GROWTH, YIELD AND OIL QUALITY

Plant growth of olives, i.e. dry weight, shoot length, total leaf area and root length, is reduced by moderate and high levels of salinity (Therios and Misopolinos, 1988). The amount of growth reduction is a function of the cultivar, salt type and its concentration and the length of exposure of plants to the salt. In the olive, leaves and their total area are more affected by salts than other plant parts. Salinity above a critical limit leads to stomatal closure, smaller leaf size, reduction in the number of perfect flowers per inflorescence and also pollen viability. Furthermore, the top:root ratio is reduced.

The effect of salinization on pollen characteristics results in a significant reduction of viability, germination capacity and fruit set (Cresti *et al.*, 1994). High salinity reduces the yield of olive trees (Gucci and Tattini, 1997); high salinity and brackish water reduce yield and oil percentage in most cases. However, some authors (Bouaziz, 1990) report that irrigation with brackish water did not affect yield and oil percentage and did not cause alternate bearing. Klein *et al.* (1992) reported for electrical conductivity (Ec) values up to 4.2 dS/m an increase or decrease in the yield observed, this being a function of planting density. However, when the Ec of the irrigation water was 7.5 dS/m oil yield and total fresh weight of fruits declined by 60–80% in comparison with the control irrigated by good-quality water. Other authors report that saline reduces oil content and that this decrease is cultivar dependent (Chartzoulakis *et al.*, 2004; Stefanoudaki, 2004).

Salinity stress affects olive oil composition by an acceleration of ripening, and its effects can be both positive and negative (Cresti *et al.*, 1994). The linoleic:oleic acid ratio sharply increases in the oil during the growing season, as the degree of unsaturation increases with the fruit ripening. Also, the activity of linoleic acid desaturase is strongly inhibited under saline conditions. Another effect is the increase of aliphatic and triterpenic alcohols, as a consequence of accelerated ripening.

Salinity results in increased phenol concentration in the oil (Stefanoudaki, 2004), and especially in the second fraction of secoiridoid derivatives. With regard to negative effects the increase in total saturated fatty acids – and especially that of palmitic acid – was included (Zarrouk *et al.*, 1996; Stefanoudaki, 2004). Furthermore, oleic acid decreases or linoleic acid increases, and the ratio of oleic:linoleic acid decreases.

WATER RELATIONS AND PHOTOSYNTHESIS

The effect of salinity can be divided into osmotic and toxic components. Due to the osmotic effect water absorption is reduced as salinity increases (Therios and Misopolinos, 1988). Upon exposure to salinity leaf water potential (Ψ_{leaf}) and relative water content (RWC) are reduced. The decrease in Ψ is followed by a decrease in osmotic potential (Ψ_s). It is important to understand that when salinized olive plants at levels of up to 100 mM NaCl are exposed again to non-salinized solutions the plants completely recover and start growing again. The olive plant has, therefore, the ability to adjust osmotically. Osmotic adjustment is attained either by ion accumulation in the vacuoles or through soluble carbohydrates. The most common osmoregulatory substance in olives is mannitol.

In experiments with the cultivars 'Leccino' and 'Frantoio', cations contributed about 40% of the total Ψ, while the contribution of mannitol and glucose was about 30%. An analysis of the water relation characteristics of olive leaves under salt stress for the salt-tolerant 'Frantoio' and salt-sensitive 'Leccino' was reported by Gucci *et al.* (1997a). The osmotic contribution of Na was 0.1–2.1% in the controls and 15–20% for the 200 mM salt-treated plants (Gucci *et al.*, 1997a).

When searching the literature on the subject of photosynthesis, all published papers to date show that Pn is reduced under salinity. The initial Pn value of about 14–15 μmol is significantly reduced. The rate of decrease of Pn depends on the salt tolerance of the olive cultivar; this decrease varies from 20% for the salt-tolerant olive cv. 'Kalamon' up to 62% for the cv. 'Amphissis', which is moderately sensitive (Chartzoulakis *et al.*, 2002). This reduction in Pn could be ascribed to Na and Cl accumulation in the leaves under salt stress (toxic effect), and also to the decrease in mesophyll conductance due to increased leaf thickness (Syvertsen *et al.*, 1995).

Salinity has a negative effect on photosynthesis (Tattini *et al.*, 1995, 1997, 1999; Vigo *et al.*, 2002; Centritto *et al.*, 2003; Loreto *et al.*, 2003; Chartzoulakis *et al.*, 2004; Chartzoulakis, 2005; Tabatabaei, 2006). Salinity stress brought about a reduction in net CO_2 assimilation and stomatal conductance in two cultivars (Tattini *et al.*, 1997), but the effect was more pronounced in the salt-tolerant cv. 'Frantoio' than in the salt-sensitive 'Leccino'. Hence, gas exchange parameters may be misleading if used to evaluate the salt tolerance of olive genotypes. Furthermore, upon relief of NaCl stress the recovery in gas exchange parameters was slower in the salt-sensitive cultivar.

ION CONTENT OF VARIOUS OLIVE TISSUES

The effect of salinity on nutrient content is evidenced by the increase or decrease in the concentration of certain micro- and macronutrients. Hence, increases in Cl^- and Na^+ concentrations in olive tissues occur with increasing NaCl in the soil (Tattini *et al.*, 1992). The most tolerant cultivars prevent the entry or translocation of Na and Cl from the roots to the parts above ground level. Therefore, salt-tolerant cultivars compartmentalize the toxic ions to root or trunk tissues. When the concentrations of NaCl are moderate, olive cultivars activate an Na^+ exclusion mechanism.

However, when the concentration of salts in the soil solution is high, sodium is transported to the tops, especially in salt-sensitive cultivars. Thus, toxicity symptoms appear in the form of peripheral necrosis of leaves. The opposite is true for the salt-tolerant cultivars, which translocate less sodium to tops. Hence, tolerant cultivars such as the Greek cv. 'Kalamon' and the Italian 'Frantoio' have more efficient mechanisms of inhibiting Na^+ transport to leaves or effecting Na^+ exclusion or retention by the roots.

The effect of salinity on P absorption depends on various factors such as plant species, cultivar, stage of growth, salt concentration and the ionic form of P in the substrate. In many cases salinity reduces P concentration (Sharpley *et al.*, 1992), while in other cultivars P concentration remains constant. Decreased P concentration, as a result of salinity, was observed in experiments conducted in the soil. The K^+ concentration of leaves was reduced when the NaCl of the nutrient solution was increased (Janzen and Chang, 1987). Furthermore, the K^+ content of leaves in four olive cvs ('Amygdalolia', 'Adramittini', 'Manzanillo' and 'Frantoio') was decreased by increasing the

NaCl concentration in the nutrient solution (Therios and Karagiannidis, 1991) due to K^+–Na^+ antagonism. The highest K^+ decrease occurs in roots and old leaves, while young olive leaves have higher K^+ concentrations, which may help osmoregulation in the plant.

Calcium plays many important roles in olive nutrition: cell division, cell wall formation, protein synthesis, competition with other elements and selectivity and semipermeability of membranes. Calcium competes with other cations like Na^+ (Marschner, 1997). Calcium is a key element in limiting the toxic effects of Na^+, increasing the Ca^{2+}:Na^+ ratio in the external solution, which alleviates the toxic symptoms of NaCl (Rinaldelli and Mancuso, 1994).

Calcium exerts a positive role concerning salt tolerance. This effect is manifested in both the trans-membrane and trans-root potentials. The presence of Ca^{2+} considerably limits the decrease in extra- and intracellular resistance due to Na^+. Therefore, the function of Ca is protective with respect to membrane integrity by both maintenance of adequate Ca^{2+} binding to the plasma membrane and prevention of K^+ leakage (Cramer et al., 1985; Mancuso and Rinaldelli, 1996).

Salinity can be reduced by soil reclamation and drainage, but the cost is high. Therefore, a more cost-effective approach is to add Ca^{2+} to the saline growth medium. Addition of calcium sulphate maintains membrane permeability and selectivity. Salinity stress has been shown to induce transient Ca^{2+} influx into the cell cytoplasm. Therefore, calcium channels responsible for Ca^{2+} influx may represent one type of sensor to these stress signals. Increases in cytosolic Ca^{2+} are perceived by various Ca^{2+}-binding proteins such as Ca^{2+}-dependent protein kinases and the SOS3 family of Ca^{2+} sensors. Genetic analysis indicates that the salt-overly-sensitive (SOS) mutants SOS1, SOS2 and SOS3 function in a common pathway in controlling salt tolerance.

CYTOKININS AND THE RESPONSE TO SALINITY STRESS

Cytokinins (CKs) are involved in the regulation of many aspects of growth and differentiation, such as nutrient mobilization and apical dominance (Hare et al., 1997). Furthermore, CKs are involved in responses to adverse conditions, like salinity. Plant hormones play a role in the transformation of stress signals into the gene expression necessary to effect adaptation to marginal environmental conditions. Among the growth regulators linked to responses to abiotic stresses ABA and ethylene are included. Cytokinins are antagonists of ABA in many processes such as stomatal opening, and in plant responses to high salinity. Among the growth regulators ABA and CKs are likely to play a crucial role in controlling the responses of plants to stressful conditions of the environment.

Endogenous and exogenous cytokinins and their effects on stress responses

Under field conditions water deficit is produced by water deprivation and soil salinity. Studies with root exudates indicate that, when sunflower plants were exposed to water stress, their root exudates contained less kinetin-like activity substances. In leaves of tomato plants during the first 2 days after the initiation of salt stress, CKs increased and this was followed by a depression in concentration afterwards. Furthermore, the levels of CKs in the shoots of sunflower plants affected by salinity were almost half of those of the unstressed plants. The decrease in CK content is an early response to salt stress, and the effects of NaCl on salt-sensitive varieties are not mediated by CKs. The exogenous application of kinetin alleviated the salinity stress effects seen during the growth of wheat seedlings. The CK thidiazuron (phenylurea-type CK) significantly increased the yield of salinity-stressed wheat.

Mineral deficiency, salinization and cytokinin levels

Salinity reduces mineral availability and this significantly affects CK levels *in vivo*. Salt stress lowers N levels in plants and leads to a decrease in CK levels (Darral and Wareing, 1981). The same is true for P deficiency and CK levels. When CKs are applied exogenously in N-deficient plants growth is stimulated, indicating that CKs change the nutrient supply, especially of N. The symptoms of Ca^{2+} deficiency are delayed by benzyladenine (BA) application.

The effect of salinity on cytokinin metabolism and transport

Cytokinin metabolism is a dynamic process between CK biosynthesis and catabolism, resulting in loss of biological activity. Due to the existence of at least 20 natural CKs in plants it is not easy to evaluate the changes in CK levels under salinity stress conditions. In fact there is not enough information on olives about the pathway of CK biosynthesis and the enzymes which regulate certain metabolic steps. The activity of CKs can be reduced by processes leading to their inactivation via the formation of CK nucleosides or N-glycosylation, conjugation or oxidation. The enzyme which degrades CKs is CK oxidase. The reversible conjugation of CKs is a possible mechanism by which olive plants achieve hormonal homeostasis and CK activity under adverse conditions. CKs are considered to be transported from the roots to the shoots via the xylem and transpiration stream, indicating the mechanism by which the root affects shoot physiology.

Many CK effects have long been known to be mediated via stimulation of ethylene production. Ethylene is induced by many stress factors such as drought and salinity. Cytokinin-mediated stimulation of ethylene production appears to be of particular importance during water and salt stress. Apart from ethylene, salicylic acid (SA) and jasmonic acid (JA) are also important signal molecules involved in the activation of plant deficiency responses to salinity (Sano et al., 1996). Cytokinin is required for the biosynthesis of SA and JA, which subsequently serve as endogenous inducers.

SALINITY AND OXIDATIVE STRESS

Salinity involves both an osmotic and a toxic effect of salts. Furthermore, salinity generates oxidative stress, thus increasing the level of reactive oxygen species (ROS) such as superoxide radical (O_2^-), hydroxyl radical (OH^-) and hydrogen peroxide (H_2O_2), all of which damage membranes, proteins, DNA and lipids (Fadzilla et al., 1997).

Determination of the malondialdehyde (MDA) content, and therefore the extent of membrane lipid peroxidation, is used to assess the degree of plant sensitivity to oxidative damage. Lipid peroxidation remains unchanged in plants tolerant to salinity and drought.

SALINITY AND MANNITOL CONTENT

In olive pulp, apart from the other components, concentrations of up to 8 mg/g d.m. of mannitol have been recorded (Conde et al., 2007). Mannitol is a polyol or sugar alcohol. In some plant species polyols are products of photosynthesis of mature leaves. Mannitol is present in more than 100 species. The most common families include the following: Rubiaceae (coffee), Oleaceae and Apiaceae. Mannitol is produced in leaves from mannose 6-phosphate via the activities of NADH-dependent mannose 6-phosphate reductase and mannitol 6-phosphate phosphatase.

Mannitol production leads to more efficient carbon use and confers salt tolerance. Mannitol content increases as the salinity of the growth medium increases. Addition of 100–500 mM NaCl to cultivated olive cells enhances the capacity of the polyol–H^+ symport system. Furthermore, cell viability is greater in mannitol-grown cells after a 250 or 500 mM NaCl pulse, while sucrose-grown olive cells have reduced viability.

TREATMENT OF NUTRIENT SOLUTIONS OR IRRIGATION WATER WITH RADIO WAVES IN THE REDUCTION OF SALT TOXICITY

A new approach to decreasing the negative effects of salinity of irrigation water is by changing the characteristics of the nutrient solution or the irrigation water. This is accomplished by using an electromagnetic field to which the nutrient solution or the irrigation water is exposed before its use. Experiments in hydroponic culture of tomatoes indicate that the electromagnetic field improves the solubility of salts and the total yield (Roberts, 2002). The equipment used produces low-frequency radio waves, i.e. 2×10^6 vibrations/s, which breaks the salts down to a size of 2–4 μ. This treatment does not modify the chemical composition, and offers the following advantages:

- Use of brackish or saline water for irrigation (Bouaziz, 1990; Benlloch *et al.*, 1991; Briccoli Bati *et al.*, 1994; Fodale *et al.*, 2006).
- Decrease in salt toxicity.
- Improvement in fruit size and yield.

Data from the cv. 'Chondrolia Chalkidikis' indicate that, even with an Ec of 30 dS/m, plants continue to grow and no symptoms of toxicity appear (Ioannis Therios, unpublished data). The number of leaves is significantly greater with the use of the electronic equipment. Potassium concentration is constant at all Ec values with the use of the equipment and decreases without it. Furthermore, Ca concentration is significantly greater with this method, but lower and decreased with increasing Ec. Phosphorus follows the same pattern as Ca.

Another way to reduce Na accumulation in olive leaves and to restore K levels similar to those of controls is the inclusion of 0.1–100 mM K in the nutrient solution. It has been suggested that K^+ is one of the osmolytes, and its accumulation in plant cells might facilitate osmotic adjustment, lower the internal osmotic potential and contribute to salt tolerance. However addition of K^+ to the irrigation water did not improve Pn in olives compared with plants receiving the same salinity treatment but with no added K^+.

SALT TOLERANCE

Olive is a moderately salt-tolerant plant (Rugini and Fedeli, 1990). At Ec values of 4–6 dS/m olive growth is reduced by 10%. Therios and Misopolinos (1988) found that 3-year-old olive plants could be grown without problems at NaCl concentrations < 80 mM (Ecw of 8.0 dS/m) during a 3-month period (see Fig. 19.1). Further studies indicated that olives can even do well under higher NaCl concentrations: experiments were conducted with NaCl concentrations of up to 150 mM NaCl. In another experiment olive trees were irrigated with Hoagland nutrient solution diluted with sea water (Ec of sea water = 58 dS/m); the maximum rate of dilution was 23% and its Ec value was 13 dS/m (Vigo *et al.*, 2005). Furthermore, the type of salt in the irrigation water affects the severity of the toxicity symptoms. In one experiment the salts NaCl, $CaCl_2$ and KCl were tested. From these, at iso-osmotic concentrations, the most toxic was KCl, reducing growth by a far greater extent than the other two (Vigo, 1999;

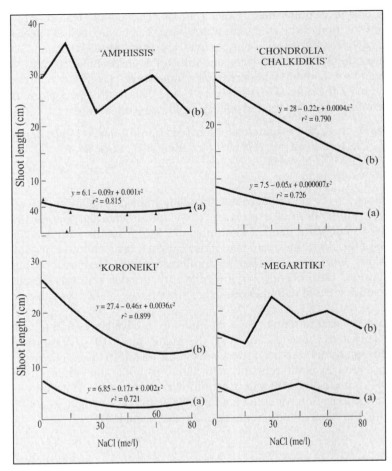

Fig. 19.1. Effects of NaCl (0–80 me/l) concentrations on shoot growth 1 (a) and 4 (b) months into an experimental trial on four olive cultivars (from Therios and Misopolinos, 1988).

Vigo et al., 2002). Furthermore, other authors (Bartolini et al., 1991) found that Na_2SO_4 was more toxic than NaCl when the cv. 'Maurino' was used. Salt toxicity, of course, is a function of the olive cultivar and there is significant genotypic variation.

Various authors provide more information on the classification of salt tolerance in olive cultivars (Therios and Misopolinos, 1988; Bouaziz, 1990; Tattini et al., 1992, 1994; Benlloch et al., 1994; Marin et al., 1995; Gucci and Tattini, 1997; Vigo, 1999; Chartzoulakis et al., 2002; Troncoso et al., 2004; Aragüés et al., 2005; Chartzoulakis, 2005; Demiral, 2005). Chartzoulakis

(2005) classified these data in the form of a table giving the resistance of cultivars, their origin and the source of the literature. Olive cultivars are classified as salt tolerant, semi-tolerant or sensitive.

- Tolerant: 'Megaritiki', Lianolia Kerkiras', 'Kalamon', 'Kothreiki', 'Frantoio', 'Arbequina', 'Picual', 'Lechin de Sevilla', 'Hamed' and 'Chemlali'.
- Semi-tolerant: 'Amphissis', 'Koroneiki', 'Mastoides', 'Valanolia', 'Adramittini', 'Coratina', 'Moraiolo' and 'Maurino'.
- Sensitive: 'Chondrolia Chalkidikis', 'Throubolia', 'Leccino', 'Nabal', 'Chetoui', 'Calego' and 'Meski'.

Salt tolerance in olives could be ascribed to certain mechanisms (Tattini *et al.*, 1994; Amane *et al.*, 1999; Al-Absi *et al.*, 2003), such as retention of Na or Cl by the roots and stem (Tattini *et al.*, 1994) or ion exclusion due to K selectivity instead of Na. Another mechanism is compartmentation of toxic ions. Salt-tolerant cultivars restrict the ingress of toxic ions such as Na or Cl into vacuoles and thus avoid toxicity (Loreto and Bongi, 1987).

FUTURE GOALS

One widely used approach to unravel olive salt-tolerance mechanisms has been to identify cellular processes and genes whose activity or expression is regulated by salt stress (Watad *et al.*, 1991; Hasegawa *et al.*, 2000). The assumption is that salt-regulated processes and genes probably function in salt tolerance. Changes in cellular activities in response to salt stress include cell wall alteration, decline in photosynthesis (Locy *et al.*, 1996), potassium content (Rains, 1972) and increases in Na^+ and organic solutes (proline, mannitol, etc.) (Greenway and Munns, 1980). Another approach is to identify genes and cellular processes that are crucial for olive salt tolerance. Research with the species *Arabidopsis* resulted in the isolation of several salt-overly sensitive (SOS) mutants such as SOS1 (Wu *et al.*, 1996), which is defective in high-affinity K uptake. This means that K uptake is essential for salt tolerance in glycophytes (Wu *et al.*, 1996) under NaCl stress. SOS1 mutants accumulate more proline (Liu and Zhu, 1997) and absorb less Na^+ and K^+.

The SOS2 mutants are specifically hypersensitive to inhibition by Na^+ or Li^+ and not hypersensitive to general osmotic stress. Furthermore, SOS2 mutants are necessary for K^+ nutrition, since they do not grow in a culture medium with low K^+ concentration. Therefore, SOS1, SOS2 and SOS3 are postulated to encode regulatory components controlling plant K^+ nutrition that, in turn, are essential for salt tolerance.

Some questions which should be asked include:

1. Do genes similar to SOS1, SOS2 and SOS3 exist?
2. Is salt sensitivity in the olive correlated with cellular Na^+ content?

3. Do genes exist in olives encoding for: (i) the enzyme PCP synthetase involved in proline biosynthesis; (ii) a protein with a potential protective function during desiccation (Yamaguchi-Shinozaki and Shinozaki, 1993); and (iii) a gene of cold tolerance, drought and salt stress (MYB)-related transcription factor and phospholipase that plays a role in osmotic signal transduction (Hirayama *et al.*, 1995)?

4. Which is more detrimental to plants under salt stress: higher tissue Na^+ or lower K^+?

5. Does the level of salt tolerance, as measured by root growth, correlate closely with K^+ content?

6. The accumulation of compatible osmolytes is often considered to be a universal protective mechanism used by many plants under salt stress. Is proline accumulation an adaptive process or a response to salt stress injury? This is an open question.

Pruning

WHAT IS PRUNING?

Pruning of olive trees has two main objectives, to improve vegetative growth and reproductive growth. The term pruning includes the techniques known as *cutting, heading, incision, inclination, twisting* and *girdling*. Successful pruning involves the knowledge and experience of olive tree physiology; the response of trees to pruning varies with the age of the tree, vigour and cultivar, and it is an excellent and very efficient means of controlling plant growth and productivity. In pruning, by cutting certain parts of the canopy, the growth of the remaining parts is stimulated; cutting is separated into the categories of thinning and heading. When we undertake thinning we eliminate shoots from their base. With such removal the entire shoots stimulate the vigour of the remaining shoots by reducing competition for light, nutrients and water. An excellent analysis of pruning and training systems for olive trees can be found in Gucci and Cantini (2004b). By the term heading we mean the cutting back of a young shoot to a bud or to shorten an old shoot to a lateral branch. With heading, the growth of shoots close to the cut is promoted.

WHY PRUNE?

Pruning is an important agricultural task. There are many reasons why, and circumstances when, olive trees should be pruned.

- Rapid development of tree skeleton and achievement of a balance between structure and productivity, and optimum light penetration.
- Achievement of a strong tree skeleton necessary to support fruit load and able to transmit the vibrations from mechanical harvesters.
- To achieve early onset of production and high yields.
- To renew the canopy in old, non-productive trees to stimulate productivity, (Cimato *et al.*, 1990).

- To create new canopy after the olive tree has been damaged by frost, fire, pests or disease.
- To give the appropriate shape to the olive canopy and to adjust the shape of the tree canopy for the dense and super-high-density planting systems.
- To improve the quality of olive fruit and oil.

RULES OF PRUNING

The style of pruning depends on olive tree age, crop load, the use of olives as table or olive oil fruits, soil fertility, environmental conditions and fruit load. Although these factors differ from one area to another and from cultivar to cultivar, certain general rules are applicable to pruning.

- It is not necessary to prune olive trees every year; in some cases pruning is conducted every 2 years. Therefore, pruning cost is a very important factor.
- Plant age is a determining factor of the type of pruning (light, medium, severe).
- The pruning method should be simple and fast; we can then check light penetration into the canopy and assess the need for more severe pruning.
- Pruning starts from the top of the tree and proceeds towards the base. Large shoots are cut first, followed by those of smaller diameter.
- In mature plants pruning is light; pruning intensity increases with age of plant.
- The cut should be executed close to the point of attachment of the lateral branch.

APICAL DOMINANCE

Growth in olives is controlled by endogenous hormones, and pruning can modify the regular growth pattern by changing the hormone balance. Auxin (IAA) is produced in the upper shoot tips, and in young leaves it is transported downward via the phloem and exerts a certain degree of control on shoots, buds and branches lying below the shoot tips. The large amount of IAA produced and transported downwards affects the growth of shoots by the following means:

- Inhibition of lateral bud growth.
- Decrease in growth rate in length of lateral branches.
- Increase in the angle between side branches and trunk.

These growth characteristics are linked to the supply of IAA and can be modified by pruning, girdling or hormone application. The movement of auxin and its asymmetric distribution affects lateral bud inhibition, cambial activity and stem elongation. The same endogenous concentration of auxin has varying effects depending on the plant tissue, i.e. it inhibits growth of lateral buds and stimulates cell division in the cambium. Therefore, when IAA produced by a vigourous stem moves to the base of the stem, this results in a wider crotch angle.

EFFECTS OF PRUNING

Photosynthesis

The rate of CO_2 assimilation per unit area of leaf is an important characteristic of higher plants and integrates all the biochemical and biophysical processes. Pruning techniques lead to removal and reduction of photosynthetic area. This decreases the efficiency of carbon assimilation and also the vigour of the remaining shoots. Large decreases in leaf area by pruning negatively affect olive tree growth. On the contrary, absence of pruning or very light pruning has, as a consequence, the shading of the olive canopy.

Shading delays fruit ripening and decreases olive oil production. Due to lack of pruning, in a dense canopy light penetration is sometimes very low and less than 10% of sunlight intensity. Due to shading, flower bud differentiation and fruit set can be low and this effect is a function of cultivar, duration of shading and time of year. Hence, shading from July to October decreases flower bud formation (Tombesi and Cartechini, 1986). Photosynthesis requires open stomata, and therefore CO_2 assimilation is concurrent with transpiration, leading to water loss. This means that pruning removes a proportion of leaf area and it reduces water consumption and improves water use efficiency.

Shape of the olive plant and its growth habit

When olive plants remain unpruned they retain a central leader, and their shape is spherical, hemispherical or ellipsoidal. The shape of the olive tree can be explained by the growth habit, which is basitonic. Therefore, the lateral shoots closer to the apex grow more than shoots located on the lower part of a branch. In addition, differences in canopy shape also arise from differences between cultivars.

Shoot length and pruning

The shortening of shoots should not be excessive, since production will be reduced, on the contrary, if the roots are not shortened, they are going to overproduce and finally become exhausted. A practical means of determining the physiological condition of the current year's growth of fruiting shoots is their length. Shoots < 20 cm are inadequate and pruning is necessary to increase their length. Shoots > 60 cm are considered very vigourous, but their productivity is unreliable. The best shoot length varies between 20 and 60 cm; such shoots produce the highest yields.

PRUNING AND HORMONES

Growth in olives is controlled by endogenous hormones, and pruning can alter the growth pattern by modifying the hormone balance. The auxin IAA is produced in vigourous shoot tips and in young leaves. Afterwards, it moves downwards in the phloem and modifies the growth habit of shoots, buds and branches below the point of auxin production (Westwood, 1978). The auxin thus transported affects the olive tree in the following ways:

- Inhibition of lateral bud growth.
- Depression of elongation of lateral branches.
- The angle between the trunk and lateral branches is increased.

These effects depend on the natural auxin supply, which can be modified by pruning or hormone application. Decapitated shoots treated with 250–1000 ppm of indolebutyric acid (IBA) resulted in wide branch angles of all the shoots below that level. In contrast, girdling stopped the downward movement of auxin and gave rise to narrow-angled branches.

PRUNING TOOLS

Essential tools are pruning shears, handsaw, chainsaw and gloves. Of these the handsaw is an efficient tool for cutting shoots or branches of diameter 30–80 cm. The chainsaw has to be light in weight, thus reducing the pruning time and its cost. Furthermore, pneumatic tools may be used. All tools should be clean and sharp and must be disinfected, especially if there is any possibility of the presence of *Pseudomonas syringae* pv. *savastanoi*. When trees are thus infected they should be sprayed foliarly with Cu or other pesticides. Large cuts should be protected by covering with pruning pastes or paints.

PRUNING AND TREE AGE

Pruning costs

Pruning represents 20–30% of the annual cultivation cost of olives. When employing mechanical pruning that cost is reduced to 7%. However, this type of pruning has certain disadvantages. The high cost of hand pruning is ascribed to the high cost of labour and the significant amount of time required per tree. The developing trend today is to reduce costs by reducing pruning to a minimum.

Pruning during the training period

The objectives of pruning during the training period are set out below:

- To create trees of good shape.
- To obtain heavy crops and to bring trees into production as early as possible.
- To achieve maximum light penetration into the canopy.

This can be managed with trees that have the following qualities:

- Appropriate tree size and good sanitary conditions.
- Freedom from pests such as *Prays oleae*, *Saissetia oleae*, *Spilocaea oleaginea* and *Verticillium dahliae*. The best size for pruning is 70–120 cm in height; smaller plants also can be used, but with no single trunk.

The following guidelines are applicable to the pruning of young trees (Gucci and Cantini, 2004):

- Avoid shortening of shoots, since this promotes the emergence of many and vigourous shoots close to the cutting surface.
- Shoot shortening should be achieved by cutting next to a lateral shoot.
- Sucker removal should be carried out during the summer, and also removal of very vigourous shoots, which antagonize the growth of other shoots.
- Pruning during canopy training should be light and be restricted only to elimination of watersprouts and suckers.

Pruning of mature olive trees

The following procedures should be followed for the annual pruning of mature olive trees (Sibbett, 1994):

- Removal of all broken shoots and branches, suckers and watersprouts, and vigourous and very weak shoots.
- Reduction of the tree to optimal height.
- Removal of shoots from the upper part of the canopy to allow better light penetration.
- Reduction of the length of secondary branches.
- Annual pruning of table olive cultivars.
- Pruning of mature trees varies according to their training system.

Rejuvenation of old trees

The olive tree is characterized by its longevity, since it has the ability to create new shoots from the various parts of its wood following frost damage. Furthermore, those trees very old and low in productivity can be rejuvenated. After de-topping, new, vigourous shoots develop and some of these are selected to create the new tree canopy, which will start to produce after no more than 3 years. In the case of frost or fire damage we wait for the tree to react and produce new shoots; from the new shoots we select those which will form the tree canopy and cut off all the dead parts.

PRUNING SHAPES

The main canopy shapes created by pruning of olive trees are the following:

- The vase shape, common in Greece, France and Italy.
- The spherical shape (trees with central leader), which is not so common since adequate light penetration is not allowed.
- The free palmette, which is not used extensively.
- The non-trunk tree, as found in Tunisia or with short trunk (vasebush).
- The double- or triple-trunk shape, common in Seville, Spain.
- The monoconical shape, i.e. the tree has one trunk and the canopy has a conical shape.
- The multiconical shape, in which each branch has a conical shape; common in certain parts of Italy.
- The two-branched shape; common in Andalucia (Spain) and appropriate for table olives.
- The candlestick shape, as found in Tunisia.

MECHANICAL PRUNING

Nowadays, it is difficult to find persons with the necessary skill to prune olive trees. Until recent times mechanization in olive pruning has been rare (Giametta, 1988; Giametta and Zimbalatti, 1993, 1997; Ferguson *et al.*, 1994, 2002). In the near future, due to increasing costs, mechanical pruning will be the only viable alternative. Mechanical pruning requires a labour input of 4 h/100 trees in comparison with 128 h/100 trees for hand-pruning.

Pruning machinery generally has a linear arrangement of five circular saws (diameter 55 cm) rotating at a speed of 2200–2300 revs/min. The blades cut the external edge of the olive canopy and the tractor works at low speed. The blades cut horizontally at the top of the canopy and vertically or with inclination at the edges. The shapes of canopies obtained are cones or pyramids. Mechanical pruning can result in 80% lower cost of pruning. However, mechanical pruning has certain disadvantages, since the cutting is indiscriminate and olive trees develop the appearance of thick, regular hedges of low productivity.

Pruning intensity depends on various factors, including cultivar, age, fruit load and soil/climatic conditions. The pruning is more severe in trees of low vigour, such as very old trees. When the trees carry a high fruit load the previous year's shoot growth is reduced. Therefore, pruning the following year should be lighter. Only watersprouts and problematic shoots should be removed.

Pruning intensity can be determined based on the grower's experience; growers observe the vegetative reaction of their trees the year following pruning. One good criterion is the length of annual shoots 2 m above ground level: shoots shorter than 15 cm indicate that the tree requires more severe pruning the following winter or spring. In contrast, too many watersprouts on old wood, or suckers from the trunk base and very long annual shoots, indicate that the next season's pruning should be very light.

TIMES AND FREQUENCY OF PRUNING

The best time for olive pruning is between the end of winter and the initiation of flowering, in order to avoid frost problems. Of course, in mild climates pruning can be started in winter and performed after harvesting. Early pruning expands the pruning period and thus spreads the timescale for input of labour. Late pruning, when inflorescences appear, is not recommended, since assimilate reserves are lost along with the cut shoots. It is possible to complete pruning in summer by removing suckers and watersprouts, as soon as they are tender. Also, with summer pruning we can correct any problems of tree shape. Furthermore, summer pruning improves the table olive quality, by reducing crop load.

In most cases olive trees are pruned every year. Pruning every year is necessary in table olive cultivars or when the annual shoot length is < 20 cm due to excessive fruit load, water stress or tree ageing. If there is no problem with the trees, pruning can be conducted every 2 years or even less frequently. However, pruning intervals of more than 3 years are not appropriate. As a rule we say that trees and training systems with a strong skeleton need annual pruning, while those with a free canopy may be pruned less frequently.

21

OLIVE RIPENING

INTRODUCTION

Olives are grown for one of two purposes: olive oil extraction and green mature olives for processing. The olive fruit is a drupe and the ripening period is determined by the time elapsing between the first purple spots and the peel turning black. By green maturation we mean the changes in fruit colour and characteristics during the period the olives are green.

Olive growers have established certain harvesting dates and other criteria such as fruit colour and fruit drop (Snobar, 1978). These criteria are used by growers as indices in order to determine the initiation of harvesting. However, in order to determine the optimum time for harvesting other parameters are also of value, such as oil content, fruit detachment force and fruit colour.

ASSIMILATE SUPPLY AND OLIVE FRUIT GROWTH

The developing fruit is a strong sink that requires a continuous supply of building materials. Fruit competes with shoot growth for newly assimilated materials, as well as for reserves accumulated in different tissues. The leaf:fruit ratio plays a significant role in fruit growth and photosynthesis (Proietti, 2003; Proietti et al., 2006). Most assimilates are supplied by the leaves on the same shoot where fruits are located, and light and shading affect fruit growth and composition (Tombesi et al., 1999), and oil synthesis (Proietti et al., 1994). The dry matter of the endocarp increases by the end of August, and afterwards it remains constant. Endocarp grows faster than mesocarp until the end of August, and afterwards its rate slows (Rapoport et al., 1990). Oil synthesis starts after 60 days of full bloom. The availability of large quantities of assimilates in July affects mostly endocarp growth. In the period August–November it affects mesocarp growth and oil synthesis, which is of cytoplasmic origin (Rangel et al., 1997). Variations in fruit load create differences in assimilate availability and influence fruit ripening. Changes in the availability of assimilates at different times during fruit development cause variations in assimilate distribution between different components of the fruit. During ripening the colour of fruit skin and flesh changes from green to reddish, red or bluish black and the skin becomes glossy. During fruit ripening the seasonal changes occurring in fruit size and chemical composition are described in the following section.

SEASONAL CHANGES IN FRUIT DEVELOPMENT

Fruit growth measured as fresh fruit weight differs for each cultivar and is probably genetically determined. During maturation fruit weight increases, with the exception of the final harvesting date, where fruit weight decreases due to water loss caused by low temperatures.

The olive fruit is a drupe, which consists of epicarp, mesocarp (flesh) and endocarp (pit or stone). The epicarp represents 1.5–3.0% of fruit weight. The greatest part of the fruit is the mesocarp which, depending on cultivar, constitutes 70–90% of the fruit and yields 90% of the oil. After pollination and fertilization of the flowers, fruit growth starts. Initially the endocarp becomes hard, but afterwards the flesh starts to grow (Rapoport et al., 1990). The fruit, as ripening proceeds, changes colour from green to yellowish green and finally blue or bluish black. Ethephon application promotes earlier ripening (Rugini et al., 1982). As the fruit grows from July to November sugar content is reduced; the remaining sugars are fermented during the processing of green olives. This decrease is followed by an increase in the synthesis of olive oil.

From fruit setting to olive fruit maturation the timespan is 6–7 months, and the fruit generally follows three growth phases in most varieties:

1. *First Phase.* Initially, fruit growth is rapid during the first 2 months (June–July). During this period the weight of the fruit increases, mostly due to endocarp growth rather than to flesh (see Figs 21.1, 21.2, 21.3, 21.4 and 21.5). Polyphenols increase and later decrease (see Fig. 21.6), while hydroxytyrosol decreases with ripening (see Fig. 21.7).
2. *Second Phase.* During the period August–September fruit growth is faster, the flesh grows and pit hardening occurs.
3. *Third Phase.* The growth of fruits is rapid (October) and the fruit starts to change colour from green to bluish, and finally to black. The changes in fresh weight and percentages of water, oil and dry weight are presented in Figs 21.1, 21.2, 21.3 and 21.4.

In the cv. 'Koroneiki' five stages of fruit growth are recorded:

1. *Stage 1.* This lasts from May to the middle of July. During this period the individual fruit parts (exocarp, mesocarp and endocarp) are obvious.
2. *Stage 2.* During this stage (middle of July to end of August) the rate of fruit growth is smaller compared with that in stage 1 but greater than that of the third stage.

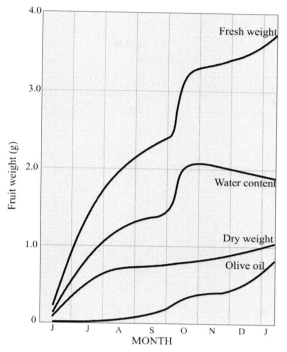

Fig. 21.1. The seasonal changes in fresh weight, water content, oil content and dry weight of olive fruits.

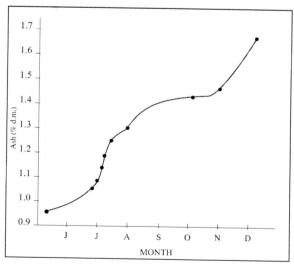

Fig. 21.2. The seasonal variation in ash content of olive cv. 'Koroneiki' during fruit growth and ripening.

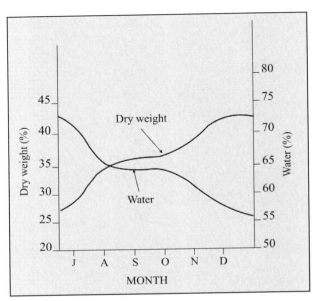

Fig. 21.3. The seasonal changes in percentage dry weight and water content in olive fruits.

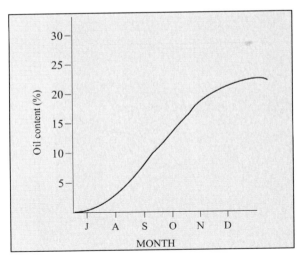

Fig. 21.4. The seasonal changes in oil content (%) of the olive cv. 'Koroneiki' during fruit growth and ripening.

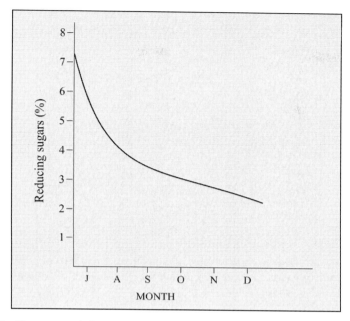

Fig. 21.5. The seasonal changes in reducing sugars (%) of the olive cv. 'Koroneiki' during fruit growth and ripening.

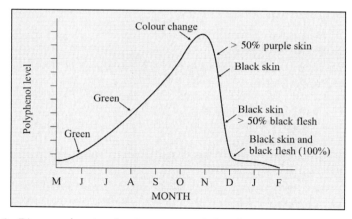

Fig. 21.6. Diagram showing the time course of olive fruit polyphenol levels and colour changes.

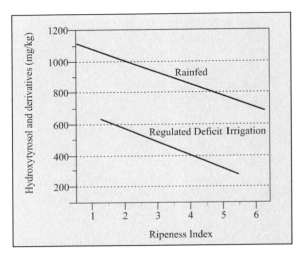

Fig. 21.7. Levels of hydroxytyrosol and its derivatives during fruit ripening as a function of the irrigation method used during the period 2003/2004 (rainfed or regulated deficit irrigation, RDI).

3. *Stage 3.* The main characteristic of this stage is the small rate of fresh weight increase, starting early in September and lasting to around 20 October.
4. *Stage 4.* The growth rate is rapid, commencing at the end of October and continuing to early December.
5. *Stage 5.* This starts in the middle of December and ends during olive fruit harvesting.

FACTORS AFFECTING OLIVE FRUIT SIZE AND PRECOCITY OF RIPENING

1. Genotype. Based on fruit size, olive varieties can be distinguished by those with small, medium or large fruits.
2. Tree age and vigour. Young trees produce larger fruits and their ripening is more precocious.
3. Soil moisture and mineral nutrition. The soil moisture should be adequate, especially during the first stages of fruit growth, when cells expand. Dry winds result in early maturation and fruit shrivelling.
4. Olive orchard orientation. Olive orchards oriented in a westerly direction are more exposed to sunshine and their fruits ripen earlier. Furthermore, from such orchards the oil produced has a higher concentration of aromatic components.

COMPONENTS OF THE OLIVE FRUIT

During the period of fruit growth and development fruit composition varies (Nergiz and Engez, 2000), and a small quantity of olive oil is produced while the maximum quantity is produced during ripening. The maximum quantity of olive oil is produced in the mesocarp (96–98%), with the remainder in the endocarp. The olive fruit flesh contains oil, water, sugars, proteins, organic acids, pectin, pigments, flavonoids (Vlahov, 1992), sterols, phenols and – especially – oleuropein, tocopherol and minerals.

Organic acids of olive oil

Most of the organic acids in olive oil are unsaturated. The most common acid is oleic acid (C18:1), followed by linoleic acid (C18:2), linolenic acid (C18:3), arachidonic acid (C20:4) and palmitic acid. Among the saturated organic acids, palmitic acid (C16:0) and steatic acid (C18:0) are the most common. The glycerides of oleic acid represent 70–80% of olive oil and are in a liquid state at room temperature. Oleic acid ranges between 56 and 83%, linoleic acid 3.5–20% and palmitic acid 7.5–20%. The percentage composition of virgin olive oil with respect to organic acids is presented in Table 21.1. Table 21.2 shows the composition of the non-saponifiable component of virgin olive oil and the oil produced from the endocarp. Table 21.3 shows the content of virgin olive and other oils with regard to organic acids. Ohbogge and Jaworski (1997) give details of the regulation of fatty acid synthesis.

Water content

Water represents 70% of the fresh olive fruit weight. However, this percentage depends on various factors such as stage of fruit growth, cultivar and water stress.

Table 21.1. Organic acid composition of virgin olive oil.

Fatty acid	Composition (%)
Oleic acid	56.0–83.0
Palmitic acid	7.5–20.0
Linoleic acid	3.5–20.0
Steatic acid	0.5–5.0
Palmitoleic acid	0.3–0.5
Linolenic acid	0.0–1.5
Myristic acid	0.0–0.1
Arachidonic acid	> 0.8
Lignoceric acid	> 1.0
Heptadecanoic acid	> 0.5

Table 21.2. Percentage composition of the non-saponifiable component of virgin olive oil and of the oil produced from the fruit endocarp.

Component	Virgin olive oil	Stone oil
Squalene and other hydrocarbons	30–50	12
Sterols	15	25
Sterol esters	–	1
Triterpenoid alcohols	10	10
Higher fatty alcohols	–	15
Waxes	–	2
Antioxidants, etc.	25–45	35

Table 21.3. Comparison of organic acid content (%) of virgin olive oil with other oils.

Fatty acids	Olive oil	Maize oil	Linseed oil	Soybean oil	Cotton seed oil	Ground-nut oil	Sun-flower oil	Sesame oil
14:0	t	–	–	t	1	t	–	–
16:0	14	13	6	11	23	6	11	10
18:0	2	4	4	4	4	5	6	5
20:0	t	t	t	t	t	2	–	–
22:0	–	t	–	t	–	3	–	–
24:0	–	–	–	–	–	1	–	–
14:1	–	–	–	–	t	–	–	–
16:1	2	–	–	–	2	t	–	–
18:1	66	29	22	25	24	61	29	40
20:1	–	–	–	–	–	–	–	–
22:1	–	–	–	–	–	–	–	–
18:2	16	54	16	51	40	23	54	45
18:3	–	–	52	9	–	–	1	–

t, traces.

Sugars

Olive fruits contain the following sugars: glucose, fructose, mannose and sucrose. The sugar concentration is high in the edible green olives. Sugars are very important during fermentation in the production of lactic acid, which gives olives their characteristic flavour and also increases their storability.

Sugar levels, standing at 10% initially, are reduced to 1–2% during the complete ripening process. The olive oil content, at 1% initially, is increased gradually with the progress of ripening. The olive fruit passes through the stages of green, red, bluish and black during ripening and ripens in 7–8 months from flowering. In the green stage, the fruit photosynthesizes, in the same manner as the leaves, and contains a high concentration of starch and tannin. Slowly, the olive oil is produced in the fruit and chlorophyll and starch disappear.

Oils

Fruit weight increases from September to December, almost solely due to oil synthesis. Olive oil contains triglycerides, carbohydrates and alcohols (0.5–1.5%). This process continues even when the day temperature falls below 13°C. The production of triglycerides and elongation of the fatty acid chain are a function of time. Through desaturation, oleic acid concentration increases. Temperature decrease is the determining factor, which triggers the synthesis of mono- or polyunsaturated fatty acids. Elongation leads to the formation of chains of fatty acids longer than C16. Desaturation and the process of elongation produce oleic acid, and the desaturation of oleic acid (C18:1) produces linolenic acid (C18:2). The total yield and harvesting time both affect oil content (Lavee and Wodner, 2004).

Oleuropein

This is the substance that gives the bitter flavour to olive oil and olives and belongs to the polyphenol group. Oleuropein levels are reduced during ripening (Tovar *et al.*, 2002). Oxidation of oleuropein gives the black colour to ripe olives. Oleuropein under alkaline pH is transformed to caffeic acid. During the processing of olives, oleuropein is removed by successive leaching with tap water or by using NaOH and subsequent leaching.

Proteins

The olive fruit contains 1.5–3.0% protein. Furthermore, the olive fruit flesh contains free amino acids such as arginine, glutamic and aspartic acid.

Chlorophyll and carotenoids

The most important pigment of olive oil is chlorophyll (Stefanoudaki-Katzouraki and Koustsaftakis, 1992; Minguez-Mosquera and Garrido-Fernández, 1989), which gives the oil its green colour. However, the presence of chlorophyll under light conditions is the causal factor of oxidative degradation of the oil. Both chlorophylls α and β are present in olive oil; the concentration of chlorophyll in virgin olive oil can be as high as 10 ppm. When olive fruits are harvested unripe, chlorophyll concentration is greater.

Three chloroplast pigments identified in olives are those of all green plant tissues and do not undergo any change or modification during the stages of ripening. The carotenoids 3-carotene, phytofluene and luteoxanthin are present in very small amounts. Both chlorophylls and carotenoids decrease as the season progresses, almost disappearing at the moment of maturity, while anthocyanins begin to appear, little by little, invading the skin and later the whole pulp. The α-chlorophyll is the major component, followed by β-chlorophyll. Carotenoids have been found to be minor components, such as lutein, which is the major xanthophyll, and β-carotene, the principal carotene. Carotenoids present in olive oil are represented by xanthophylls, carotenoids a, b and γ and lycopene. β-Carotene comprises 85% and α-carotene 15% of carotenoids. As olive fruit ripens, photosynthetic activity decreases and chlorophyll disappears. As a consequence, the colour of the skin changes from green to yellow, reddish or red. During this period the concentration of carotenoids and chlorophylls diminishes, while the proportion of xanthophylls increases. Chlorophyll degradation is accompanied by the synthesis of other compounds, anthocyanins, because the carotenoids do not produce the final pigmentation of the ripe fruits, i.e. reddish or purple.

Sterols

Olive oil contains various sterols such as campesterol (12%), stigmasterol (0.5%), β-sitosterol (89.5%), avenasterol (8.0%) and traces of cholesterol. The sterol concentration of olive oil is 180–265 mg/100 g. Olive oxidation reduces the concentration of sterols. The most common sterols found in oils of plant origin are campesterol, stigmasterol and β-sitosterol. Determination of sterol concentration provides a means of determining the authenticity of olive oil.

Aromatic compounds

Of the aromatic substances giving olive oil its characteristic aroma are included 40 aldehydes (C_7–C_{12}), monosaturated aldehydes and terpenes (see Table 21.4).

Tocopherols

These are present in all oils of plant origin, and also in animal lipids. The α-tocopherol comprises 88.5% of all tocopherols in olive oil. The concentration of α-tocopherol in olive oil is 12–150 ppm. Oils derived from the fruit stone contain higher concentrations of tocopherols, which play the role of antioxidants. The concentration of tocopherols provides a means of determining the authenticity of olive oil (see Table 21.5).

Table 21.4. Some of the aromatic components of olive oil.

Hydrocarbons	Aldehydes	Aliphatic alcohols
Isopentane	Acetaldehyde	Methanol
2-methyl-pentane	Propanol	Ethanol
Hexane	2-methyl-butanol	Isopropyl alcohol
Octane	Butanol	1-pentanol
Napthalene	Pentanol	3-methyl-butanol
Ethyl napthalene		1-penten-3-ol
		1-exanol
		1-heptanol
		1-octanol
		Monoenole
		2-phenylethanol

Table 21.5. Tocopherol concentrations of certain oils of plant origin.

Oil	Tocopherols (mg/g)			
	α	$\beta + \gamma$	δ	Total
Olive oil	0.24	t	t	0.25
Cotton seed oil	0.56	0.38	t	0.94
Maize oil	0.26	0.92	t	1.18
Soybean oil	0.07	0.78	0.24	1.09
Groundnut oil	0.23	0.31	t	0.54

t, traces.

Phenols

These are antioxidative substances and are present in both leaves and fruit (Ryan *et al.*, 1999; Gomez-Rico *et al.*, 2006). Important phenols include tyrosol and hydroxytyrosol, which are produced after oleuropein hydrolysis. A significant portion of the phenols from the fruit flesh is removed with the olive mill wastes. Figures 3.6 and 3.7 present the mean ripening periods of 25 Greek olive cultivars and 36 cultivars cultivated on a worldwide basis.

Mineral analysis of olive fruits

The overall composition of minerals varies between olive cultivars. Of the various elements K is the most abundant element in the fruit, followed by Mg, Ca, Na and Fe (Nergiz and Engez, 2000).

RIPENING-RELATED CHANGES

Softening of the cell wall

The study of ripening in the olive is of great interest, because the ripening stage of the fruit influences the industrial production of olive oil and olive processing (Jimenez et al., 2001). Ripening is usually characterized by softening due to modification and/or degradation of cell wall components. Changes in cell wall polysaccharides (Mafra et al., 2001) and cell wall-associated enzymes (Fernández-Bolanos et al., 1995; Mazzuca et al., 2006) play a significant role in determining olive fruit texture during ripening. During this process the changes occurring are characterized by an increase in the solubilization of pectic and hemicellulosic polysaccharides, i.e. an increase in the amount of arabinose in pectic polysaccharides and a decrease in the degree of methylesterification of pectic polysaccharides. With regard to cell wall phenolics, p-coumaric acid levels increase at harvesting time (Briante et al., 2002a).

Changes in chloroplast pigments

The olive fruit during the ripening process becomes purplish black at the same time as the oil content increases. During ripening, photosynthetic activity decreases and both chlorophylls and carotenoids decrease progressively (Stefanoudaki-Katzouraki and Koustsaftakis, 1992). At maturation the purple colour of the olive fruit can be ascribed to the formation of anthocyanin. These changes with ripening are expressed in pigment composition of the olive oil. However, during crushing and malaxation, chlorophylls are transformed into their Mg-free derivatives. The pigment content of olive fruits is a function of their genotype.

Changes in catechol oxidase

Catechol oxidase is the main enzyme causing browning in fruits and other tissues, and is located mainly in the chloroplasts (Ben-Shalow et al., 1977). The amount of o-diphenols, which can serve as substrates for catechol oxidase, often declines during olive fruit development. The total o-diphenol content in the olive continues to rise parallel with fruit growth. The phenolic content can be one of the factors determining the rate of enzymatic browning, and the browning potential of olives will not change during development. In green olives there is a tight binding of the enzyme to chlorophyll-containing membranes. During the final stage of fruit development, when the olive blackens, the enzyme appears predominantly in the soluble fraction and these changes coincide with the appearance of anthocyanins. The properties of catechol oxidase are essentially the same in both green and black olives.

PHENOLIC AND VOLATILE COMPONENTS OF EXTRA VIRGIN OLIVE OIL

Ripening level of olives could be evaluated by using any of the following methods: harvesting day, ripening index (RI), fruit skin colour, fruit firmness and amount of chlorophylls and carotenoids in the oil. Oil quality is evaluated by using the parameters of acidity, K_{232}, K_{270}, peroxide index and panel test (Yousfi et al., 2005). The degree of ripening of the fruit and irrigation management significantly affect the amounts of secoiridoid phenolic compounds (Tovar et al., 2002; Prenzler et al., 2003). These compounds greatly decrease upon increase of both ripening and irrigation. Furthermore, secoiridoid derivatives of hydroxytyrosol decreased more than those of tyrosol. Hydroxytyrosol has a protective role in the human erythrocyte against oxidative damage (Lavee and Wodner, 1991). Oil quality evaluated using the parameters of acidity, K_{232}, K_{270}, peroxide index and panel test was not affected by fruit ripening or by increased rainfall (Yousfi et al., 2005).

FRUIT RIPENING AND OLIVE OIL QUALITY

Harvesting of olives at the green stage of ripening results in a reduced yield. However, this permits easier handling of harvested olives, since green olives are more resistant to mechanical damage and fungal infection than the more ripe fruit. Soft olives should be processed as soon as possible, because they lose the initial quality of their oil very rapidly (Garcia et al., 1996). The oil content is more or less stable during the different ripening stages. Furthermore, the oils obtained from olives at more advanced ripening stages have an increased content of conjugated fatty acids; the bitterness index of the oil decreases as ripening progresses. In general, the parameters that are used to measure the oxidation of the oils (K_{232}, K_{270} and stability to oxidation) show a progressive deterioration of oil quality as the ripening process proceeds. As the olive fruit ripens, a percentage decrease in palmitic and oleic acids is observed, while that of linoleic acid increases.

PHENOLS IN OLIVE OIL

Oleuropein, the bitter substance of olives and olive oil, is present in the aglycone form. As maturity proceeds oleuropein is hydrolysed, producing other substances which participate the characteristic taste of olive oil. Another process taking place is oxidation of phenolic compounds and the production of substances contributing to the aroma and flavour of olive oil. The presence of phenols gives a bitter and pungent taste, this being due to inactivation of ptyalin and to various interactions between phenolic compounds and the human sensors of taste.

In olive oil the bitter taste is due to phenolic compounds and cinnamic acid, while the pungency is due to oleuropein. The amount of phenolic compounds in olive oil cultivars varies from 150 to 700 ppm and is affected by various factors such as cultivar, degree of maturation, climatic conditions and pest damage to fruit. The best stage for picking is when the colour changes from pale green to brown. At this stage the olive oil has a high phenolic content. Furthermore, the best way of harvesting is hand-picking, and fruit processing should be conducted at temperatures < 30°C in order to maintain its high phenolic content. Green olives give oil high in phenolic content and with a more aromatic and fruity flavour.

MINOR CONSTITUENTS OF OLIVE OIL

Among the minor constituents of virgin olive oil the most important contributing to oil colour are α- and γ-tocopherol (200 mg/kg) and β-carotene and chlorophyll, which give the oil its colour (yellow or green). Other minor constituents include phytosterols, squalene, pigments, terpenic acids, flavonoids (luteolin, quercetin) and phenolic compounds known also as polyphenols. Other minor components present are coumaric, ferulic and caffeic acids.

EFFECT OF STONE REMOVAL ON VOLATILE COMPOUNDS OF OLIVE OIL

Olive oils produced after stone removal have a greater concentration of C6 volatile compounds. This greater C6 concentration in de-stoned fruits is related to the release of the membrane-bound enzymes participating in the lipoxygenase (LOX) pathway. The reason for this is more effective pulp grinding, which causes the cellular disruption that releases the enzymes. De-stoning increases the hexanol content of olive oil, and this has positive effects on flavour, especially the fruity aroma.

METHODS FOR EVALUATION OF MATURATION STAGE

An acceptable method for maturity index (MI) evaluation of olives should be objective, cheap, non-destructive and easy to apply. Various methods have been used in order to achieve this, including: (i) measurement of contents of soluble solids, starch or organic acids; (ii) measurement of respiration rate or ethylene production; (iii) fruit size; or (iv) skin colour.

The most common method is the ripening index (RI), a subjective method that takes into account the changes in the skin and flesh colour during the process of fruit ripening. This method is not valid for the olive varieties that undergo a range of colour changes as maturity proceeds. Some methods are based on the fruit's resistance to compression, which is a destructive method. Other methods, non-destructive, include the use of a hand densimeter or those based on transmission of acoustic waves through the fruit.

A maturity index (MI) has been developed to help producers categorize their fruit's maturity level. The MI will help determine when each variety should be harvested in order to achieve the desired product quality. For MI calculation a sample of 100 olive fruits is collected and separated into eight colour categories, as follows (UC Cooperative Extension, 2006):

- 0, fruit surface is deep green and the fruit is firm.
- 1, fruit surface is yellowish green and fruit starting to soften.
- 2, < 50% of the fruit surface is red.
- 3, > 50% of the fruit surface is red.
- 4, 100% of the fruit surface is purple.
- 5, 100% of the fruit surface is purple or black; 50% of the flesh is purple.
- 6, 100% of the fruit surface is purple or black; > 50% of the flesh is purple.
- 7, 100% of the fruit surface is purple or black; 100% of the flesh is purple.

For MI calculation we multiply the number of fruits of each colour category by the number with that colour category (0–7):

$$MI = \frac{A \times N_1 + B \times N_2 + C \times N_3 + D \times N_4 + E \times N_5 + F \times N_6 + G \times N_7}{100}$$

CHARACTERISTICS OF TABLE OLIVES AND CRITERIA FOR RIPENING

Table olives should have the following characteristics:

- High flesh:pit ratio, ranging from 5 to 12.
- High sugar concentration, 5–6%.
- Relatively small oil content – around 12–20% of the fresh weight of the pulp. A greater oil content is disastrous for the storage of table olives, since product taste deteriorates and the fruit softens.
- The flesh should be firm. Hence, the fruit can maintain its good characteristics during both the various stages of processing and the subsequent storage of the processed olives.

- Easy removal of the pit from the flesh.
- Fruit epidermis should be thin, elastic and tolerant during processing.

The criteria for ripening are the following:

- Olives destined for oil production are harvested when the fruit is ripe. This means that a period of 7–8 months should elapsed after flowering.
- Table olives are harvested when fruit growth is complete and the colour starts to change, for Californian-type olives. For green olive processing harvesting starts earlier, before fruit softening commences. Fruit size is a good criterion for harvesting for some table olive cultivars.

22

OLIVE FRUIT HARVESTING

INTRODUCTION

This chapter outlines the various methods employed in the harvesting of olive fruits, from the simple and traditional to the modern and sophisticated.

HAND-HARVESTING

Most olive harvesting around the world still involves the traditional methods, i.e. picking by hand (Martin *et al.*, 1994b). Labourers pick the fruits one by one or beat the tree, with poles, causing fruit drop. Under the trees, nets or canvases are placed to collect the fallen fruits. This method of harvesting is time-consuming, involves a high input of labour and a high percentage of fruit damage. Also, the olive fruit may simply drop through natural forces on to the harvesting nets. The cost of labour at harvesting represents 55–80% of the total labour cost of olive culture.

The use of poles for olive harvesting increases the tendency of trees to biennial bearing, because it damages a significant number of annual shoots necessary for the following year's crop.

Harvesting of olive fruit by hand, while the fruit is attached to the tree, gives a high-quality fruit, while harvesting of olives allowed to drop naturally to the ground produces olive oil with a high acidity and unpleasant taste.

HAND-HARVESTING WITH ABSCISSION CHEMICALS

In more recent times growers have started to promote abscission by the use of chemicals such as Ethrel or Alsol, which reduce fruit detachment force (FDF). The classical harvesting method for olive oil varieties is by beating the shoots and collecting the fruits one by one from the soil surface, but this method incurs a high labour cost. Furthermore, labourers are not available during the period of unfavourable climatic conditions (low temperature and precipitation). Today, new methods involving tree-beating and harvesting machinery are used. However, their suitability depends on the cultivar (Al-Jalil *et al.*, 1999).

PNEUMATIC BEATING POLES

These poles should have minimum weight, and therefore they are constructed from plastic or aluminum material. This method is suitable for the harvesting of low-fruited and shorter varieties. Tools previously used had a low harvesting cost, short harvesting time and did not damage the tree. Pneumatic beating poles are used when the olive fruit is ripe, since the FDF is reduced; they are recommended only for varieties destined for olive oil production.

The best beating poles revolve, since they have a greater efficiency and do not damage the trees. In the commercial market two types are available, i.e. the linear and the T-shaped (see Fig. 22.1). Each pneumatic beating pole has a vertical axis revolving with the aid of a motor. At various locations on the vertical axis are attached easy-to-bend accessories, four to six at each level. The fruits drop on to plastic nets or canvases placed on the soil surface below the tree canopy. Workers should use the pneumatic beating pole in a rapid fashion, since beating the same shoot for a relatively long time damages the shoots and causes their defoliation.

The linear beating pole is recommended for most varieties, with the exception of 'Koroneiki'. This machinery is highly efficient – one operative

Fig. 22.1. Three versions of olive-beating machines: (a) with horizontal branches; (b) with T-shape; and (c) with a vertical axis.

can harvest around 1000 kg/day. The T-shaped beating pole can harvest 1500 kg/day and is appropriate for all varieties. This version incorporates six to eight flexible beating poles, which revolve and harvest the fruit. The percentage leaf drop with this device is lower in comparison with the linear beating pole. The pneumatic beating pole is powered from a 12 v battery or from an air compressor of 6–8 atm pressure, adapted to a tractor.

All types of machinery above should be tolerant to stress, have a high degree of efficiency and not be too heavy in use. After completion of the beating of shoots, the olive fruits are harvested from plastic nets or canvases, or the collection may done by hand if no nets are used. The soil should be flat and without stones; otherwise, the soil should be levelled off. Harvesting from the ground is accomplished by a rotating cylinder with needles, which collect the fruits. This method, however, damages the fruits and results in deterioration of oil quality.

HARVESTING NETS AND CANVASES

The use of plastic nets and canvases is very common in Mediterranean countries as an aid in olive fruit harvesting, in conjunction with the use of pneumatic beating poles. Nets and canvases should be tolerant to mechanical and light damage and not heavy (around 100 g/m^2). After harvesting, nets and canvases are cleaned and dried under shade and stored in warehouses for the next harvesting period. The most common colours of nets available are white, black and green, and their sizes vary from 3.3×5.0 to 6.5×12.0 m. The nets are manufactured from polyethylene, improved to resist UV radiation. In certain heavily sloping olive groves with large trees that are not regularly pruned the nets are placed in their position permanently and the olives simply drop onto the nets. From time to time the fruits are collected from the nets. With this method the fruits may be damaged by insects and diseases, while the oil loses its aromatic components and fine taste.

After harvesting the fruits are carried in plastic bags or containers, each containing 210–420 kg, to the oil-processing unit. These containers are manufactured from high-quality plastic, not wood, for better aseptic conditions.

HARVESTING AND GATHERING TOOLS

Scissors-type harvester

This tool consists of two combs with prolongations made of hard plastic material. When the harvester is closed and pulled, it harvests the fruits from small twigs without removing the leaves.

Tweezers-type harvester

The main parts of this tool comprise two rotating cylinders. When the device is closed the small shoots pass through the two cylinders and the fruits are detached. Afterwards, collection of the detached fruits from the ground takes place.

Manually operated cylinder harvester

This tool has one cylinder with plastic needles that rotate; fruits dropped on the ground are harvested, then collected in a metal basket.

Plastic nets

Plastic nets are placed below each tree canopy before harvesting. These have the disadvantage of their significant cost and short duration of life. In order to avoid olive drop on the ground, the nets are placed under the trees early and remain there until the end of the harvesting period. Therefore, around 32% of the olives are collected in the nets by natural drop and the remaining portion after shoot-beating with pneumatic poles. Beating of olive trees increases the tendency for biennial bearing. When using mechanical vibrating harvesters, certain factors such as fruit tolerance to abscission, fruit weight, tree shape and the power of the harvester may affect harvesting success. For reduction in FDF, various substances have been tried, with varied results.

MECHANICAL HARVESTERS

The conventional methods of olive growing and the popularity of tall trees both increase the cost of harvesting. Recent decades have seen the introduction of harvesting for olives and other fruits by mechanical harvesters, which can vibrate either only one branch or the whole tree The efficiency of vibration and its transmission is affected by the cultivar (Antognozzi et al., 1990a, b; Visco et al., 2004). Each mechanical harvester consists of the vibration system, the collection system (umbrella) and the fruit suction system (see Fig. 22.2). The method of fruit collection consists of two semicircular pieces of cloth attached to a metal skeleton, placed under the olive tree canopy.

This is an innovative type of harvester, incorporating an excavator that controls the shaker and the gathering umbrella. It can harvest 15 or more olive trees/h and has a harvesting efficiency of 90.5%. The force necessary to vibrate one branch is 8–10 Hp, and the trunk 40–120 Hp; the maximum vibration

Fig. 22.2. Mechanical trunk vibrator with a reverse umbrella harvesting an olive tree (from Tombesi, 2006).

power is 4000 cycles/min, suitable for the trunk – 1200 cycles/min are recommended for branches (see Fig. 22.3). Only single-trunked trees are suitable for harvesting by this device. The fruit suction component consists of a rotating axis, which creates the suction force. Furthermore, the air current produced removes the leaves and other foreign material from the harvested olives.

Over-row harvester

This harvester is derived from grape or coffee harvesters and vibrates the olive fruit branches by means of ten to 12 shaking bars (see Fig. 22.4). The fruits are collected by a chain, with baskets running at the same speed as the machine. Fruit removal percentage is high (90–95%), even in small-fruited varieties. The harvester moves with a speed of operation between 0.3 and 1.0 Km/h. An alternative system consists of a vertical axis with lateral sticks that vibrate.

Trees suitable for harvesting by this type of harvester should be lower than 2.5 m and no wider than 1.5 m, with elastic branches; therefore, dwarf and very productive varieties are most suitable. For the SHD orchard (1600 trees/ha) 'Arbequina' and 'Arbosana' are the best varieties. 'Arbequina' showed the highest total production between the third and sixth years following planting, followed by 'Arbosana' and 'Koroneiki'. 'Arbequina' has a semi-erect habit, 'Arbosana' an open habit. In SHD orchards the objective is to attain

Fig. 22.3. Trunk shaker for olive harvesting. The fruits drop and are collected on a net or canvas placed under the tree canopy.

Fig. 22.4. An over-row olive harvester harvesting a super-high-density olive grove (from Tous *et al.*, 2006).

production by the third or fourth year, to use mechanical pruning and to use efficient harvesting machines, in order to reduce costs. However, it is necessary to determine for how many seasons the plantation remains economically productive.

FEATURES AND ADVANTAGES OF MECHANICAL HARVESTING

There are many reports in the literature dealing with mechanical harvesting of olives and its limitations and advantages (Mannino and Pannelli, 1990; Martin, 1994b; Tombesi *et al.*, 1998; Giametta, 2001; Ravetti, 2004).

- Harvesting cost is one-third to one-half of that of hand-harvesting.
- The speed of harvesting is greater; therefore, pests and diseases are avoided.
- The olive fruits are not in contact with the soil surface. Therefore, they avoid infection by microorganisms.
- Harvesting requires very few operatives. One mechanical harvester can harvest ten to 15 trees/h or more with only two or three operatives.

Mechanical harvesting demands groves of limited slope and adequate planting distances, in order to permit the movement of machinery. Furthermore, the tree canopy should be trained appropriately. Trees intended for harvesting by a trunk vibrator have the following characteristics:

- The trees should have only one trunk, 15–25 cm or more in diameter and trunk height ≥ 1 m in order to allow easy attachment of the vibrator to the trunk.
- The shoots located at the lowest part of the tree canopy should be cut, since they represent obstacles to the use of a mechanical harvester.
- The trees should have no more than three branches; if not, these are removed. Experience has shown that branches of mature trees < 15 cm in diameter are not suitable for vibration.

In the case of branch vibrators the main branches should be arranged in such a way that the harvester can vibrate them without the need to change position. This reduces the harvesting time required for each tree, and therefore harvesting cost.

The proportion of fruits harvested by the mechanical harvester depends on the following criteria.

Vibration characteristics

The greater the acceleration the more easily the fruits are harvested. The acceleration produced by a mechanical harvester is compared with the natural forces required by the fruit in order to drop, and this is described by the following equation:

$$\text{Fruit acceleration} = \text{Fruit weight}/\text{Required force for fruit removal}.$$

Force of fruit detachment

This force is measured by dynamometer, and decreases significantly after spraying the fruits with one of the following chemical loosening agents (Martin et al., 1981; Ben-Tal and Wodner, 1997; Metzidakis, 1999):

- Cycloheximide.
- Ethrel (ethephon, 2-chloroethyl-phosphonic acid), producing ethylene (Hartmann et al., 1970; Ben-Tal and Lavee, 1976b; Ben-Tal et al., 1979; Lavee and Martin, 1981).
- CGA (2-chloroethyl-tris-2-methoxy-ethoxy silane, or Alsol).
- Chloroethyl-sulphonic acid.

These substances are sprayed 5–7 days before the date of mechanical harvesting. The use of ethephon in order to enhance fruit ripening is very common. Ethephon is used commercially in order to facilitate mechanical harvesting or hand-harvesting of olive fruits (Ben-Tal and Lavee, 1976a; Ben-Tal, 1987, 1992; Denney and Martin, 1994; Gerasopoulos et al., 1999). Environmental factors affect ethephon efficiency in the olive (Klein et al., 1978). From the applied ethephon only an amount lower than 3% penetrates the olive pedicels and releases ethylene to promote abscission (Martin et al., 1994b). Ethephon is not translocated from leaves and fruits to the fruit pedicels.

When we use looseners we attempt to achieve maximum fruit removal with minimum leaf abscission. Ethephon application induces the onset of leaf abscission 36–60 h after treatment and, in the following 8–12 h, separation is evident (Martin et al., 1994b) in the abaxial cortical cells close to the vascular system. Separation in the adaxial side then proceeds. Finally, separation takes place in the vascular region and the epidermal cells. During the abscission process, starch is accumulated near the vascular region.

The abscission of olive fruit occurs in two abscission zones, one between the peduncle and pedicel and the other between the pedicel and fruit. Application of ethylene-releasing compounds results in abscission mostly in the pedicel/fruit zone. However, fruit abscission may be observed in the peduncle/pedicel zone. Under field conditions, ethylene action is sometimes variable due to variations in temperature and moisture conditions. High relative humidity (RH) reduces the speed of ethylene decomposition, while low (35–70%) RH speeds up ethylene decomposition, especially when the temperature is high. Furthermore, both the duration and concentration of ethylene-releasing compounds affect fruit abscission. An optimum fruit abscission is achieved by applying about 3 ppm ethylene for 28 h. In olive fruits, ethylene production is low, i.e. there is no ethylene autocatalytic enhancement. However, olive leaves show autocatalytic enhancement 120 h after external application (Martin et al., 1994b).

Date of harvesting

As maturation proceeds the FDF is reduced and the fruits are harvested more easily. The ratio of FDF:fruit weight is reduced with the progress of ripening, but starts to increase again after the middle of the growth period.

Olive varieties with small fruits have a high F:W ratio (F, force, W, weight). In contrast, large-fruited varieties with high FDF have a low ratio. The F:W ratio may constitute a criterion for ease of fruit removal (Tsatsarelis, 1987). In olives the maximum fruit abscission with a low leaf drop, by use of ethylene-releasing chemicals, is of great significance. Once ethephon is applied, fruit abscission starts within 36–60 h; even after 8–10 h separation in the abscission zone is obvious. Histochemical examination of the fruit abscission zone indicates that considerable starch is accumulated close to the vascular bundles, and its content does not change during abscission.

Efficiency of mechanical harvesting

Shaking tests have shown that the percentage harvesting of olives without application of a loosener was < 50%; application of abscission chemicals reduced the FDF:W ratio and thus increased this figure. The highest fruit removal (96%) was obtained by using a concentration of 12.5 ml/l of ethephon. The frequency used for the individual fruit–stem system is a key criterion in the design of a mechanical harvester that uses vibration. The detachment of fruits is mainly a function of the applied frequency, the total time of shaking and the relative proportions of mature and immature fruits on the tree (Sessiz and Özcan, 2006).

Variety in tree size and agronomic parameters affect harvesting efficiency (Dias *et al.*, 1999; Proietti *et al.*, 2002b). Of the training systems in use, the vase form showed a slightly greater adaptability to mechanical harvesting as compared with the monocone form. This is most probably due to the higher deviations and distance from the main branch of fruiting branches in the monocone system. The oil obtained during the early periods of harvest had a higher polyphenol content and was more fruity, bitter and spicy than that obtained from the later harvesting periods.

Factors affecting efficiency of trunk shakers

The most important factors affecting the efficiency of mechanical trunk shakers are:

1. Planting distance: planting distances of 6 × 6 or 6 × 7 m are ideal for the reverse umbrella technique.

2. Training system: olives should be trained in such a way that there is only one trunk, 1.0–1.2 m in height.
3. Cultivar: the variety affects the shaking efficiency, since that determines fruit size, detachment force, maturity pattern and branch elasticity.
4. Fruit size and its FDF: fruits > 2 g are harvested easier than small fruits. Fruit size also determines the fruit detachment force – the FDF should be around 4N (Newton), in order to facilitate the efficiency of the harvester.
5. The optimum canopy volume is 40–50 m^3 (Tombesi, 2006).

MOLECULAR STRATEGIES FOR IMPROVING OLIVE HARVESTING

The cost of hand-harvesting olives represents more than 50% of the total production cost. Mechanical harvesting, therefore, has many advantages in comparison with traditional hand-harvesting. Furthermore, mechanical harvesting efficiency is increased by ethylene-releasing chemicals, which induce olive fruit abscission. However, mechanical harvesting cause 15–20% leaf abscission and affects the following year's yield.

Olive fruit is not a climacteric one and produces negligible amounts of ethylene during maturation; such fruits produce very little ethylene. Production of ethylene can be increased by exogenous application of ACC. This means that the lack of significant ethylene production by olive fruit is likely to be ascribed to low levels of ACC synthesis in the tissue. Therefore, ethylene production in olives can be achieved by using transgenic plants or after insertion of an active ACC synthase into the trees. Today, transformation methods can be used and such a strategy could be utilized to solve the problem of the high labour cost of hand-harvesting by producing a transgenic olive plant with ACC synthase expressed specifically in the fruit at the correct maturation stage (Ferrante *et al.*, 2004).

23

OLIVE VARIETIES

INTRODUCTION

The olive tree has been cultivated for thousands of years and includes many varieties, creating problems in their classification (Barranco et al., 2000). The problem becomes more acute because, very often, the same cultivars at various locations are known under different names (Therios, 2005b). The criteria for classification of olive cultivars are the following:

- Tree height.
- Leaf and inflorescence characteristics.
- Fruit characteristics (shape, colour, size, percentages of flesh and pit, flesh:pit ratio, shape of pit and seed, percentage of oil).
- Precociousness and productivity.
- Resistance to pests and diseases.
- Adaptability to soil and climatic conditions.
- Use of DNA markers to separate cultivars.

THE USE OF MOLECULAR MARKERS TO CHARACTERIZE AND CLASSIFY OLIVE (*OLEA EUROPAEA*) GERMPLASM

The considerable diversity within olive and the presence of a great number of synonyms require the development of efficient and rapid methods to determine cultivars (Pontikis *et al.*, 1980; Belaj *et al.*, 2001, 2004). Although molecular analysis is able to characterize olive cultivars it is not a substitute for morphological description: it is only a complementary tool to study olive germplasm.

Among the molecular techniques applied for cultivar identification there are four major ones:

- RFLP (Restriction Fragment Length Polymorphism).
- RAPD (Random Amplified Polymorphic DNA).
- AFLP (Amplified Fragment Length Polymorphism).
- SSR (Simple Sequence Repeats) (Martin-Lopes et al., 2007).

Other markers used, but to a lesser extent, to study genetic diversity in olives include SCAR (Sequence Characterized Amplified Region) and ISSR (Inter-Simple Sequence Repeats) (Martin-Lopes et al., 2007).

The RFLP markers are the molecular technique that was first applied to olives, the aim being to discriminate between wild and cultivated olive varieties. The RAPD markers are used for genetic characterization in olives (Bogani et al., 1994; Fabbri et al., 1995; Belaj et al., 2001; Besnard et al., 2001; Hagidimitriou et al., 2005; Martin-Lopes et al., 2007). AFLP is used to study the genetic relationship between cultivated olives, wild olives and related species. AFLP also tests for genetic interrelationships between olive cultivars (Bandelj et al., 2004; Hagidimitriou et al., 2005; Montemurro et al., 2005).

SSR markers are used to study variability in olives and are also used for variety identification. Other markers used to study olive germplasm variability utilize chloroplast DNA variations by comparing (i) RFLP and PCR polymorphisms (Besnard and Berville, 2002; Intrieri et al., 2007); (ii) microsatellite markers (Bandelj et al., 2002; Diaz et al., 2006a); (iii) polymorphic and single nucleotide polymorphism markers (Reale et al., 2006); (iv) fractals and moments to describe olive cultivars (Bari et al., 2003); and (v) genetic distances to study the history of olive cultivars (Loukas and Krimbas, 1983).

The number of cultivars grown and their synonyms, their separation into type (olive oil, table olives and dual-purpose cultivars) and their distribution throughout various countries are given in Table 23.1. The main olive cultivars grown in various countries are presented in Table 23.2, and in Table 23.3 the taxonomy of 24 Greek cultivars is presented.

SMALL-FRUITED OLIVE CULTIVARS

'Ladolia' (*Olea europaea* var. *mastoides* or *microphylla*)

It is considered that this cultivar belongs to the species *O. mastoides* or *O. microphylla* and is known also under the names 'Patrini', 'Kurelia' or 'Kutsurelia'. The growth habit is erect and tree height at maturity is 5–7 m. The leaves are small and the leaf blade is broader at the middle and the top. The length of the blade is 3.3–5.3 cm and the length:width ratio is 3.66:1. The fruit is spherical in shape, with a characteristic mastoid protuberance. The fruit ripens late, from November onwards, and its weight is 1.1–1.62 g. The ratio of

Table 23.1. Number of local cultivars and synonyms by type (olive oil, table olives or dual-purpose) grown throughout various countries worldwide.

Country	Cultivars	Synonyms	Olive oil	Table olives	Dual-purpose
Albania	21	28	14	1	6
Algeria	51	33	37	1	13
Argentina	6	2	5	1	–
Australia	2	–	1	1	–
Brazil	–	–	–	–	–
Chile	1	–	–	–	1
China	9	–	9	–	–
Cyprus	1	–	1	–	–
Egypt	6	–	1	5	–
France	99	435	73	7	19
Greece	60	67	46	3	11
USA	3	–	3	–	–
India	–	–	–	–	–
Iraq	10	2	8	–	2
Israel	46	13	13	17	16
Italy	476	1599	390	13	73
Jordan	3	–	1	1	1
Lebanon	13	17	4	1	8
Libya	6	13	5	–	1
Morocco	6	8	2	–	1
Mexico	1	–	–	1	–
Pakistan	1	–	1	–	–
Peru	1	–	–	1	–
Portugal	19	132	6	20	11
Spain	196	513	135	25	36
Tunisia	50	42	27	4	19
Turkey	34	46	7	18	9
Former Yugoslavia	47	70	20	18	9
Total	1188	512	812	117	259

flesh:pit is 3.58–5.00:1 and the oil content of the fruit is 16–22%. It is considered as a good cultivar for olive oil production.

'Koroneiki' (*Olea europaea* var. *microcarpa alba*)

Other names of this cultivar are 'Psilolia', 'Lianolia' or 'Korani'. It is widespread in the main olive-producing districts of Greece (i.e. Crete, Peloponnese, etc.) and is expanding in other areas of the world too. In recent years this cultivar

Table 23.2. The main olive cultivars grown in various countries.

Country	Cultivars
Spain	'Picual', 'Verdal', 'Arbequina', 'Manzanillo', 'Sevillana'
Italy	'Frantoio', 'Moraiolo', 'Leccino', 'Coratina', 'Ascolano'
Portugal	'Galega', 'Verdial', 'Radondil', 'Gordal'
Tunisia	'Chemlali'
Algeria	'Sigoise', 'Azeradj', 'Harma'
Morocco	'Picholine marocaine'
France	'Picholine', 'Anglandaou', 'Salonenque', 'Tanche'
Turkey	'Milas', 'Kilis', 'Cakir', 'Ismir'
Argentina	'Arbequina', 'Leccino', 'Frantoio', 'Manzanillo'
Israel	'Souri', 'Malissi', 'Barnea'
Lebanon	'Chami', 'Souri'
USA	'Mission', 'Manzanillo', 'Sevillana', 'Ascolano'
South Africa	'Sigoise', 'Barouni', 'Meski'
Greece	'Koroneiki', 'Megaritiki', 'Kalamon', 'Amphissis', 'Kolovi', 'Adramytini', 'Lianolia', 'Chondrolia Chalkidikis'

Table 23.3. Twenty-four Greek olive cultivars of *Olea europaea* and their taxonomy.

Cultivar	Taxonomy
'Konservolia'	*O. europaea* var. *rotunda*
'Karydolia'	*O. europaea* var. *med. maxima*
'Gaidourolia'	*O. europaea* var. *major macrocarpa*
'Amygdalolia'	*O. europaea* var. *amygdaliformis*
'Kalamon'	*O. europaea* var. *ceraticarpa calamata*
'Adramitini'	*O. europaea* var. *media subrotunda*
'Kothreiki'	*O. europaea* var. *minor rotunda*
'Karolia'	*O. europaea* var. *oblonga*
'Kolymbada'	*O. europaea* var. *uberina*
'Tragolia'	*O. europaea* var. *minor oblonga*
'Mastoidis'	*O. europaea* var. *mamilaris*
'Throumbolia'	*O. europaea* var. *media oblonga*
'Valanolia'	*O. europaea* var. *pyriformis*
'Megaritiki'	*O. europaea* var. *argentata*
'Lianolia Kerkiras'	*O. europaea* var. *craniomorpha*
'Maurolia'	*O. europaea* var. *nigra microcarpa*
'Koroneiki'	*O. europaea* var. *microcarpa alba*
'Vasilikada'	*O. europaea* var. *regalis*
'Aguromanacolia'	*O. europaea* var. *ovalis*
'Kalokaerida'	*O. europaea* var. *precox*
'Myrtolia'	*O. europaea* var. *microcarpa subrotunda*
'Dafnolia'	*O. europaea* var. *clavata*
'Asprolia'	*O. europaea* var. *alba*
'Chrysolia'	*O. europaea* var. *chrysophylla*

has started to be grown in the form of super-dense plantings all over the world (e.g. Australia, Italy, Spain). The leaves are thick, with a small leaf blade. The length of the blade is 4.5–5.2 cm and the ratio of length:width is 4.2–5.5:1. The fruit is very small (0.5 g), with a mastoid shape and ending in a teat. The fruit ripens from mid- to late season and turns black at full ripening. The pit is small and cylindroconical in shape. The ratio of flesh:pit is 1.63–4.06:1. The medium yield per tree is 50–60 kg. This cultivar is resistant to water stress and wind, but sensitive to *Dacus oleae*, *Euphyllura olivina*, *Pseudomonas savastanoi* and attacks from rhynchites. The olive oil content is 27%. Its tolerance to cold is low and its rooting ability from leafy cuttings is constant and medium.

'Lianolia Kerkiras' (*Olea europaea* var. *craniomorpha*)

Other names of this cultivar include 'Suvlolia', 'Nerolia', 'Striftolia' and 'Prevezana'. This is the main cultivar in Kerkira, a Greek island in the Ionian Sea. The tree is very tall, 12–14 m, with a maximum of 20–25 m. The tree trunk is crooked, but the trees are erect and the leaves are big, with a deep green colour on their upper surface. The ratio of length:width is 5:1. The average number of flowers/inflorescence is 23–30. The fruit is small, with a cylindroconical shape and ends in a teat. The fruits are formed in clumps of two to six. The pit is relatively large in comparison with fruit size, and of cylindroconical shape. The number of fruits/kg is 580 and the ratio of flesh:pit is 6.21:1. The oil content is 20–22% and of excellent quality. The tree starts to produce late, and biennial bearing is common due to late harvesting (sometimes in spring). This is an oil-producing cultivar with high oil content. It is partially self-fertile, with stable and high productivity but with a low rooting potential from leafy cuttings. Its tolerance to drought, wind and cold is high, while the tolerance to *Bactrocera oleae*, *Cycloconium oleaginum* and *Armillaria mellea* is low. This cultivar is cultivated in the Ionian Sea on the Greek islands of Kerkira, Kefalonia, Zakinthos and Paxoi.

'Tsunati' (*Olea europaea* var. *mamilaris* subvar. *minima*)

Other names of this cultivar are 'Muratolia' and 'Athinolia'. It is cultivated on the Greek island of Crete. The fruit is oval in shape, with a characteristic teat and it looks like a lemon in its shape. The fruits are borne two to three together, ripening during the period November–December and their average weight is 1.3 g. The flesh:pit ratio is 4.86:1.

'Lianolia Patron'

Other names of this cultivar are 'Ladolia', 'Lianolia', 'Kothrelia', 'Kutrelia', 'Kutsurelia' or 'Kutsuliera'. This variety is cultivated in the Greek area of Peloponnese. The tree height is 5–7 m, the leaves are small and the fruit is spherical. The mean weight of fruits is 0.9–1.7 g (600–1100 fruits/kg). Its oil content is 15–28%.

'Mastoidis' (*O. europaea* var. *mamilaris*)

Other names of this cultivar are 'Mastolia', 'Athinolia' and 'Pitsadeiki'. It is cultivated in Crete and Peloponnese. The leaves are 6–7 cm in length and 1.1 cm wide; their colour is light green to green. The fruit is cylindroconical in shape, weighing 1.4–2.2 g. Its oil content is 28–35% and the flesh:pit ratio is 6.0:1. It tolerates low temperatures and can be grown in areas up to 1000 m in elevation.

MEDIUM-FRUITED OLIVE CULTIVARS

'Megaritiki' (*Olea europaea* var. *argentata*)

Other names for this cultivar are 'Ladolia', 'Perachoritiki' and 'Vododitiki'. This is a Greek cultivar cultivated in the area of Megara and part of Peloponnese. The length:width ratio of leaves is 6.4:1. The average fruit weight is 3.5–4.5 g. The fruit has a characteristic heteromorphism in its shape. The fruits are borne singly or in pairs and have characteristic grey spots on their epidermis during ripening. The ratio of flesh:pit in the fruit is 8.6–10.1:1. The oil content is 22–25%. This cultivar requires pollination since it is partially non-self-fruiting with a tendency towards alternate bearing and medium rooting ability of its cuttings. It has low tolerance to *Aspidiotus hederae*, *Parlatoria oleae* and *Saissetia oleae*, medium tolerance to cold, salinity and *Verticillium dahliae* and high tolerance to drought.

'Galatsaniki'

Other names of this cultivar are 'Agiou Orous' or 'Galatistas'. This is a medium-sized cultivar with cylindroconical fruit, ending in a teat. Its oil content is 20%. It ripens early (September) and its fruit at ripening becomes completely black. This cultivar is sensitive to *Dacus* and *Cycloconium*, but tolerant to cold.

'Thasitiki'

Other names of this cultivar are 'Thasou', 'Throumba' or 'Throumbolia'. It is a Greek cultivar grown on the Greek island of Thasos (northern Greece). Its fruit is cylindroconical in shape, ending in a teat. This is a dual-purpose cultivar with 20% oil content. The leaves are dark green and the shoots are hanging. It is resistant to cold, *Cycloconium* and *Bacterium tumefaciens*.

'Kothreiki' (*Olea europaea* var. *minor rotunda*)

This is also known under the names 'Manaki', 'Manakolia', 'Glykomanako' or 'Korinthiaki'. It is cultivated in central Greece and Peloponnese. The leaves are broad, with a length:width ratio of 4.0:1 and the fruit is spherical, without a teat. The mean fruit weight is 3 g and the pit is cylindroconical. The ratio of flesh:pit is 4.70:1. This cultivar is resistant to cold and can be grown at up to 800 m elevation. It is a very productive dual-purpose olive cultivar.

'Kalamon' (*Olea europaea* var. *ceraticarpa calamata*)

This is known also by the names 'Kalamatiani', 'Aetonichi', 'Tsigeli' and 'Karakolia'. It is an excellent cultivar for both table olives and oil production, and has now begun to be expanded from Greece to other countries (see Fig. 23.1). Its leaves are distinctively large, slightly twisted from end to end and their length:width ratio is 4.11:1. The fruit has a mean weight of 2.6–5.5 g and it is lengthy and pointed, with a distinctly bent point at its tip. The fruits are presented singly or in pairs. The fruit starts to ripen in November and it turns black when

Fig. 23.1. The world-famous Greek olive cv. 'Kalamon'.

fully ripe, and it has a mean oil content of 17%. The flesh:pit ratio is 6.72:1. This is partially non-self-fruiting and its best pollinizers are the cultivars 'Manzanillo' or 'Gordales'. The tree obtains a height of 7–10 m; its growth is erect, with broad dark green leaves (8–9 cm × 1.5–2.0 cm). The fruit has a cylindroconical shape and the ratio of flesh:pit is 8.3:1. The flesh of the fruit is compact and the fruit ripens in late November–December. It is resistant to *Dacus* and needs medium-textured soils with an optimum pH of 7.0. During the summer it requires at least three irrigations in order to produce fruit of adequate size. Since it is a vigourous cultivar N fertilization should be carefully adjusted. This cultivar does not root easily with leafy cuttings under mist propagation, and it is the best known Greek olive cultivar, with a worldwide reputation. It is an excellent table variety producing a high-quality product as regards colour, texture and taste. Although it originated in the region of Kalamata in south-east Peloponnese, it has proved its adaptability as an excellent cultivar in both warm and cold areas worldwide.

'Kolovi' (*Olea europaea* var. *pyriformis*)

This is known also by the names 'Valana', 'Milolia' or 'Mytilinia'. It is cultivated mostly on the Greek island of Lesbos (70%), and less so on other Greek islands. The tree is vigorous and its height is 6–8 m. The leaves have a large dark green blade and their length:width ratio is 4.33:1. The fruit (3–6 g) has the shape of an acorn and matures late (end of December). One to three fruits are presented together. The pit has six to seven carvings. The oil content of this cultivar is 25% and is of excellent quality. This cultivar is used mostly for olive oil production and only partly for pickling of green and black table olives. It is sensitive to *Cycloconium*.

'Adramitini' (*Olea europaea* var. *med.* subvar. *otunda*)

Known also under the names 'Aivaliotiki' and 'Fragolia', it represents 20% of olive culture on the Greek island of Lesbos. The leaves are of average size, with a length:width ratio of 4.16:1. The fruit is round to slightly oval in shape. The mean fruit weight is 2.17 g, the ratio of flesh:pit is 3.82:1 and the month of ripening is December. The oil content of this cultivar is 22% and the oil quality is excellent. However, this cultivar is sensitive to *Dacus*.

'Frantoio'

Other names for this cultivar include 'Paragon' (in Australia), 'Frantoijano', 'Corregiolo', 'Razzo' and 'Gentile'. These five are all considered to belong to the same family or population. The cv. 'Paragon' of Australia, after DNA analysis, has been shown to be similar to the cv. 'Frantoio' grown in the Tuscany region of Italy. This is the most productive cultivar in central Italy, is very adaptable to various conditions and is very cold resistant. The oil quality and flavour are both excellent. The productivity is light to medium and this is one of the most important olive trees in New Zealand. It does not require chilling for flowering bud differentiation. Although this cultivar is considered self-fruiting, the use of the cv. 'Pendulina' as pollinator may increase the yield by up to 10%. The proportion of pollinators should be 5–10% of the trees in an orchard. 'Frantoio' is considered as being sensitive to peacock spot (*Cycloconium oleaginum*).

'FS-17'

This cultivar originated from free pollination of the cultivar 'Frantoio' and it is a widespread variety throughout Italy. The cultivar 'FS-17' is highly suited to dense planting and mechanical harvesting (Fontanazza *et al.*, 1998). Nowadays this cultivar is grown in many parts of the world – Italy, Spain, South Africa, Argentina, Chile, Greece and Australia. It is an excellent variety for oil production and has low vigour, the tree having a drooping appearance. The leaves are medium-sized and elliptical. The fruit weight is 2–4 g and the flesh:pit ratio 9.73:1, with an oval pit. 'FS-17' roots very easily under mist propagation and is self-fruiting. The flowering period and ripening season are both average. The tree starts producing early and its productivity is stable and high. It can be used as rootstock for the cultivar 'Giaraffa', reducing the vigour of the latter by 50%.

'Manzanillo'

This is also known as 'Manzanilla', 'Manzanilla de Sevilla' or 'Fina'. The tree is about 7 m in height at maturity. The canopy is spreading, is highly productive and early bearing. The tree grows better in warm locations without heavy frosts, and it is considered to be the world's most popular table olive, especially in California and Spain.

The size of the fruit is medium (4.8 g, 200–280 fruits/kg) and the flesh:stone ratio is 8.2:1 (see Fig. 23.2). The shape of the fruit is oval, with a thick skin and good texture. The fruit at maturation has a slightly violet–black colour. It is a cultivar sensitive to extended cold periods, and represents 60 and 80% of green pickling fruit for Mediterranean countries and California, respectively. It is considered as the best dual-purpose olive cultivar in the world. 'Manzanillo' is susceptible to *Verticillium* wilt. For good production and cross-pollination the cvs 'Sevillana' and 'Santo Agostino' are used as pollinators.

Fig. 23.2. The olive cv. 'Manzanillo'.

The tree is very productive. It produces every year and its propagation by leafy cuttings under mist is difficult. The fruit contains 20% oil and has a small pit. The flesh of this cultivar contains also 5.65% total sugars (5.41% reducing). Examination by RAPD (Random Amplified Polymorphic DNA) revealed that 'Manzanillo' consists of 14 genotypes, as shown in Table 23.4.

'Meski'

'Meski' is a table olive cultivar that is self-sterile, with medium rooting ability and low oil content. It has low tolerance to cold, *Cycloconium oleaginum*, drought, iron chlorosis and lime, medium tolerance to *Pseudomonas savastanoi* and high tolerance to salinity and *Veticillium dahliae*. This cultivar is distributed throughout Tunisia, Argentina, Turkey and the USA. Synonyms of this cultivar are 'Gerbide', 'Getlet' and 'Gherlide'.

'Mission'

This cultivar is dual-purpose, i.e. for both table and olive oil production. It is self-fertile, with high rooting ability of leafy cuttings under mist, and its tendency for alternate bearing is intermediate. The tolerance of this cultivar is high to cold, medium to drought and *Pseudomonas savastanoi* and low to *Aspidiotus hederae*, *Cycloconium oleaginum*, *Gloesporium olivarum*, *Parlatoria oleae*, *Saissetia oleae*, *Verticillium dahliae* and salinity. 'Mission' is distributed throughout Argentina, Australia, Azerbaijan, Egypt, Iraq, Japan, the USA and other areas.

Table 23.4. The 14 genotypes of olive cv. 'Manzanilla'.

Genotype	Origin
'Manzanilla sevillana'	Spain
'Manzanilla de Carmona'	Spain
'Manzanilla de Tortosa'	Spain
'Manzanilla de Jaen'	Spain
'Manzanilla de Sevilla'	Spain
'Manzanilla dos Hermanas'	Spain
'Manzanilla Hermanas'	Spain
'Manzanilla del Piquito'	Spain
'Manzanilla de Almodovar'	Italy
'Manzanilla Italy'	Portugal
'Manzanilla Carrasquena'	Argentina
'Manzanilla Commune Argentina'	Argentina
'Hass improved Manzanilla'	USA
'Manzanilla Israel'	Israel

'Barnea' (or 'K-18')

This is a new cultivar selected in Israel by Professor Shimon Lavee. It was found in the Kadesh Barnea region, located between the Sinai Desert and Israel. It is an early producer, giving third-year yields up to 10 t/ha with an oil content of 20%. It is suitable for mechanical harvesting and more than 40 trees/1000 m^2 can be planted.

The tree is vigourous, with thin fruiting shoots. The canopy is open. The leaves are medium to large in size. The fruit is elongated and pointed to the apex; the stone is banana-shaped. The tree is sensitive to water stress. It is propagated by leafy cuttings that root easily. The fruits mature in mid-season. This cultivar is used for oil extraction and for black pickles, and has some resistance to leaf spot (*Cycloconium oleaginum*). It is prolific in growth and fruiting, very productive and with little alternate bearing. The average production is 1.5 t/1000 m^2. The most appropriate pollinator is the cv. 'Manzanillo'. 'Barnea' trees can be adapted to heavy soils.

'Arbequina'

'Arbequina' is an oil-producing cultivar with high oil content, very productive, producing every year and with a high rooting ability for its leafy cuttings. 'Arbequina' is very tolerant to cold, salinity and high atmospheric moisture, moderately tolerant to drought, *Pseudomonas savastanoi* and poorly tolerant to *Bactrocera oleae*, *Cycloconium oleaginum*, *Gloesporium olivarum*, iron chlorosis, *Meloidogyne arenaria*, *Meloidogyne incognita*, *Meloidogyne javanica*, *Pratylenchus penetrans*, *Pratylenchus vulnus*, *Saissetia oleae* and *Verticillium dahliae*. Synonyms of this cultivar include the following: 'Arbequin', 'Alberchina', 'Catalana, 'Blancal', 'Oliva de Arbela', 'Manglot', 'Blancas' and 'Oliva de Borjas'. The cultivar is distributed throughout Spain, Algeria, Argentina, Australia, Bolivia, Brazil, Chile, China, France, Israel, Peru and the USA. It is used for super-high-density plantings, together with 'Arbosana' and 'Koroneiki' (Larbi *et al.*, 2006). The tree's requirements for chilling time are low, and therefore this cultivar can be expanded in southern areas with very mild winters.

'Saloneque'

This variety has medium vigour and adequate productivity. The inflorescences have many flowers; the fruit is oval and of medium size. The green ripe fruits are used for canning.

'Tanche'

The tree is very vigourous and very tolerant to cold. Its inflorescences are short and compact. The productivity of the tree is good and constant. The fruit is of medium size, heart-shaped and it is used for canning.

'Coratina'

This cv. originated in Italy. The tree is of medium size, spherical and with long shoots. The leaves are elliptical in shape, the fruit weighs 4.5 g and is oval in shape. The season of ripening is average and its oil content is 21–26%. The cultivar is non-self-fruiting, with high and constant productivity. The best pollinizers for this cultivar are 'Frantoio' and 'Moraiolo'. It is tolerant to water stress and cold and it has adapted to varying soil conditions.

'Verdale', 'S.A. Verdale' and 'Wagga Verdale'

A synonym of this cultivar is 'Verdial', and it originated in the southern part of France. Today there are known many different 'Verdales' all over the world, differing in fruit size, oil content, flesh:pit ratio, hardiness and other characteristics. The South Australian 'Verdale' has a larger oval fruit (7–10 g) than the normal 'Verdale'. The 'Wagga Verdale' has smaller fruit like the common 'Verdale', but yields more than the Australian clone. These three clones are well adapted to South Australian conditions. The oil content of these clones is low, not exceeding 20%.

'Aglandau'

This is an oil-producing cultivar with medium oil content, self-fertile, an intermediate tendency for alternate bearing and a high rooting ability for leafy cuttings. The cultivar is considered as being very tolerant to cold and *Pseudomonas savastanoi*, moderately tolerant to *Bactrocera oleae*, drought, *Prays oleae*, *Saissetia oleae* and wind and poorly tolerant to *Cycloconium oleaginum*. This cultivar is distributed throughout France, Azerbaijan, Australia and Ukraine and is also known by various synonyms such as 'Aglandaou', 'Aglando', 'Airane', 'Argental', 'Blancane', 'Blanchet', 'Blanquet', 'Blanquette', 'Olivier Commun', 'Verdale', 'Verdaou' and 'Luzen'.

LARGE-FRUITED OLIVE CULTIVARS

'Konservolia' (*Olea europaea* var. *med. rotunda*)

Known also under the names 'Voliotiki', 'Piliou' and 'Amphissis', this is cultivated in various regions of Greece such as Agrinio, Arta, Amphissa, Pilio, Stylis, Atalanti and Magnisia. This cultivar is self-fruiting, although planting of pollinizers such as 'Chondrolia Chalkidikis', 'Adramytini', 'Gordales' and 'Kalamon' increases productivity. The mean tree height is 6–8 m. The mean fruit weight is 6.5 g and is mostly used for olive pickling and for black ripe olives. The fruit is spherical and the flesh:pit ratio is 8.28:1. It prefers soils of medium texture or fertile ones. It can be cultivated at up to 600 m elevation and it is sensitive to *Dacus* infection and to frost damage.

'Gaidourolia' (*Olea europaea* var. *major macrocarpa*)

Known also under the names 'Koromilolia' and 'Damaskinati', the tree has a height of 5–6 m, leaves of light green colour and has an elongated fruit with an average weight of 10–12 g. The surface of the fruit has many lenticels, whitish to light green in colour. The ratio of flesh:pit is 9.7:1 and its oil content is 17%. The main use of this cultivar is in the pickling of green table olives. When the fruit load is average, the fruit size is very large – up to 30 g. This is a promising variety for pickling green olives.

'Karydolia' (*Olea europaea* var. *makima*)

Other names of this cultivar are 'Kolymbada', 'Kolymbati' and 'Karydorachati'. This cultivar is cultivated in some regions of Greece such as Chalkidiki, Fokis and Attiki. The tree height is 5–8 m, with light green leaves and cylindroconical fruits of average weight 5–8 g, with a teat. The ratio of flesh:pit is 6.6:1 and its oil content is 14%. This is a processing cultivar appropriate for green and black ripe olives.

'Amygdalolia' (*Olea europaea* var. *amygdaliformis*)

Other names of this cultivar include 'Stravomita' or 'Kurunolia'. It is grown in some regions of Greece such as Amphissa and Thiva. Its canopy is spherical, the leaves are broad and lengthy and the fruit weight is 5–10 g. The ratio of flesh:pit is 5.0–6.5:1 and its oil content is 8–20%.

'Chondrolia Chalkidikis'

This variety is cultivated mostly in the region of Chalkidiki (northern Greece). Its fruit is very large and can exceed 10 g in weight. It is sensitive to frost and non-self-fruiting. The best pollinizers of this cultivar are 'Amphissis', 'Megaritiki', 'Koroneiki', 'Manzanillo' and 'Gordales'. The main use of this cultivar is for either pickling of green olives or stuffed (with almond or chili) green olives. The remaining product inappropriate for processing is used for oil production. It is very sensitive to *Dacus oleae*.

'Sevillano' (syn. 'Gordal', 'Gordal Sevillano' and 'Sevillana')

This is a large-fruited variety with 14% oil content (see Fig. 23.3). The trees are less productive than 'Manzanillo', with a spreading habit of the shoots. The cultivar is difficult to adapt and is not easily propagated by cuttings under mist. The tree requires fertile soils and the fruit weight is 12–14 g, with irregular shape. The flesh is not easily separated from the pit. It requires 2000 h at < 7°C to differentiate flowering buds. The best pollinizers are the cvs 'Chondrolia Chalkidikis', 'Kolovi', 'Adramytini' and 'Manzanillo'.

The flesh:pit ratio is 7.3:1. The fruits are oval in shape, with an indent at the stem, which can give it a heart-shaped appearance. The skin is thin and speckled, with white markings. The fruit flesh is green, turning purplish black when ripe and with a good texture for pickling. The pit has deep grooves.

'Sevillano' occupies second place after 'Manzanillo' in both the Spanish and Californian table fruit market, where it obtains good prices. Early harvesting can reduce processing problems, such as bruising and split pits.

'Sevillano' is resistant to cold, although it requires 2000 h at < 7°C in Spain, and in Italy it is cultivated in warmer regions without problems. Currently, this cultivar covers 20,000 ha in Spain. Due to high levels of soil N

Fig. 23.3. The olive cv. 'Sevillano'.

fertilization, Ca uptake is reduced and a problem known as 'soft nose' appears, i.e. the bottom part of the fruit becomes soft and deteriorates.

This is a table olive cultivar characterized by self-sterility and shows an intermediate tendency for alternate bearing. 'Gordal Sevillana' shows low tolerance to *Bactrocera oleae*, cherry leaf roll virus (CLRV), *Camarosporium dalmatica*, drought, *Gloesporium olivarum*, *Meloidogyne arenaria* and *javanica*, *Pratylenchus vulnus*, *Pseudomonas savastanoi*, strawberry latent ringspot virus (SLRV) and *Saissetia oleae* and medium tolerance to cold, *Cycloconium oleaginum*, salinity and soil moisture.

This cultivar is distributed throughout many regions such as Argentina, Australia, Israel, Cyprus, Croatia, Italy, Lebanon, Morocco, Spain, France, Portugal, Chile, the USA, Algeria, Japan, South Africa and Ukraine.

'Ascolano'

This is a large-fruited variety (8–10 g) with low oil content (13%). The flesh has a pleasant taste and is easily separated from the pit. The flesh:pit ratio is 8.2:1. The period of ripening is during the middle of October. The tree is vigourous, with a spreading habit. The cv. 'Ascolana tenera' originated from 'Ascolano', which is self-fruiting and tolerant to *Cycloconium*, *Bacterium tumefaciens* and low temperatures.

'Ascolana tenera'

The tree is vigourous, with a very dense canopy and grows with a spreading habit. The leaves are medium-sized, elliptical in shape and dark green. The fruit is ellipsoid and asymmetrical. The mean fruit weight is 8–10 g and is used for pickling table olives. The flesh comprises 86% of the fruit and its oil content is

16–18%. 'Ascolana tenera' is a non-self-fruiting cultivar with high percentage of imperfect flowers, average tendency for alternate bearing and a high rooting ability of leafy cuttings. The best pollinizers for 'Ascolana tenera' are the cvs 'Lea', 'Rosciola', 'Leccino', 'Frantoio' and 'Pendolino'.

'Ascolana tenera' is a very tolerant cultivar to cold, cucumber mosaic virus (CuMV), iron chlorosis, *Meloidogyne arenaria* and *Meloidogyne hapla*, moderately tolerant to drought, *Gloesporium olivae*, *Parlatoria oleae* and SLRV and poorly tolerant to *Bactrocera oleae*, *Cycloconium oleaginum*, *Gloesporium olivarum*, *Meloidogyne incognita*, *Meloidogyne javanica*, *Pratylenchus vulnus*, *Prays oleae* and *Saissetia oleae*. This cultivar is known also by various synonyms such as 'Ascolano', 'Askolano', 'Asiolani', 'Noce', 'Nociola', 'Nociu', 'Oliva grossa' and 'Spanish Queen'. 'Ascolana tenera' is distributed throughout Argentina, Albania, Croatia, Cyprus, Egypt, France, Italy, Israel, Lebanon, Pakistan, Slovenia, Serbia, Montenegro, Turkey, Brazil, Chile and the USA.

'Santo Agostino'

The region of origin of this cultivar is the area of Puglia, in Italy. The tree is vigourous and the fruit is large (8–9 g), asymmetric and oval. The fruit is harvested during September for production of green canned olives and the period of ripening is November. The fruit flesh represents 90% of its weight. The cultivar is self-sterile and the best pollinizers are the cultivars 'Moraiolo' and 'Corregiolo'. It resists *Cycloconium oleaginum*.

'Santa Caterina'

This originated from Italy. The tree is vigourous and spherical, with light green leaves. The fruit weighs 7–9 g, is elliptical and asymmetric in shape. It ripens early and harvesting starts in September. Its tolerance to cold is adequate and moderate to *Cycloconium oleaginum*.

'Grossa di Spagna'

This also originated in the Puglia region of Italy. The vigour of the tree is average, with dense shoots. The leaves are of medium size, elliptical in shape and dark green. The fruit weight is 9–12 g and its shape is elliptical and asymmetric. The period of ripening is early and the fruit flesh comprises 75% of the fruit, containing 16–19% olive oil. Its productivity is medium to high. The cultivar is self-sterile and the best pollinizers are the cultivars 'Maurino', 'Coratina', 'Pentolino', 'Frantoio' and 'Leccino'. It is suitable for green olive canning and requires fertile and irrigated soils.

TABLE OLIVES

INTRODUCTION

Worldwide interest exists in using processing methods for olives such as brine, fermentation and salt-/heat-drying methods rather than processing with lye (caustic soda). Primary processing involves any of several operations (soaking, fermentation, lye treatment or heat). Secondary processing involves increasing the organoleptic value of the olive by, e.g. addition of herbs, spices or vegetables, de-stoning and stuffing with peppers, cheeses, almonds, garlic or onion.

The most important table olive varieties internationally are 'Kalamata', 'Conservolia', 'Manzanillo', 'Sevillana', 'Hojiblanca' and 'Ascolana tenera'. The Californian table olive industry is based on five cultivars; in order of volume sales they are 'Manzanillo', 'Sevillano', 'Mission', 'Ascolano' and 'Barouni' (Ferguson, 2006). Many olive products have now attained a worldwide commercial base, i.e.:

- 'Kalamata' type.
- Green ripe Californian type.
- 'Castelvetrano' type (Romeo *et al.*, 2006).
- 'Throumbes'.
- Sicilian-type green olives.
- 'Picholine' type.
- Spanish green olives.

A detailed description of table olives is given by Zervakis (2006) and Kyritsakis (1998, 2007).

BLACK OLIVES IN BRINE

The olives are subjected to a primary process for de-bittering, and this is followed, after de-stoning, by addition of other materials to the brine such as olive oil, vinegar, herbs and spices or pimiento and almonds.

The optimum quality of table olives is determined by many factors, in particular the olive grove conditions and the selected variety. The oleuropein concentration is markedly reduced during processing, leading to de-bittering. The soluble sugars of olive flesh are essential for supporting fermentation during processing.

Most consumers prefer medium- to large-sized olives, with a flesh:stone ratio 5–6:1 or more. Freestone varieties are best, since the flesh separates easily from the stone. Olive size, shape, flesh:stone ratio, ease of de-stoning, colour and texture are the most important selection criteria for table olives. Most olive trees suitable for table olive processing are clonally propagated, with the exception of difficult-to-root varieties like 'Kalamata'. Some of the most important varieties for table olive processing include 'Manzanillo', 'Sevillana', 'Barouni' and 'Hojiblanca' for green olives and 'Conservolia' and 'Kalamata' for black ripe olives. Other varieties well known internationally are 'Chondrolia Chalkidikis', 'Nocellara del Belice' and 'Picholine'.

Very few chemicals are required for successful olive cultivation; correct pruning will help to control many problems; Cu sprays following harvesting and pruning can be used as a general antifungal treatment.

Table olives should be harvested by hand in preference to mechanical harvesting, in order to reduce the risk of damaged fruit that, when processed, produces an inferior product (Balatsouras, 1990). Olive harvesting by hand (or mechanical) rakes, tree shakers or overhead harvesters increases the risk of bruising, leading to soft fruits after processing. Mechanical harvesting with shakers or vibrators has limited application for table olive production. Ethylene-releasing compounds have also been used to improve fruit abscission, but their use is limited since that leads to substantial leaf loss. Table olives should be harvested in three stages: green ripe, turning colour and black ripe. Harvesting time depends on the variety, growing conditions and crop load. Green ripe olives are picked from late summer to early autumn, and black ripe olives from late autumn to late winter.

Green ripe olives are processed as green table olives, the period when they have attained maximum size. Naturally black ripe olives (those that fully ripen on the tree) have adequate flesh colour and are hand-harvested. To avoid over-ripening, the optimum stage for harvesting is when the pigments are present in around two-thirds to three-quarters of the flesh.

Raw olives should be processed as quickly as possible after picking, i.e. 1–2 days, to avoid damage; damaged fruits yield a product of poor quality and taste. Storage of raw olives at low temperatures decreases the danger of deterioration. A temperature of 5–10°C may preserve the quality of raw olives for 4–8 weeks. However, temperatures < 5°C may cause fruit browning. A controlled atmosphere with low oxygen (2%) and 90–95% air humidity prolongs the storage period. The nutrient levels in the flesh of raw olives are as follows: K, 0.53–3.39%; P, 0.02–0.25%; Ca, 0.02–0.16%; Mg, 0.01–0.06%; S, 0.01–0.13%; B, 4–22 mg/kg; Fe, 3–95 mg/kg; Mn, 0.91–55.00 mg/kg; and Zn, 1.4–33.0 mg/kg.

PROCESSING

Raw olives are bitter and require processing in order to become suitable for consumption. Processing should be conducted under good sanitary practices in order to maintain all ingredients and all the necessary chemical and microbiological standards. Processing affects the concentration of the major compounds, depending on the type of olive (see Table 24.1; Bianchi, 2003).

For high-quality table olives the following requirements are important:

- Good-quality water.
- Excellent quality of raw olives.
- Excellent quality of the chemicals and additives used.

The flavour and taste of processed olives depend on the variety, fermentation, processing solution and packing solutions such as vinegar, olive oil and flavourings.

The equipment required for table olives includes sorters, graders, tanks, pumps and packing equipment. At every stage in the process sanitation is very important in safeguarding consumers' health, with all statutory food and safety standards complied with. Table olives are derived from good-quality raw olive fruits harvested at the appropriate stage of ripening and processed to give a tasty and safe product. Various table olive processing methods are used, depending on cultivar, ripeness, cultural condition and technology of processing.

Among the various types of olives destined for the table, the most popular are the following:

- 'Kalamon', used for 'Kalamata'-type olives.
- Spanish-type green olives ('Sevillana' or 'Manzanillo').
- Greek 'Conservolia' for naturally black ripe olives.
- 'Throumbes' – best for production of dried olives.
- Californian black olives – 'Mission'.
- Sicilian green olives.

Table 24.1. The effect of processing on the major compounds in three types of processed table olives (from Bianchi, 2003).

Compound	Olive type		
	Spanish style	Greek style	Californian style
Triglycerides	No change	No change	No change
Phenols	No change or decrease	No change or decrease	Decrease
Triterpene acids	No change or decrease	No change or decrease	No change or decrease
Glucosides	Decrease	Decrease	Decrease
Sugars	Decrease	Decrease	Decrease
Proteins	No change	No change	No change

Green ripe olives

These olives are harvested at the yellowish green to straw colour; the fruit should be firm. Green ripe olives are treated with a lye solution (caustic soda, NaOH), washed to remove the lye and stored in brine to undergo complete lactic acid fermentation.

Black ripe olives

Black ripe olives are harvested and processed by placing them into brine for spontaneous fermentation. Consequently, they are preserved in brine, by sterilization or pasteurization or by preservatives. Due to processing these olives lose their intense black colour, which can be restored by exposure to air after processing.

Shrivelled black olives

This product is achieved with or without lye treatment in brine or by addition of dry salt. With this treatment the olive flesh is dehydrated, resulting in shrivelled olives. For this process containers allow the liquid from the processed olives to drain away.

Olives darkened by oxidation

Green ripe olives are de-bittered by an alkaline solution, exposed to the air to turn black by oxidation, washed free of lye and then packed in containers containing brine.

Olive products

Such products are marketed as green olive paste, olive rollers, stuffed olives and marinaded olives.

Bruised olives

These are derived from green ripe olives or those turning colour. They are crushed in such a way that the flesh is exposed, without breaking the stone. The olives are then de-bittered in water and packed into containers containing brine, to undergo natural fermentation.

Split olives

Black ripe or green ripe olives may be used. The olives are split longitudinally with a knife or other device, to penetrate the flesh. They are then de-bittered with or without a lye treatment and packed into containers with brine to undergo natural fermentation. After fermentation, oil, lemon or vinegar can be added.

Water-cured olives

This method involves immersion in fresh water every day for 10 days, to achieve de-bittering. With every water change the bitter substance oleuropein is leached out of the olives and the leaching solution is discarded. Subsequently, the olives are stored in 10% brine. This method is fast, but it requires large quantities of water, which create disposal problems with regard to contamination of the environment. After de-bittering, vinegar or oil may be added. With this process the traditional 'Kalamata' variety of olive (naturally black) is produced.

Brine-cured olives

The raw olives are put into containers containing 10% (w/v) sodium chloride (NaCl). The olives absorb the NaCl and a slow fermentation takes place (Division of Agricultural Science, University of California, 1975). Over the following 8 weeks, the concentrations of salt in the fruit and the brining solution equilibrate. Due to salinity effects the water-soluble oleuropein, phenolics and minerals leach out from the olive fruits. During this period fermentation occurs, with the sugars being transformed into lactic and acetic acids, and other substances. The entire process lasts from 3 to 12 months depending on the variety, ripening, salt concentration and pH of the brine.

Slicing the olives speeds up the de-bittering process. Ready-for-consumption naturally black olives in brine have a pH of 4.5–4.8, free acidity of 0.1–0.6% w/v (lactic acid) and NaCl of 10% (w/v).

Spanish-type green olives

With this method Spanish-type green olives are produced. Green Spanish-style table olives contain 9–28% lipids, 1.0–1.5% proteins, 1.5–2.5% fibre and 4–6% ash. Furthermore, these olives contain Na, P, K, Mg, Fe, Mn, Zn, Cu and vitamins (Garrido-Fernández *et al.*, 2006). Raw green ripe olives are soaked in tanks containing 1–2% NaOH (lye) for 8–12 h for de-bittering. When the NaOH has penetrated through three-quarters of the flesh the lye is removed. The remaining flesh not de-bittered supplies sugars necessary for fermentation. Penetration of the lye is checked by observing the colour changes following slicing. Also, by application of phenolophthalein solution to the cut flesh the colour changes from green to red. After de-bittering, the NaOH is removed by several washings of good-quality water. Subsequently, the tanks are filled with 10% NaCl for lactic acid fermentation. These olives are suitable for human consumption within 5 weeks. The characteristics of this ready-to-eat product are: pH 3.8–4.2, free acidity 0.8–1.2% w/v (lactic acid) and NaCl 7–8% (w/v).

The varieties processed by this method include 'Manzanillo', 'Sevillana' and 'Chondrolia Chalkidikis'.

Californian-type black ripe olives

The raw product for this purpose is olives that are green or turning in colour. The fruits are treated with NaOH solutions of different concentrations until the lye penetrates the flesh and the olives are de-bittered. During this process polyphenols are oxidized and change the fruit colour to brownish black. Subsequently, the NaOH solution is removed and the treated fruit are immersed in tanks containing 0.1% ferrous gluconate, to enhance the black colour. After leaching to remove excess Fe, the fruits are placed in containers containing 2–3% NaCl and sterilized. The characteristics of ready-to-eat black olives are pH 5–8 and NaCl 1–5% (w/v). With this method the fruit is ready within 1–2 weeks. This method is very fast and does not require fermentation, but it does require large volumes of water for both the lye phase and subsequent washings.

Raw olives have a population of natural microflora that includes: Gram-negative bacteria, lactic acid bacteria (homofermentative and heterofermentative), yeast, oxidative yeast, *Clostridium*, *Propionibacteria*, etc. The Gram-negative bacteria predominate and produce CO_2, which dissolves in the NaCl solution producing H_2CO_3, thus lowering pH (see below). This process creates anaerobic conditions in the brine.

FERMENTATION OF OLIVES

Aerobic fermentation

Aerobic fermentation of olives is based on the action of lactic acid, which is produced from lactic acid bacteria starter cultures (Marsilio *et al.*, 2005), or from *Lactobacillus* and yeasts grown in the sugars released from olives during processing (Marsilio, 2006). In the process of olive fermentation lactic and acetic acids are produced; these acids lower the pH of the substrate. The dual action of low pH and high salt concentration significantly reduces microbial contamination problems in olives. With fermentation at temperatures of 20–25°C, flavour-giving compounds are produced.

Anaerobic fermentation

Another process used in table olive production is anaerobic fermentation, either as an initial process or after de-bittering of the olives by the application of lye. With this process the organic substrates of the fruits are broken down to produce lactic acid. Mature olives are habituated by homofermentative and heterofermentative lactic acid bacteria and yeasts.

The following Gram-negative bacteria can be found in olives:

- Lactic acid bacteria.
- *Clostridium*.
- *Propionibacterium*.
- *Bacillus*.

These bacteria ferment sugars and liberate CO_2; the CO_2 dissolves in the brine and produces H_2CO_3, which lowers the pH to about 5. The duration of this process is 3–4 days. The same chemical process is achieved by the addition of lactic acid. The lowering of pH and the established anaerobic conditions prevent the further growth of Gram-negative bacteria, and the product is therefore stabilized. Processed olives should maintain some bitterness and fruit flavour. Yeasts and moulds are also natural inhabitants of the olive fruit.

SECONDARY PROCESSING, PACKING AND LABELLING OF OLIVE PRODUCTS

Secondary processing involves the stuffing of olives with various food products to achieve specific flavours. Some of these additional flavours include vinegar, olive oil, herbs and spices (oregano, garlic, etc.). Furthermore, olives can be stuffed with chili, peppers, cheeses, pimiento, almonds, fish, etc. Among these the most common use is that of vinegar, which lowers the pH and thus increases shelf life. Various types of vinegars are used, such as wine vinegar, cider vinegar or balsamic vinegar. In green olives we prefer the light-coloured vinegars, while in black olives darker vinegars are used.

For packing processed olives a brine solution of 8–10% is used. In olives containing vinegar the brine solution contains three parts of brine plus one of vinegar. To avoid spoilage the surface of olive containers is covered with olive oil.

Table olives must be labelled appropriately. Labelling information should include caloric value and the content of various substances such as carbohydrate, oil, protein and minerals.

RULES FOR THE PRODUCTION OF HIGH-QUALITY TABLE OLIVES

- Use of good-quality water, for both processing and washing.
- Use of high-quality raw olives, chemicals and stuffing materials.
- Adherence to all hygiene rules in the processing unit.
- The processed olives should be of the recommended cultivar, with no defects and of appropriate marketable size.

The table olive processing facility should follow all statutory regulations regarding food, health and protection of the environment. The processing unit is comprised of the following areas:

- Reception and grading of fresh olives.
- De-bittering of olives.
- Primary and secondary processing.
- Packing of the processed product.
- Storage of chemicals.
- Storage of packed products.
- Analytical laboratory.
- Loading, marketing and offices.

STORAGE OF FRESH OLIVES

After harvesting, olives should immediately be processed, in order to avoid deterioration. After 1 week's storage at room temperature olives start to deteriorate quickly due to the presence of bacteria, yeasts and certain enzymes present in the flesh. Therefore, if processing is not to be carried out immediately, fruits should be refrigerated. During storage olives lose weight, and shrivelling may occur. Storage at 5–10°C can extend the olive life for 4–8 weeks. However, temperatures < 5°C cause flesh browning. Ideal conditions for olive storage are under a controlled atmosphere of 5.0–7.5% CO_2, 2% O_2 and 90–95% relative humidity.

Olive Oil

EXTRACTION

For olive oil extraction three systems are available:

- The traditional discontinuous processing system.
- Continuous-cycle centrifugation.
- The percolation–centrifugation system.

Olive production involves the twin stages of milling and separation of the oil from water, resulting in three constituents – oil, husk (solid waste) and waste water. In the traditional method milling is conducted with a millstone or hammer stone, and oil extraction is achieved by hydraulic pressure.

In the continuous cycle, grinding of the olives is achieved by metal crushers, and subsequently the olive paste is centrifuged with a horizontal centrifugal decanter. Afterwards, the oil is separated from waste water by a vertical centrifuge. Extra virgin olive oil is produced by cold pressing of the olive paste. The process of oil extraction influences the phenolic content of the oil. Oils obtained by centrifugation have a lower phenolic content, since this method uses large quantities of water, thus removing a significant proportion of the phenols. Furthermore, the type of mill used for oil production exerts a significant influence, and hammer mills are more efficient in extraction of phenol.

The phenolic composition of olive oils is the result of a complex interaction between several factors, which include cultivar, stage of maturation, climate, mill as crushing machine and malaxation conditions. Malaxation is the action of slowly churning or mixing milled olives in a specially designed mixer for 20–40 min. This churning allows the smaller droplets of oil released by the milling process to aggregate and be more easily separated. The paste is normally heated to around 27°C during this process.

CATEGORIES OF OILS AND STONE OILS

The characteristics of olive oil are described in EU Regulation 2568/91. These characteristics are given in Table 25.1.

Virgin olive oil

This is produced by mechanical or physical processes and relatively low temperatures that cause no deterioration in the oil. This product is exposed to no other processes than those of fruit cleaning, centrifugation and filtration.
Virgin olive oil is divided into the following categories:

Extra virgin olive oil
This is a virgin olive oil containing free fatty acids (expressed as oleic acid) ≤ 1 g/100 g (see Table 25.1). Its other characteristics are those of virgin olive oil.

Virgin olive oil
The free fatty acid content (expressed as oleic acid) is ≤ 2 g/100 g. Its other characteristics (colour, flavour, taste, etc.) are those as in the other virgin olive oil categories.

Curante virgin olive oil
The free fatty acid content is ≤ 3.3 g/100 g. All the other characteristics are those of virgin olive oil.

Lampante virgin olive oil
The free fatty acid content is > 3.3 g/100 g.

Table 25.1. The main characteristics in the categorization of olive oil, according to EU Regulation 2568/91.

Oil category	Free acidity (%)	Peroxide index	Waxes (mg/kg)
Extra virgin olive oil	≤ 1.0	≤ 20	≤ 250
Virgin olive oil	≤ 2.0	≤ 20	≤ 250
Curante virgin olive oil	≤ 3.3	≤ 20	≤ 250
Lampante virgin olive oil	> 3.3	> 20	≤ 350
Refined olive oil	≤ 0.5	≤ 5	≤ 350
Olive oil	≤ 1.5	≤ 15	≤ 350
Non-processed pomace oil	≤ 0.5	–	–
Refined pomace oil	≤ 0.5	≤ 5	–
Pomace olive oil	≤ 1.5	≤ 15	> 350

Refined olive oil
Produced from virgin olive oil in refineries, and its free fatty acid content (as oleic acid) is ≤ 0.5 g/ 100 g.

Olive oil
This is a combination product of the mixing of refined olive oil with virgin olive oil (not lampante) with free acidity ≤ 1.5 g/100 g.

Crude pomace oil
Produced after extraction of olive pomace with an organic solvent, and not fit for human consumption. EU law defines any oil containing 300–350 mg/kg of waxes, and aliphatic alcohols > 350 mg/kg, as crude pomace oil.

Refined pomace oil
The pomace oil after refining, giving a product not fit for human consumption, with free acidity $\leq 0.5\%$; its other characteristics must conform to the standards for this category.

Pomace olive oil
This is a mixture of refined stone oil and virgin olive oil (except lampante), fit for human consumption; its free acidity level is $\leq 1.5\%$

The categories of olive oil that can be sold in the retail market are extra virgin olive oil, virgin olive oil, olive oil and pomace olive oil.

STORAGE AND BOTTLING

The high-quality oils are stored in stainless steel containers at a constant temperature of 7–19°C. Both storage period and temperature affect olive oil quality (Pereira *et al.*, 2002; Clodoveo *et al.*, 2007), and especially oil and residual DNA (Luna *et al.*, 2006; Spaniolas *et al.*, 2007). After extraction the oil is stored in large stainless steel tanks for more than 1 month, in order to settle out the water and pulp. Decanting follows, in order to remove sediments that otherwise would appear in the bottled product; it also eliminates the presence of processing water, which could give rise to 'off' flavours.

As soon as the producer has an order for a certain number of bottles the oil is bottled and sent to the market, and should be consumed within a short period in order to avoid deterioration of quality within the bottle.

SENSORY EVALUATION OF QUALITY

Olive oil should have a fruity flavour that characterizes the variety or the mixture of varieties from which that oil was produced. The quality can be evaluated for certain characteristics with equipment or by sensory analysis, which could be much more accurate. With sensory evaluation the aroma, taste and textural changes are easily evaluated. Sensory evaluation determines not only defects due to improper storage or processing but also the positive characteristics of the oil. Among these characteristics, bitterness, pungency and fruity olive flavour are included. Bitterness and pungency are frequent in olive oils following fruit processing. These are not defects, but important characteristics that are related to the long shelf life of olive oil. Bitter and pungent oils have greater polyphenol content and can be stored for a longer period than sweet oils.

CRITERIA OF QUALITY AND STANDARD VALUES

Certain laboratory measurements are used to determine adultery by seed oils, and also the levels of free acidity and peroxide that indicate the degree of oxidation in an olive oil. Due to oxidation olive oil develops off-flavours and turns rancid. Table 25.2 shows the quality criteria and standards for olive oil (according to EU Regulation 2568/91) and Table 25.3 defines the 100C purity and authenticity criteria (sterols, fatty acids, saturated fatty acids) for olive oil.

CERTIFICATION

A great number of certifications exist, the two most important of which, in the context of olives, are described below.

Protected Designation of Origin (PDO)

Determination of the geographical origin of an olive oil can be conducted using ^{13}C nuclear magnetic resonance (NMR) (Vlahov et al., 2003). NMR

Table 25.2. Quality criteria and standards for olive oil, according to EU Regulation 2568/91.

Oil category	Free acidity (%)	Peroxide index	K_{270}	K_{232}	ΔK	Mean defects
Extra virgin oil	≤ 0.8	≤ 20	≤ 0.22	≤ 2.5	≤ 0.01	0
Virgin oil	≤ 2.0	≤ 20	≤ 0.25	≤ 2.5	≤ 0.01	≤ 2.5
Ordinary virgin oil	≤ 3.3	≤ 20	≤ 0.25	≤ 2.60	≤ 0.01	2.5–6.0
Lampante oil	> 3.3	–	–	–	–	> 6.0
Refined olive oil	≤ 0.3	≤ 5	≤ 1.10	–	≤ 0.16	–
Olive oil	≤ 1.0	≤ 15	≤ 0.90	–	≤ 0.15	–

Table 25.3. Purity and authenticity standards for olive oil (from IOOC data, www.cesonoma.ucdavis.edu/hortic).

Content and composition	Statutory level
Individual sterols (mg/kg)	
Cholesterol	Max. 0.5[a,b]
Campesterol	Max. 4.0[a,b]
Stigmasterol	
D-7-stigmasterol	Max. 0.5[a,b]
β-sitosterol	Min. 93[a,b]
Total sterols (mg/kg)	Min. 1600[b], 1000[c], 2500[d], 1800[e]
Fatty acids composition[f]	
Myristic acid, C14:0	Max. 0.05[a,b]
Palmitic acid, C16:0	7.5–20[a,b]
Palmitoleic acid, C16:1	0.3–3.5[a,b]
Heptadecenoic acid, C17:0	Max. 0.3[a,b]
Heptadecenoic acid, C17:1	Max. 0.3[a,b]
Steatic acid, C18:0	0.5–5.0[a,b]
Oleic acid, C18:1	55.0–83.0[a,b]
Linoleic acid, C18:2	3.5–21.0[a,b]
Linoleic acid, C18:3	Max. 1.0 (EU = 0.3)[a,b]
Arachidic acid, C20:0	Max. 0.6[a,b]
Gadoleic acid, C20:1	Max. 0.4[a,b]
Behenic acid, C22:0	Max. 0.2[a,b]
Saturated fatty acids (%)	Max. 1.5 (EU = 1.3)[a,c,e]
	Max. 1.8 (EU = 1.5)[e]
	Max. 1.8 (EU = 1.5)[a]
	Max. 2.2 (EU = 2.0)[d]
	Max. 2.2 (EU = 2.0)[b]

[a]Olive oil; [b]refined pomace oil; [c]virgin oil; [d]crude pomace oil; [e]refined olive oil; [f]% of methylesters.

spectroscopy (Mannina *et al.*, 2001) or sterol composition (Alves *et al.*, 2005). This definition refers to olive oil produced in a specific region using traditional production methods from the defined area, of whose name the product bears. The area of origin affects virgin olive oil quality (Ranalli *et al.*, 1999).

Protected Geographical Indication (PGI)

This indicates that the olive oil is produced in the geographical region whose name is written on the bottle, allowing for a more flexible link to certain regions having a specific characteristic.

CRITERIA FOR DETERMINATION OF QUALITY

The International Olive Oil Council (IOOC) has established certain criteria in determining oil quality, such as acidity, oxidation, colour and organoleptic characteristics.

Free acidity

Free acidity is an indicator of the quality of olives and of the procedures of harvesting, handling, transportation and storage prior to olive milling. Free acidity measures the hydrolytic breakdown of triglycerides to di- and monoglycerides, leading to fatty acid liberation. It is expressed as percentage free fatty acids – especially on the basis of oleic acid, which is the main fatty acid in olive oil.

Peroxide index

Peroxide index is an indicator of oxidation, the means of creating peroxides in olive oil. High peroxide values mean that the olive oil is defective and that the olives were not correctly treated before oil extraction. Peroxide values are expressed in milliequivalents (meq)/kg of free O_2 in oil. Measurement is conducted by titration with potassium iodide, which liberates iodine.

Ultraviolet (UV) light absorbance

This is also an indicator of oxidation, and in particular for oils that have been heated during the process of refinement in the processing plant. With this process the absorbance of certain oxidized compounds, at 232 and 270 nm, is measured. The value $(\Delta)K$ detects treatment by decolorizing agents and also adulteration by refined or pomace oil Optical Density (OD_{270} and OD_{266}–OD_{274}).

Flashpoint

The parameter of flashpoint measures the temperature at which the oil begins to burn. The flashpoint in refined olive oil, pomace oil and seed oils is 190–200°C, while in virgin oils it is 210–220°C.

Insoluble impurities

These provide an index of the presence of impurities, such as alkaline soaps of palmitic and stearic acids, proteins suspended in the oil, dirt, oxidized fatty acids, resins and mineral elements. Insoluble impurities are expressed as a percentage and are determined by dissolving oil in petroleum ether, and subsequently by filtration of impurities.

Colour

The type of lipophilic pigments present in the olive oil at harvesting, such as chlorophylls or carotenoids, determines its colour.

Flavour

The flavour is due to the presence of volatile substances, organic acids such as oleic and linoleic, and also polyphenols.

The quality of oil depends on both the microclimate of the olive grove and anthropogenic factors. The organoleptic characteristics (flavour, aroma) and the colour depend on soil and climatic conditions and also on the olive cultivar. The time of olive harvesting also determines oil quality. Harvesting should be completed quickly, and so too should processing for oil production. In that way, any potential problems due to *Dacus oleae*, fruit deterioration or oxidation are reduced. Reduction in the use of phytoprotective chemicals and replacing them by biological means improves oil quality. Finally, significant improvement in olive oil quality has been achieved by reducing the time from harvesting to extraction in the mills.

Olive oil has a high biological and nutritional value. However, it can be polluted with certain chemicals dangerous to human health during the stages of extraction, refining, bottling and marketing. These include the following:

- Agricultural phytoprotective chemicals.
- Volatile substances and Freon due to its leakage from refrigerators located adjacent to olive mills or bottling units.
- Heavy metals, due to contact of oil with the metal parts of machinery or storage tanks.
- Polycylic or aromatic hydrocarbons, in the case of pomace oil.
- Environmental pollutants such as dioxins and toluene.

AUTHENTICITY

Authenticity means that the oil purchased by the consumer matches its description. Undeclared addition of vegetable oils and false statements regarding the source of the oil (geographic origin, cultivar, organic olive oil) are examples of adulteration (Bazakos, 2007). Growers prefer the production of superior-quality olive oil, since they can market it more easily and at higher prices. Food distributors also want their products to comply with current legislation. Furthermore, most consumers are ready to pay more for olive oil they consider of high quality derived from cultivation practices, method of processing, cultivar or geographic origin.

DETECTION OF ADULTERATION

Detection of adulteration is possible by the use of certain analytical methods. However, adulteration methods are constantly changing and there is always a need for reassessment of these techniques. Nowadays, various conventional or non-conventional methods have been developed for detection of adulteration. Conventional methods are those not based on DNA analysis.

Conventional methods

These methods are based on the analysis of various constituents and properties of olive oil. Olive oil could be adulterated by other vegetable oils such as pomace oil, maize oil, groundnut oil, sunflower oil, soybean oil, sesame oil, animal lipids or other sources.

Conventional control of olive oil adultery includes the following methods.

Determination of physical and chemical constants and chemical compounds of olive oil (Aparicio *et al.*, 1997)

These constants include specific gravity, iodine number, refraction index and chemometric procedures (Bucci *et al.*, 2002). They are increased by any adulteration by groundnut oil, cotton oil, soybean oil, sunflower oil, maize oil or sesame oil.

Changes in squalene concentration

The accepted level of squalene (a natural organic compound) is 136–708 mg/100g of olive oil. Adulteration shows values outside this range.

Ultraviolet (UV) absorption

Ultraviolet absorption in vegetable oils, at 208–210 nm, is three times greater than that of olive oil.

Gas or HPLC and infiltrated spectroscopy

These methods are based on a carrier gas movement through a column (gas) to generation of a gradient of two solvents (HPLC) and to spectroscopy.

Sterol measurement

High concentrations of the sterol stigmasterol indicate adulteration of olive oil with soybean oil. The following techniques are available.

NEAR-INFRARED (NIF)   This method utilizes the NIF part of the light spectrum. NIF causes inter-atomic vibrations of molecules at a frequency of 10^{13}–10^{14} Hz. The NIF measurement is a fast, continuous and non-destructive process.

NUCLEAR MAGNETIC RESONANCE (NMR)   This high-resolution method provides information on liquid or semi-solid materials, involving the use of ^{1}H and ^{13}C (Vlahov et al., 2003).

ULTRAVIOLET (UV), VISIBLE (VIS) AND INFRARED (IR) OR FTIR SPECTROSCOPY   Ultraviolet, VIS, IR and Fourier Transform Infrared (FTIR) electromagnetic radiation all interact with various materials which, upon excitation, emit light that can be measured.

The techniques described above represent non-destructive spectroscopic methods used to distinguish authentic and non-authentic olive oil samples (Tapp et al., 2003).

CHROMATOGRAPHY   Various forms of chromatography are used, such as gas chromatography (GC), liquid chromatography (LC), thin-layer chromatography (TLC) and high-performance liquid chromatography (HPLC). The stationary phase consists of a thin layer of liquid held in place on the surface of a porous inert solid. A wide variation of liquid combinations are suitable. The movement of a mobile phase results is a differential migration and separation of the sample components. The stationary phase can be either a solid or liquid and the mobile phase can be liquid or gas (liquid–solid, LSC, or liquid–liquid, LLC).

These methods achieve a rapid and reliable separation of molecules with similar characteristics, based on partition of the molecules between a mobile and a stationary phase. They are used for the detection of certain adulterants or to determine oil authenticity. The principles of these methods are:

- GC: there is a stationary solid phase and a mobile gas phase.
- TLC: there is a solid planar surface and the liquid phase is mobile through capillary forces.
- HPLC: there is a column filled with the stationary phase and a mobile liquid phase.

ENZYMATIC METHODS   These include assessment of enzyme activity by measuring the rate of consumption of the reactants or the formation of products. These methods allow quick and accurate determination of certain compounds.

Within the enzymatic methods are included immunoassay, which is based on the ability of antibodies (immunoglobulins) to interact with various substances. This method involves at least one antibody having specificity for a particular antigen. When the antigen is immobilized a detection antibody is added, which forms a complex with the antigen.

Non-conventional methods

The residual DNA of olive oil can be used to determine olive oil authenticity (Cresti *et al.*, 1996; Muzzalupo *et al.*, 2007; Spaniolas *et al.*, 2008b).

DNA polymorphism
This is defined as genetic changes in the DNA of certain materials, i.e. the genome of the nucleus, mitochondria or ribosomes, and also of other organelles. Molecular markers provide information about variations in the DNA sequence. The main molecular markers are RAPD, SCAR, AFLP, SSR, SNP and intron polymorphism (Spaniolas *et al.*, 2008a). These markers can be used to determine olive oil authenticity. A short description of these markers is now given.

RAPD (Random Amplified Polymorphic DNA)
This marker provides a fast and low-cost method, and also requires only a small amount of DNA. It uses one or two primers in order to provide information on the multi-banded fingerprints of a genome.

SCAR (Sequence Characterized Regions)
A method involving two specific primers that amplify a well-defined genetic locus, which is derived by sequencing a single RAPD amplicon. The polymorphism between SCAR bands is determined by variation in length of the sequence between the two primes.

AFLP (Amplified Fragment Length Polymorphism)
The origin of polymorphism in AFLP is the same as that of RFLP, i.e. changes in restriction site sequence, deletions or insertions between two adjacent restriction sites.

SSR (Simple Sequence Repeats)
These markers are also known as microsatellites and consist of 1–10 bp, which are repeated in tandem; the number of repetitions is variable (Carriero *et al.*, 2002).

SNP (Single Nucleotide Polymorphism)

A technique involving changes of a single base in the DNA sequence that can be identified by polymerase chain reaction (PCR) (Lavee and Haskal, 1993; Marmiroli et al., 2003; Spaniolas, 2007; Spaniolas et al., 2007). This method does not provide a lot of information. Therefore, several adjacent SNPs can be studied in one sequence. Many types of SNPs are known, such as exon, intron and promoter, based on the SNP position in a gene. Other types include regulatory SNPs (rSNPs). Single-nucleotide polymorphism is the most common DNA polymorphism, which offers the possibility of developing genetic maps to study various genes.

ISSR (Inter-simple Sequence Repeats)

This technique does not require any prior knowledge of the genomic sequence. It is a PCR system making use of only one primer, which is complementary to the 5′ or 3′ end of a microsatellite region and further extends into the flanking region, thus allowing the amplification of regions found between inversely oriented repeats. It is a multilocus marker system useful for fingerprinting, diversity analysis and genome mapping (Godwin et al., 1997).

Of the markers above, RAPD, SCAR, AFLP, SSR, ISSR and SNP can be used in tandem with PCR analysis.

ADULTERATION OF OLIVE OIL

The high price of olive oil in comparison with that of vegetable oils increases the likelihood of adulteration, which takes on various guises.

1. Mixing of extra virgin olive oil with low-quality olive oil or vegetable oils.
2. Exposure to treatments other than those required for its production. Processes such as decolorization, de-odouring or neutralizing acidity are examples.
3. Designation of geographic origin. Olive oils from designated geographical areas are, by definition, of high quality; designation of the area of origin commands higher prices. Details of the particular cultivar from which the oil has been produced will be written on the bottle label. Olive oils labelled as carrying designation of origin (PDO) and/or geographical indication (PGI) are protected by EEC regulation 510/2006. Any oil sourced from outside these criteria but purporting to be either PDO or PGI is classified as having been adulterated.

Many chemical and biochemical methods have been developed in order to determine the authenticity of olive oil, and today the methods of DNA analysis are more important, since the classical techniques cannot easily distinguish between closely related materials. Furthermore, conventional chemical methods are not always able to detect the region of origin of olive oil, since environment affects chemical composition and phenotype. Molecular markers can distinguish or identify cultivars, estimate germplasm variability and trace olive origin. Molecular markers are not affected by the environment and can help resolve problems such as nursery certification, variety identity, geographic origin and identification of germplasm collections; many olive cultivars are screened by Random Amplified Polymorphic DNA (RAPD). The SSRs (Simple Sequence Repeats) are of small size and can be detected in degraded DNA, as can DNA extracted from olive oil. Amplified fragment length polymorphism and microsatellites have also been used. However, a reliable and promising molecular marker that can significantly contribute to olive oil characterization is the single nucleotide polymorphism (SNP). In olives, SNPs have been used to classify olive cultivars and can differentiate 72% of the olive samples tested.

From the 60 Greek varieties 43 have been sequenced. According to the SNP database of the Greek olive varieties, an initial differentiation was carried out. The SNP data resulted from lupeol synthase and cycloartenol synthase targets. However, more genes are required to be sequenced and analysed. By this technique 19 of the 43 varieties can efficiently be tested for adulteration. These varieties are given in Table 25.4: 'Amfissis', 'Asprolia Lefkados', 'Adramytini', 'Agouromanakolia', 'Gaidourelia', 'Kalokairida', 'Koutsourelia', 'Koroneiki', 'Kothreiki', 'Klonares Koropiou', 'Kerkyras', 'Mastoides', 'Rachati', 'Thiaki', 'Tragolia', 'Throumbolia', 'Strogylolia', 'Valanolia' and 'Vasilikada'. In some varieties only one SNP is enough for their differentiation. Furthermore, classification can be performed in some varieties in relation to the region of origin. Hence, 'Adramytini' from the Greek island Lesbos can be differentiated from that from Agios Mamas (northern Greece) through the SNP(13) of lupeol synthase. Also, 'Koroneiki' and 'Mastoides' from Chania can be differentiated from 'Koroneiki' and 'Mastoides' from Agios Mamas through SNP(6) and SNP(2) of lupeol synthase, respectively. The same differentiation is possible in many other varieties, such as 'Gaidourelia', 'Agouromanakolia', etc. Some groups of varieties have the same SNPs and cannot be differentiated, e.g. 'Chondrolia Chalkidikis' and 'Karydolia Chalkidikis'.

The varieties 'Amygdalolia' and 'Kalamon' exhibit no difference in SNP positions, but they can be differentiated by SNP(1), SNP(2), SNP(2+) of lupeol synthase and SNP(6), SNP(8), SNP(9) and SNP(10) of cycloartenol synthase.

Table 25.5 presents the Greek SNP database of lupeol synthase (lupeol 2F-lupeol 2R).

Table 25.4. Differentiation of olive varieties according to their SNP position (from Bazakos, 2007).

Olive variety	SNP position
'Amfissis'	1,2,3 (lupeol2) and 15 (cyclo3)
'Asprolia Lefkados'	1 (lupeol2) and 1 (cyclo2)
'Adramytini' (Lesbos)	1,5,8 (lupeol2) and 3 (cyclo2)
'Agouromanakolia' (NAGREF Chania)	2,7,8 (lupeol2)
'Gaidourelia' (NAGREF Chania)	1,2,7,8 (lupeol2)
'Gaidourelia' (Kostelenos)	2,2+,3−,7 (lupeol2) and 6,7 (cyclo2)
'Kalokairida'	1,2,2+,7,8 (lupeol2) and 5,6,7 (cyclo2)
'Koutsourelia' (NAGREF Chania)	6,7,8 (lupeol2)
'Koutsourelia' (Kostelenos)	6,7 (cyclo2)
'Koroneiki' (NAGREF Chania)	6 (lupeol2)
'Koroneiki' (Kostelenos and NAGREF Agios Mamas)	6,3 (lupeol2) and 4,9 (cyclo2)
'Kothreiki' (NAGREF Agios Mamas)	6 (cyclo2)
'Klonares Koropiou'	1,2,2+ (lupeol2) and 9 (cyclo2)
'Kerkyras'	1,2,2+ (lupeol2) and 8,9,10 (cyclo2)
'Mastoides' (Metzidakis)	2 (lupeol2)
'Mastoides' (Kostelenos and NAGREF Agios Mamas)	1,2,2+ (lupeol2) and 5 (cyclo2)
'Rachati'	1,6 (lupeol2) and 3 (cyclo2)
'Thiaki'	1 (cyclo2)
'Tragolia'	1,2,7,8,9 (lupeol2) and 4,5,6 (cyclo2)
'Throumbolia'	1,3− (lupeol2)
'Strogylolia'	1,2,7,8,9 (lupeol2) and 4,5,6 (cyclo2)
'Valanolia' (Lesbos)	5,6 (lupeol2)
'Valanolia' (Kostelenos and NAGREF Agios Mamas)	1,6 (lupeol2) and 3 (cyclo2)
'Vasilikada'	1,2,7,8,9 (lupeol2) and 4,5,6 (cyclo2)
'Chondrolia Chalkidikis', 'Karydolia Chalkidikis'	2,3 (lupeol2) and 4 (cyclo2)
'Konservolia', 'Dafnelia'	1,2,2+ (lupeol2) and 5 (cyclo2)
'Zakinthou', 'Lianolia Kerkyras'	6,7,8 (lupeol2)
'Amygdalolia', 'Kalamon'	1,2,2+ (lupeol2) and 6,8,9,10 (cyclo2)
'Galatistas', 'Dopia Spetson'	7,8,9 (lupeol2) and 1,2 (cyclo2)
'Throumbolia' (NAGREF Chania), 'Kolireiki', 'Karolia', 'Matolia',	1,2,3−,6 (lupeol2), 6,9 (cyclo2) and 15 (cyclo3)
'GPL', 'Kolympada', 'Karydolia	1,2,2+ (lupeol2) and 6,8,9,10 (cyclo2)
Agouromanakolia' (Kostelenos), Adramytini' (Kostelenos, NAGREF Chania, NAGREF Agios Mamas), 'Manaki' (Kostelenos, NAGREF Agios ios Mamas), 'Smertolia', 'Throumbolia' (NAGREF Chania)	All SNP positions

Table 25.5. Greek SNP database of the lupeol synthase (lupeol2F-lupeol2R) fragment (from Bazakos, 2007).

A/A	1	2	3	4	5	6	7	8	9	10	11	12	13	14
	1	2 3 4 5	6 7	8	9 10 11	12 13 14	15	16 17	18	19	20	21	22 23	24 25

Variety	SNP values (SNP No. 1, 2+, 3−, 3, 4, 5, 6, 7, 8, 9)
'Agouromanakolia' (NAGREF Chania)	CC, CC, —, AA, AG?, AA, TT, —, GA, —
'Agouromanakolia' (Kostelenos)	—, —, —, AA, AA, GG, AA, TT, GG, —
'Adramytini' (Lesbos)	CC, CC, —, AA, AG, —, AA, TT, GA, —
'Adramytini' (Kostelenos)	CC, CC, CC, AA, AA, GG, AA, TT, GG, AA
'Adramytini' (NAGREF Chania)	CC, CC, CC, AA, AA, GG, AA, TT, GG, AA
'Adramytini' (NAGREF Agios Mamas)	CC, CC, CC, AA, AA, AA, GG, AA, TT, GG, AA
'Amygdalolia' (Kostellenos)	CC, CC, —, AA, AA, GG, AA, TT, GG, AA
'Amygdalolia' (NAGREF Agios Mamas)	CC, CC, —, AA, AA, AA, GG, AA, TT, GG, AA
'Amfissis' (NAGREF Agios Mamas)	CG?, CT, —, AA, AA, AA, GG, AA, TT, GG, AA
'Amfissis' (Kostelenos)	CC, CC, CC, AA, AA, AA, GG, AA, TT, GG, AA
'Asprolia Lefkados' (Kostelenos)	CC, CC, CC, AA, AA, AA, GG, AA, TT, GG, AA
'Chondrolia' (Polygyros)	CC, CT, CA, AA, AT, AA, GG, AC, TC, GG, AA
'Chondrolia' (NAGREF Chania)	CC, TC, —, AA, AA, —, GG, AC, TC, GG, —
'Chondrolia Chalkidikis' (NAGREF Agios Mamas)	CC, CT, CA, AA, AT, AA, GG, AC, CT, GG, AA
'Dafnelia' (Kostelenos)	CC, CC, —, AA, AA, AA, GG, AA, TT, GG, AA
'Gaidourelia' (NAGREF Chania)	CC, TC, —, AA?, AA, AG?, AC, TC, GA, —
'Gaidourelia' (Kostelenos)	CC, TC, CA, AA, AT, AA, GG, AC, TC, GG, —
'Galatistas' (Kostelenos)	TC, CA?, GA?, AT?, AA, GG, AC, TC, GG, AG
Genotype not yet named 'GPL'	CC, CC, —, AA, AA, AA, GG, AA, TT, GG, AA
'Koutsourelia' (NAGREF Chania)	CC, CC, —, AA, AA, AG?, AA, TC, GA, —
'Konservolia' (NAGREF Chania)	CC, CC, —, AA, AA, AA, GG, AA, TT, GG, AA
'Kerkyras' (NAGREF Chania)	CC, CC, CC, —, —, —, —, —, —, —
'Koroneiki' (NAGREF Chania)	CC, TC, —, AA, AG, —, AG, CC, CC, GA, —
'Koroneiki' (Kostelenos)	CC, TC, CC, —, AA, AT, AA, GG, AC, CC, GG, AG
'Koroneiki' (NAGREF Agios Mamas)	CT, AC?, AG, AT, AA, GG, CC, GG, AG

Table 25.5. Continued.

A/A		Variety	SNP No. 1	2+	3-	3	4	5	6	7	8	9
15	26	'Kalamon' (Kostelenos)	CC	CC	CC	AA	AA	GG	TT	GG	AA	
15	27	'Kalamon' (NAGREF Agios Mamas)	CC	CC	CC	AA	AA	GG	TT	GG	AA	
16	28	'Manaki' (Kostelenos)	CC	CC	CC	AA	AA	GG	TT	GG	AA	
16	29	'Kothreiki' (NAGREF Agios Mamas)	CC	CT	CA	AG	AT	GG	AC	CT	GG	AG
16	30	'Manaki' (NAGREF Agios Mamas)	CC	CC	CC	AA	AA	GG	TT	GG	AA	
17	31	'Kolymbada' (Kostelenos)		CC		AA	AG	GG	TT	GG		
18	32	'Kalokairida' (Kostelenos)	CC	CC		AA	AA	GG	TT	GG	AA	
18	33	'Karydolia' (Kostelenos)	CC	CC		AA	AA	GG	TT	GG	AA	
19	34	'Karydolia Chalkidikis' (NAGREF Agios Mamas)	CC	CT	AC	AA	AT	AA	GG	AC	CT	GG	AA
19	35	'Karydolia Chalkidikis' (Kostelenos)	CC	CT	CA	AA	AT	AA	GG	AC	CT	GG	AA
20	36	'Klonares' (Kostelenos)	CG	CC		AA	AA	GG	TT	GG	AA	
21	37	'Karolia' (Kostelenos)	CC	CC	CC	AA	AA	GG	TT	GG	AA	
22	38	'Kolireiki' (Kostelenos)	CC	CC	CC	AA	AA	GG	TT	GG	AA	
23	39	'Lianolia Kerkyras' (Kostelenos)				AA	AA	GG	AA	TC	GG	AA
23	40	'Lianolia Kerkyras' (NAGREF Agios Mamas)	CC	CC	CC	AA	AA	GG	CT	GG	AA	
24	41	'Mastoides' (NAGREF Chania)	CG	TT		AA	AA	AG?	AA	TT	GA	
24	42	'Mastoides' A (Kostelenos)		CC	CC							
24	43	'Mastoides' B (Kostelenos)	GC	CC	CC	AA	AA	GG	TT	GG	AA	
24	44	'Mastoides' (NAGREF Agios Mamas)	CC	CC	CC	AA	AA	GG	TT	GG	AA	
25	45	'Matolia' (Kostelenos)	CC	CC	CC	AA	AA	GG	AA	TT	GG	AA
26	46	'Megaritiki' (Kostelenos)	CC	CT	CC	AG	AT	GG	AC	TC	GG	AG
26	47	'Megaritiki' (NAGREF Agios Mamas)	CC?	CT	CA	AG?	AT	AA	GG	AC	CT	GG	AG
27	48	'Dopia Spetson' (Kostelenos)						GG	AC	TC	GG	AG

Table 25.5. *Continued.*

N/A	28	29	30	31	32	33	34			35	36	37			38		39
	49	50	51	52	53	54	55	56	57	58	59	60	61	62	63	64	65
Cultivar	'Nissiotiki A' (Kostelenos)	'Patrini' (NAGREF Agios Mamas)	'Pikrolia' (Kostelenos)'	'Rachati' (Kostelenos)	'Smertolia A' (Kostelenos)	'Strogylolia' (Kostelenos)	'Throumbolia' (NAGREF Chania)	'Throumbolia' (Kostelenos)	'Throumbolia' (Agios Mamas)	'Tragolia' (Kostelenos)	'Thiaki' (Kostelenos)	'Valanolia' (Lesvos)	'Valanolia' (Kostelenos)	'Valanolia' (NAGREF Agios Mamas)	'Vasilikada' (Kostelenos)	'Vasilikada' (NAGREF Agios Mamas)	'Zakynthou' (Kostelenos)

SNP No.	49	50	51	52	53	54	55	56	57	58	59	60	61	62	63	64	65
1	CT		CC	CC?	CC	CC?		CG	CG	CC	CC	CG	CG	CG	CC	CC	CC
2	CA?	CC	CC	CG?	CC	CC?	CC	CC	CC	CC	CC	TC	TC	CT	CC	CC	CC
2+	AG	AA	AA	CC	AA	CA?		CC	CC	CC			CA	CA	CC	CC	CC
3–	AT	AA	AA	AA	AA	AA	AA	AA	AA	AA	AA		AA	AA	AA	AA	AA
3	AA	GG	AG?	AA?	AA	AA	AA	AA	AA	AA	AA	AT	AT	AT	AA	AA	AA
4	GG	CC	GG	AC?	GG	GG	GG	GG	GG	GG	GG	AA	AA	AA	AA	AA	AA
5	AC	GG	AA		AA	AC?	AA	AA	AA	AA	AA	AG	GG	GG	GG	GG	GG
6	CC		CC		TT	TC	TT	TT	TT	TC	AA	AC	AC	AC	AA	AA	AA
7	GG		GG		GG	GG	GG	GG	GG	GG	TT	TC	TC	CT	TC	CT	TC
8	AG		AA		AA	AA	AA	AA	AA	AA	GG	GA	GG	GG	GG	GG	GG
9													AA	AA	AA	AA	AA

26

OLIVE MILL PRODUCTS AND ENVIRONMENTAL IMPACT OF OLIVE OIL PRODUCTION

OLIVE GROWING AND ENVIRONMENTAL PROTECTION

Olive cultivation has a positive impact on the environment and maintenance of landscape. Olive culture also helps combat desertification, one of the biggest problems in the Mediterranean area. Furthermore, the olive tree gives shelter and food to wildlife. There are, however, instances where olive growing damages the environment. The intensification of olive growing is accompanied by increased input of fertilizers, insecticides, herbicides and irrigation water. Such systematic removal of vegetation by various means has an adverse effect on biodiversity and increases the loss of soil organic matter. This leads to environmental deterioration and erosion or desertification.

The environmental problems associated with olive oil extraction mills are related to water consumption in regions where its supplies are limited. In oil extraction systems, where the oil is extracted by either pressure or centrifugation with a three-phase process (oil, liquid extract and residue) – found mostly in Italy and Greece – environmental problems lie in both the large volumes of water required and the removal of liquid extracts. To avoid damage to the environment liquid extracts should be treated and purified before discharge into water courses. Furthermore, the liquid extracts should be applied to the soil, as liquid fertilizers. However, the quantities applied should be low in order to avoid pollution of groundwater. In the two-phase process significantly less water is required, and this is the reason for its use in various countries. Furthermore, with two-phase centrifugation a significant amount of residue is produced, requiring drying at high temperatures to extract the remaining oil. Furthermore, the legacy of table olive production is a highly polluting liquid waste. This waste has high organic and sodium contents, rendering it unsuitable for application as fertilizer on olive groves.

UTILIZATION OF OLIVE MILL BY-PRODUCTS

After the virgin olive oil has been extracted from olives, a liquid and solid residue remains, consisting of the following: washing water, olive mill waste water (OMWW), leaves and olive pomace. From these residues the pomace was traditionally used in order to extract pomace oil, by solvent extraction; the remainder of the liquid and solid residues were disposed of into the environment, creating a pollution problem. However, in today's more enlightened world all waste products require recycling, to both increase revenue and conserve energy. The use of centrifugal decanters of either two or three phases results in the production of olive oil with a higher content of antioxidants. Below, we analyse the various by-products individually.

Washing water

The washing water used following olive processing attains a brown colour and contains organic and inorganic materials that can cause pollution. Washing water represents 10% (v/w) of the total weight of olive production waste.

Olive leaves

The leaves are collected, together with the fruit, and are transported to the processing unit. The percentage of olive leaves is a function of the method used for olive harvesting: a greater proportion from harvesting by the use of shakers. A proportion of up to 5–8% is common. These leaves should be removed before processing for oil extraction, since too many leaves give a very bitter taste to the oil. The separated olive leaves could be used in the following ways: (i) after drying, as a fuel in power plants, together with other plant material; (ii) as a fertilizer, by incorporating them in the soil of the olive orchard, intact or following crushing (leaves are a rich source of organic matter and mineral nutrients); or (iii) for phenol production (1–2% of fresh weight). The most important phenol is oleuropein, which gives olive oil its bitter taste and has important properties for human health.

Olive pomace

Olive pomace is the final product of the separation of olive oil, leaves, washing water and OMWW. Its quantity and its chemical composition depend on the type of decanter used for oil extraction, i.e. triphasic or biphasic. Olive pomace has the following two uses: (i) as a fuel in power plants; and (ii) production of compost by mixing with other plant materials (Cegarra *et al.*, 1996); this compost mixture improves soil texture and nutrient composition.

Extraction of pomace oil with solvents

Table 26.1 presents the characteristics of olive pomace after processing by three different techniques.

The endocarp is a fuel, since it produces only small amounts of ash and an exhaust that is low in nitrogen gas (N_2) and sulphurous gases. Therefore, separation of stones from the wet olive pomace is a technique routinely used in olive oil processing plants.

Preparation of compost from moist pomace

The moist pomace is mixed with other plant materials in order to supply cellulose; it is then mixed with bovine manure in order to supply N; after a period of 6 months a stable compost is produced. Furthermore, moist pomace on its own could be used as a soil fertility additive.

Olive mill waste water (OMWW)

The OMWW is the liquid obtained after olive processing. It has a high content of K, organic matter and phenols with antibacterial properties (Capasso *et al.*,

Table 26.1. Characteristics and quantities of olive pomace produced by three different extraction methods: (i) traditional pressing; (ii) triphasic decanting; and (iii) biphasic decanting (from Di Giovacchino and Preziuso, 2006).

Parameter	Traditional pressing	Triphasic decanting	Biphasic decanting
Total pomace (kg/t olives)	250–350	450–550	800–850
Moisture (%)	22–35	45–55	65–75
Oil (% f.w.)	6.0–8.0	3.5–4.5	3.0–4.0
Fibre (%)	20–35	15–25	10–15
Endocarp (%)	30–45	20–28	12–18
Ash (%)	3–4	2–4	3–4
N (mg/100g)	250–350	200–300	250–350
K (mg/100g)	40–60	30–40	40–50
P (mg/100g)	150–200	100–150	150–250
Total phenols (mg/100g)	200–300	200–300	400–600

1995). The quantity of OMWW is a function of the type of decanter used, i.e. biphasic or triphasic. OMWW is a potential polluter, but it is also a source of valuable materials such as oleuropein and hydroxytyrosol (Capasso et al., 1999; Visioli et al., 2001; Allouche et al., 2004). This by-product can be sterilized by various processes such as:

- Chemical oxidation or Fenton's reaction (Beltran-Heredia et al., 2001).
- Photo-oxidation.
- Electrolysis (Israilides et al., 1997).
- Electrochemical oxidation.
- Anaerobic degradation.
- Aerobic degradation, by use of biofilm or activated sludge.
- Membrane technology, utilizing reverse osmosis or ultrafiltration. This gives a liquid with low organic content and a more concentrated liquid or sludge, with many organic compounds that can be separated and used.
- Evaporation and distillation. Evaporation, involving formation of pools of waste liquid, has been used in many countries (Di Giacomo et al., 1991). Evaporation, however, is an expensive process in terms of thermal energy, and in practice solar energy sometimes is adequate, especially in the southern areas of Mediterranean countries.

The controlled spreading of OMWW on the soil of the olive orchard is a common method for utilization of this liquid waste, since it achieves the recycling of both organic and inorganic materials (see Table 26.2).

With regard to the numbers of microorganisms, after application of OMWW their population increases but after 2 months decreases. These microorganisms include nitrifying bacteria, *Actinomycetes*, yeasts, etc.

Table 26.2. Nutritional parameters of the effects of controlled spreading of OMWW on the olive orchard (from Di Giovacchino and Preziuso, 2006).

Parameter	Control	OMWW		
		5 l/m^2	10 l/m^2	30 l/m^2
Productivity (kg/tree)	4.3	4.6	5.4	5.2
Moisture of olive fruits (%)	51.8	52.0	52.0	52.6
Oil (%)	17.3	17.2	16.6	17.4
Free fatty acids (%)	0.50	0.50	0.50	0.54
Peroxide index (meq O$_2$/kg)	8.1	6.3	7.9	7.3
K$_{232}$	1.98	2.09	2.0	2.01
K$_{270}$	0.16	0.17	0.16	0.17
Soil pH	7.15	7.20	7.32	7.27
Organic matter (%)	1.97	2.02	2.04	2.08
N (%)	0.12	0.12	0.13	0.13
Reducing substances (mg/100 g)	0.24	0.38	0.39	0.48

ANTIOXIDANTS AND OTHER BIOLOGICAL BY-PRODUCTS

Olive mill waste water is the major by-product of olive oil production. At present olive mills have to discard it, which increases the cost of disposal, contaminates soil and creates ecological problems. Olive mill waste water contains hydroxytyrosol, which is amphiphilic and acts as an oil–water interface; it also contains antioxidants (Visioli et al., 1995b; Allouche et al., 2004). The ability of OMWW extracts to scavenge superoxide (Visioli et al., 1999), as happens with hydroxytyrosol and oleuropein, suggests that OMWW can be used in situations where Fenton and Haber–Weiss reactions take place. In such places we observe contaminant products of superoxide and nitric oxide, which yield peroxynitrile, a powerful oxidant. Since foods very often come in close contact with chlorine bleaches used as disinfectors, the use of hypochlorous acid (HOCl) scavengers may provide additional protection against reactive chlorine species. The major phenolic compounds identified in OMWW extracts are listed in Table 26.3.

USE OF OLIVE MILL WASTE WATER AS FERTILIZER

Huge amounts of OMWW are produced in the process of olive oil extraction, and these quantities represent a serious environmental problem. Therefore, strenuous efforts are now being made to reduce pollution due to their detrimental effect or to recycle them into valuable products.

Some of the techniques used include thermal concentration, treatment with physicochemical methods or with microorganisms and their use directly on agricultural soils in the form of liquid fertilizer (Tomati and Galli, 1992; Ben Ruina et al., 1999; Di Giovacchino et al., 2005; Papadopoulos, 2006). Application of OMWW, as a fertilizer, has many advantages for soil fertility, but also several disadvantages and potential problems.

Table 26.3. Major phenolic compounds identified in OMWW extracts.

Phenolic compound	Concentration (g/100 g d.w.)
Hydroxytyrosol	1.56
Tyrosol	0.85
Elenolic acid	4.3
Oleuropein	0.5
Luteolin-7-glucoside	0.22
Quercetin	0.13
Cinnamic acid	0.55
Total phenols	8.4

Advantages

- Addition of extra water to the soil, especially given the low precipitation levels and lack of irrigation water seen today in many olive-growing areas.
- Improvement in the physical and chemical properties of the soil through the addition of organic matter (OM) and enhancement of microbial activity.
- Contains large quantities of K, N, P and Mg (see Table 26.4).

Disadvantages

- The presence of fatty acids and polyphenols, which reduce soil fertility.
- It is difficult to store or dispose of the large amounts of OMWW, since it is produced in a relatively short time, and particularly during a period of high rainfall.
- Direct application should be at a level no greater than 30 m^3/ha/year for OMWW from traditional olive mills.
- Its application should be carried out at least 1 month before the time of sowing of annual crops.
- It should be applied at a minimum distance from trees.

Table 26.4. Characteristics of mature OMWW composts (d.w.) (from Cegarra et al., 1996).

Parameter	SCO	MOS
pH (H$_2$O)	7.84	8.73
Ec (1:10) (mS/cm)	7.66	5.03
Organic matter (%)	56.43	37.40
Organic carbon (%)	29.37	18.78
Nitrogen (%)	3.11	1.44
NO$_3$-N (%)	0.36	t
C:N ratio	9.44	13.04
Phosphorus (%)	0.87	0.22
Potassium (%)	2.61	3.25
Iron (%)	0.50	1.30
Copper (mg/kg)	52	62
Manganese (mg/kg)	241	268
Zinc (mg/kg)	245	73
CEC (me/100 g)	109.8	96.2

SCO, sewage sludge and cotton waste mixtures watered with fresh OMWW; MOS, mixing of 11.1% of manure with 88.9% of OMWW sludge; t, traces.

No data are available for the use of OMWW as compost for olive production. Composting is probably a good method for recycling liquid wastes, leading to the production of CO_2, water, mineral salts and stabilized OM, which contains humic-like substances. The process of composting is thermophilic and leads to release of phytotoxins. Furthermore, OMWW may be mixed with plant waste materials to be transformed into organic fertilizers or composts, which have no phytotoxicity.

HYDROGEN PRODUCTION

Hydrogen is a clean energy source for both industrial and domestic consumption. Biological production of hydrogen gas (H_2) utilizes biophotolysis of water and photo-fermentation of organic materials, usually carbohydrates, by bacteria. Carbohydrate-containing solid wastes, such as olive mill waste, can be used for hydrogen production by using suitable bio-process technologies (Kapdan and Kargi, 2006). Utilization of such wastes for hydrogen production provides inexpensive energy generation with simultaneous waste treatment.

Olive mill waste water can be used as a raw material for bio-hydrogen production. It will require pretreatment to remove undesirable components. The carbohydrate content is converted to organic acids, and subsequently to H_2, by the use of certain bio-processing technologies.

Glucose is an easily biodegradable carbon source. The bio-conversion of 1 mol of glucose yields, theoretically, 12 mol of H_2 and conversion of 1 mol of glucose to acetate yields 4 mol H_2. However, when butyrate is the end product only 2 mol H_2/1 mol glucose is formed. In practice the yield of H_2 does not exceed 17 mol H_2/1 mol glucose. In the case of cellulose, 2.18 mol H_2/1 mol glucose is produced when the concentration of cellulose in the waste is 12.5 g/l.

Fig. 26.1 is a schematic diagram showing the pathways involved in bio-hydrogen production from agricultural wastes containing cellulose and starch by two-stage anaerobic dark photo-fermentation.

Agricultural wastes such as olive pomace contain starch and cellulose (Kapdan and Kargi, 2006). Starch can be hydrolysed to glucose and maltose by acid hydrolysis, and this is followed by conversion of carbohydrates to organic acids and H_2. Cellulose and hemicellulose content also can be hydrolysed to carbohydrates, and subsequently to organic acids, which also produce H_2.

Hydrogen is considered a viable alternative fuel of the future. Furthermore, H_2 is widely used in the production of chemicals, hydrogenation of fats and oils in the food industry and desulphurization and reformulation of petrol in refineries. Hydrogen has a high energy yield that is 2.75 times greater than that of hydrocarbon fuels.

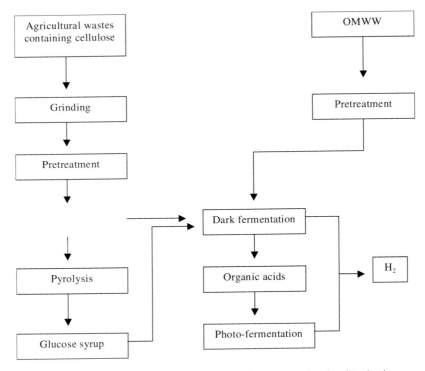

Fig. 26.1. A schematic diagram showing the pathways involved in bio-hydrogen production from agricultural wastes containing cellulose by two-stage anaerobic dark photo-fermentation.

27

OLEUROPEIN, OLIVE LEAF EXTRACT, OLIVE OIL AND THE BENEFITS OF THE MEDITERRANEAN DIET TO HUMAN HEALTH

OLEUROPEIN

Oleuropein as an antioxidant

Oleuropein is the bitter compound of olives that enhances nitric oxide production and has free-radical scavenging properties (Visioli et al., 1998). This is commonly found in the Mediterranean diet, which is rich in fresh fruits, vegetables and olive oil and has been associated with a lower incidence of cardiovascular disease and cancer. This is ascribed to its high proportion of vitamins, flavonoids and polyphenols (Ho et al., 1992). Oleuropein is the most abundant of the minor components of olive oil; it has potent antioxidant and anti-inflammatory properties and enhances nitric oxide (NO) production in lipopolysaccharide (LPS)-challenged mouse macrophages. The promotive effect is blocked by the inductible nitric oxide synthase (iNOs) inhibitor N(G)-nitro-L-orginine methyl ester (L-NAME). This indicates increased iNOs activity and that oleuropein potentiates the macrophage-mediated response, leading to higher NO production, which is considered to be beneficial in cellular protection.

Antimicrobial activity of oleuropein and hydroxytyrosol

Secoiridoids, one of the major classes of polyphenols contained in olives and olive oil, have recently been shown to inhibit or delay the rate of growth of a range of bacteria and microfungi, but there are no data in the literature concerning the possible involvement of these secoiridoids as agents against pathogenic bacteria in man (Juven et al., 1968, 1972; Juven and Henis, 1970; Ruiz-Barba et al., 1991; Aziz et al., 1998; Bisingnano et al., 1999). Oleuropein inactivates bacteria by dissolving their outer lining; the presence of 0.1% (w/v) of oleuropein delayed the growth of *Staphylococcus aureus*. Intermediate concentrations (0.2%) of oleuropein prevented growth, while concentrations of oleuropein > 0.2% inhibited growth and production of enterotoxin B in artificial nutrient media.

Caffeic and protocatechuic acids (0.3 mg/ml) inhibited the growth of *E. coli* and *Klebsiella pneumoniae*. The same compounds, apart from syringic acid (0.5 mg/ml), completely inhibited the growth of *Bacillus cereus*. Oleuropein and p-hydroxy benzoic, vanillic and p-coummaric acids (0.4 mg/ml) completely inhibited the growth of *E. coli*, *K. pneumoniae* and *B. cereus*. Vanillic and caffeic acids (0.2 mg/ml) completely inhibited the growth of aflatoxin production by both *Aspergillus flavus* and *A. parasiticus*. *Olea europaea* can therefore be considered a potential source of promising antimicrobial agents for the treatment of intestinal or respiratory tract infections in man.

Effect of oleuropein on the protection of low-density lipoprotein (LDL) from oxidation

Oleuropein (10^{-5} M) effectively inhibited $CuSO_4$-induced LDL oxidation (Visioli and Galli, 1997). Polyphenolic components of the Mediterranean diet interfere with biochemical events that are implicated in atherogenetic disease. Therefore, there is a link between the Mediterranean diet and the prevention of coronary heart disease. Oleuropein seems to both increase the high-density lipoprotein (HDL) (Mensik and Katan, 1987) and lower the bad cholesterol (LDL) levels by 30% or more without exercise or dietary changes (Baggio *et al.*, 1988; Aviram and Elias, 1993; Visioli and Galli, 1994; Visioli *et al.*, 1995a).

Blood pressure-lowering effect of oleuropein and olive leaf extract

Oleuropein has been tested for its blood pressure-lowering activity in hypersensitive rats by daily oral doses of L-NAME 50 mg/kg) for at least 4 weeks. Oral administration of the leaf extract at different doses and concurrently with L-NAME for 8 weeks reduced the rise in blood pressure at a dose of 100 mg/kg. These confirm the hypotensive effects of oleuropein. This antihypertensive effect may be related to various factors causing reversal of the vascular changes produced during hypertension (Khayyal *et al.*, 2002).

Effect of oleuropein on inactivation of cytochrome P450

Oleuropein is an inhibitor of the CYP3A-mediated androstenedione 6β-hydroxylation in human hepatic microsomes. Furthermore, oleuropein is also a potent mechanism-based inactivator of the CYP3A enzymes.

Oleuropein was clearly identified as a mechanism-based inhibitor, whereas hydroxytyrosol was not. The possible mechanism of oleuropein inhibition is associated with either its ester oxygen linkage or its non-terminal exocyclic alkene bond. The inhibition of CYP3A by oleuropein may explain the *in vivo* protective effects reported for oleuropein in LDL oxidation (Murray, 1999; Coni *et al.*, 2000).

OLIVE LEAF EXTRACT

Olive leaf extract (OLE) is a natural herbal antibacterial/antiviral extract obtained from olive leaves, which is safe and non-toxic to humans (Cherif *et al.*, 1996; Walker, 1996a,b; Briante *et al.*, 2002b; Khayyal *et al.*, 2002; Lee-Huang *et al.*, 2003). The use of OLE started as early as 1800 when a tea made from OLE was used for malaria, and in the early 1900s OLE was found to be far superior to quinine in the treatment of malaria. In 1957, the active ingredient, oleuropein, of OLE was isolated. Scientists reported that oleuropein could be used effectively to lower blood pressure (Fehri *et al.*, 1994), in the prevention of intestinal spasms, the relief of arrhythmia and to increase blood flow to the coronary arteries.

The product thus obtained is called Olivir, which is both virucidal and bactericidal, manufactured from oleuropein and a blend of OLEs. Olive leaf extract in a capsule form is a fairly recent advance, but in the early 1900s a bitter compound was found in the leaves of certain olive trees, called oleuropein, which is part of the olive tree's disease resistance system. A Dutch researcher found the chemical agent within oleuropein that could be medically important, a chemical now known as elenolic acid. During the late 1960s/early 1970s a new multifunctional monoterpene was isolated from various parts of the olive tree. This compound is calcium elenolate and is a crystalline salt form of elenolic acid. It will act effectively at low concentrations, without any harmful influence on host cell mechanisms. When this compound was injected directly into the bloodstream it was bound quickly to the blood serum protein and became ineffective.

Calcium elenolate kills viruses by interfering with the process of production of certain amino acids. This prevents virus assembly at the cell membrane and therefore the penetration of infected host cells, and irreversibly inhibits viral replication. In retroviruses, calcium elenolate neutralizes the production of the reverse transcriptase (RT) and protease enzymes. These enzymes in retroviruses (e.g. HIV) alter the RNA of healthy cells.

Calcium elenolate in animals is not toxic, even in doses several hundred times higher than those recommended. However, no studies have yet been conducted concerning the use of OLE during pregnancy and the response to its interaction with other human medicines.

Sometimes, symptoms such as fatigue, diarrhoea, headache, muscle/joint aches and flu-like symptoms may appear when using OLE. In such a case it is recommended to discontinue taking the extract until the body can eliminate the toxic substances accumulated.

Anti-HIV activity of olive leaf extract

Dried olive leaves (*O. europaea*) were rinsed in sterile distilled water to remove dust, insecticides and contaminants. The leaves (1 g leaf/40 ml water) were ground finely and extracted twice with sterile distilled water for 12 h at 80°C; centrifugation followed at 20,000 g for 30 min. The clear supernatant was concentrated by lyophilization, reconstituted with water to 0.1 g starting material/ml (1 g to 10 ml H_2O) and sterilized by millipore filtration with a 0.45 μm filter, and stored at −20°C until use. Olive leaf extract has been used for medicinal purposes for centuries, and substances contained in OLE have activity against bacteria, mycobacteria and fungi. Furthermore, OLE lowers blood pressure and inhibits lipid oxidation. The biological activity of OLE is due mainly to polyphenolic compounds, especially oleuropein and hydroxytyrosol. OLE is used in the treatment of HIV for various reasons, such as increasing the tolerance of the immune system, relief of chronic fatigue and reduction of the side effects of anti-HIV medication (Lee-Huang et al., 2003).

The mechanism of action of OLE in HIV is not known (Lee-Huang et al., 2003). However, OLE increases the activity of the HIV-RT inhibitor 3TC. OLE inhibits acute HIV-1 infection, cell-to-cell transmission and HIV-1 replication. These effects occur at 0.2 μg/ml OLE. Analysis of OLE indicates the presence of three major compounds, oleuropein (MW 539), elenolic acid (MW 377) and hydroxytyrosol (MW 153). Oleuropein accounts for 85–90% of anti-HIV activity. Treatment of HIV-1-infected cells with OLE up-regulates the expression of the apoptosis inhibitor proteins. Treatment with OLE reverses many of the changes induced by HIV-1 infection.

The polyphenolic compounds found in OLE are absorbed by the gastrointestinal tract and can be metabolized. Hence, oleuropein is metabolized to D-elenolate and hydroxytyrosol to tyrosol. Therefore, OLE is both an anti-HIV agent and it modulates the host's response to infection. HIV-1 infection triggers the following changes:

- Modulation of the expression profile of cellular genes involved in survival, stress, apoptosis, calcium and protein kinase C-signalling pathways.
- Up-regulation of the expression of heat shock proteins.

- DNA damage in inducible transcript-binding proteins.
- Down-regulation of the expression of the anti-apoptotic BC12-associated protein.

OLIVE OIL

Olive oil and human cancer

Greece, Italy and Spain obtain 71, 42 and 37%, respectively, of their total dietary fat from olive oil. Olive oil and its minor constituents have beneficial effects on human health (Visioli and Galli, 1998a,b; Visioli et al., 2006). Several epidemiological studies have evaluated the association between olive oil consumption, human health and the occurrence of cancer (Visioli and Galli, 1998a; Tripoli et al., 2005, 2006; Gaddi et al., 2006; Tripoli et al., 2006). A minimum level of essential fatty acids (4% as linoleic acid) is required before a saturated fat can display its full cancer-promoting potential. However, cancer promotion depends not only on the amount but also on the fatty acid composition of the fat (Cohen and Wydner, 1990). Olive oil-fed rats had high levels of oleic acid in total serum lipids, as well as decreased tumour prostaglandin levels. It was proposed that high-fat diets, rich in oleic acid, fail to promote mammary tumorigenesis. This is due to competition with enzymes in the linoleic acid–arachidonic acid and prostaglandin-metabolizing enzyme.

Cancer development is associated with uncontrolled cell proliferation. For signals of this proliferation, G proteins and the protein kinase C are involved; these signals control the cell division cycle. Hence, lipid alterations might be participating in the development of certain cancers; circumstantial evidence for this lies in the fact that membrane lipids were altered in cell membranes of patients suffering from cancer.

Several types of cancer cells show alterations in the membrane lipid layers, and it is feasible to apply lipid therapy to treat certain cancers. Certain anticancer drugs alter the lipid membrane structure; lipid therapy is relevant to the development of anticancer drugs. Furthermore, high olive oil intake and the use of Minerval, which is structurally and chemically close to oleic acid, have been associated with a reduced risk of human cancer.

Breast cancer

The highest rates of breast cancer occur in the most industrialized regions of the USA and Europe. Hence, the fat consumption per capita is highly correlated with breast cancer and mortality (Howe et al., 1990; Trichopoulou et al., 1993; Martin-Moreno et al., 1994; Fortes et al., 1995; La Vecchia et al., 1995; Lipworth et al., 1997). For example, today, the incidence of breast cancer in Spain is about 40% lower in comparison with the USA and northern Europe.

Furthermore, Greek women, with 71% total fat intake from olive oil, have significantly lower mortality from breast cancer than women from the USA.

The Mediterranean countries offer an ideal testing ground for the role of olive oil in breast cancer (Menendez et al., 2005), since olive oil in these countries is the main oil used for cooking. The intake of olive oil is inversely associated with the risk of breast cancer.

Ovarian cancer

In studies assessing changes in diet and cancer mortality in the Mediterranean region, it was found that Greece and Spain had the lowest rates of ovarian cancer and, at the same time, the lowest animal fat intake and the highest olive oil intake. Ovarian cancer mortality is positively associated with total fat intake in various countries (Lipworth et al., 1997).

The epidemiologic evidence concerning olive oil intake with respect to ovarian cancer risk is limited, and there has been only one study conducted in Greece. This study evaluated the association between ovarian cancer and monounsaturated fat consumption, and indications from it are that there is an inverse relation of monosaturated fat intake from olive oil and the risk of ovarian cancer.

Colonic cancer

Studies in animals have indicated that certain fats in fatty acids may affect colonic carcinogenesis (Bartoli et al., 2000; Stoneham et al., 2000), while olive oil and fish oils have the lowest carcinogenic potential. The mechanism of action is through cholesterol biosynthesis, which converts cholesterol to primary bile acids. Many studies point to the relationship between monounsaturated fats and colonic cancer. Several studies on colorectal cancer have been undertaken in Mediterranean countries (Bartoli et al., 2000; Stoneham et al., 2000).

Endometrial cancer

In a hospital-based case-control study in women with confirmed cancer of the endometrium, the only statistically suggestive association was an inverse one with monounsaturated fat, which reduced the risk of cancer by 26% (Lipworth et al., 1997).

Various studies indicate that olive oil may have a potential effect in lowering the risk of this type of cancer. However, it should be established whether the protective effect of olive oil derives from the monounsaturated fatty acids of olive oil or whether it is related to its antioxidant properties.

Other cancers

An inverse relationship exists between olive oil consumption and various cancers such as stomach, lung, bladder, urinary tract, oral cavity and pharynx.

Olive oil, membrane lipid composition and cardiovascular diseases

Cardiovascular diseases constitute the major cause of death in developed countries, and they are responsible for about 40% of all deaths. As a result of alteration in the membrane lipid composition various cardiovascular disorders have evolved, such as hypertension, atherosclerosis, coronary heart disease, sudden cardiac death, blood vessel integrity, thrombosis, etc. The type of fats consumed has been associated with cardiovascular health. Diet modifies and modulates the physical properties of membranes. Furthermore, membrane lipid composition regulates localization and function of several membrane proteins. One such case is the membrane concentration of various G proteins (transducer signals from G protein-coupled receptors (GPCRs)) involved in the control of blood pressure (Vigh *et al.*, 2005). A relationship exists between nitric oxide synthase (NOS) activity and membrane fluidity during hypertension. Protein kinase C and other protein kinases also participate in the development of hypertension. One important function regulated by GPCRs is the control of blood pressure. Significant lipid alterations occur in cell membranes of patients with cardiovascular disease. Therefore, lipid therapies should be designed for treatment of high blood pressure.

Oleic acid is a free fatty acid that regulates GPCR-associated signalling. The effect of oleic acid is exerted through its effect on protein–lipid interaction, and not modification of the activity of pure G proteins. An oleic acid diet can be considered as a form of 'lipid therapy'. Experience shows that olive oil consumption is associated with both reduced risk of development of cardiovascular disease and reduced antihypertensive medication. Apart from oleic acid, olive oil also contains tocopherol and phenols, which are believed to exert a protective function (see below). Furthermore, daily inclusion in the diet of 1 g omega-3 fatty acids reduces the likelihood of death due to cardiovascular disease.

A compound analogous to oleic acid is the synthetic substance 2-hydroxy-9-cis-octadecenoic acid (Minerval), which reduces blood pressure and regulates G proteins in a similar way to oleic acid.

Olive oil and obesity

A high body mass index (BMI) is associated with increased incidence of cardiovascular disease, diabetes and cancer. The BMI is determined as the ratio weight:height^2 and is expressed in kg/m^2. A normal BMI ranges from 25 to 30 kg/m^2, and anyone with a BMI over 30 kg/m^2 is considered obese. Obesity is associated with the accumulation of triacylglycerol, alteration in the membrane lipid composition and variation in the physical properties of membranes. Some of these changes are related to adrenergic receptors, which are involved in the control of body weight. Consumption of unsaturated fatty acids (olive oil and fish oil) induces loss of weight and thereby improves health.

The use of clinical drugs designed to interact with proteins is usually referred as *chemotherapy*. Therapies involving nucleic acids have developed in recent years, and therapies for direct treatment of cells are known as *cell therapies*. Another type of therapy is known as *lipid therapy* (see above), which involves regulation of membrane lipid structure/composition.

Olive oil and diabetes

The use of olive oil activates insulin and controls the problem of diabetes. The levels of heat stress proteins (Hsp) in diabetes are the result of reduced membrane fluidity. Due to oxidative stress and insulin deficiency, diabetes and insulin resistance are associated with stiffer and less fluid membranes. Furthermore, diabetes induces massive changes in specific lipid molecular species.

Olive oil and Alzheimer's disease

This disease is a neurodegenerative disorder, characterized by the formation of amyloid plaques and neurofibrillary tangles in the brain. The pathology of this disease may be the end result of abnormalities in lipid metabolism and peroxidation. Antioxidants of olive oil and omega-3 lipids have shown benefits in reducing the effects of Alzheimer's disease. Neural tissue has the highest lipid content in the body, after adipose tissue. Neurodegeneration in Alzheimer's disease is accompanied by alterations in lipids, such as phosphatidylcholine hydrolysis.

Antioncogenic action of the oleic acid of olive oil

Dietary fatty acids can interact with the human genome by regulating the amount and/or activity of transcription factors. This has opened up a whole new line of research for the anticancer and antioxidant benefits of the olive oil-based Mediterranean diet (Galli and Visioli, 1999; Benavente-Garcia et al., 2000; Owen et al., 2000). Recent findings reveal that oleic acid, the main monounsaturated fatty acid in olive oil, can suppress the overexpression of HER2 (erbB-2), a well-characterized oncogene, playing certain key roles in the aetiology, progression and metastasis of several human cancers:

- An exogenous supply of oleic acid significantly down-regulates HER2-coded neuroncoprotein in human cancer cells.
- Exposure to oleic acid represses the transcriptional activity of the human HER2 gene promoter in tumour-derived tissue.

- Treatment with oleic acid induces the up-regulation of the Ets protein PEA 3 (a transcriptional repressor of HER 2 gene amplification).
- Treatment with oleic acid efficiently blocks fatty acid synthase (human gene) (FASN) activity.
- Malonyl-coenzyme A (CoA) decreases HER 2 promoter activity, while oleic acid or CoA similarly up-regulate PEA 3 gene promoter activity.

Antioncogenic and anticholesterol actions of the squalene of olive oil

Squalene protects from cancer (Newmark, 1999) and atherosclerosis, since it strongly inhibits the activity of 3-hydroxy-3-methylglutaryl-CoA reductase (HMG-CoA reductase), a key enzyme in cholesterol synthesis. The HMG-CoA reductase inhibitors are now effectively being employed as cholesterol-lowering drugs.

Effects of the phenolic compounds of olive oil

Anticancer potential

The Mediterranean diet, which is high in fruits, vegetables, fibre, fish and olive oil, is a healthy and disease-preventive diet. It has chemoprotective effects against cancer and reduces heart disease problems and mortality percentages. One significant component of olive oil is *oleic acid* (C18–1), which is a monounsaturated, long-chain fatty acid, while seed oils such as sunflower are rich in polyunsaturated linoleic acid (C18–2), which can more easily be oxidized by various means.

Olive oil also contains minor components, which represent 1% of virgin olive oil. These components, and especially phenols, have anticancer potential. They number many hundreds and are classified within the following categories:

- Acids (phenoxy acids, triterpenic acids).
- Phenols (tocopherols, epoxyphenols).
- Chlorophyll.
- Alcohols (mono- or triterpenic).
- Aldehydes.
- Esters (derivatives of fatty acids with alcohols).

These compounds contribute to olive flavour, but some of them also give bitterness or a burnt taste. The concentration (mg/kg) of individual phenolic compounds in olive oil is: hydroxytyrosol, 14.42; tyrosol, 27.45; total simple phenols (TSP), 41.87; total secoiridoids (TSID), 27.72; and lignans, 41.53. The sum of TSP + TSID + lignans is 111.12.

Inhibition of enzymatic activity

Most of the phenolics in olive oil are amphiphilic, and they partition between the olive oil and waste water phases. The lipid-soluble antioxidants (tocopherols) do not affect enzymes such as lipoxygenases, NADH oxidase or NOS. Hydroxytyrosol inhibits platelet aggregation and the accumulation of the pro-aggregant agent thromboxane. However, oleuropein increases the activity and the expression of the inducible form of the enzyme NOS (iNOS). Nitric oxide inhibits platelet aggregation. Furthermore, inhibition of nitric oxide synthesis increases cellular damage and animal mortality.

Furthermore, olive oil phenols inhibit human hepatic microsomal activity. Olive oil phenols tested positive as being inhibitors of human hepatic microsomal androstenedione-6b-hydroxylation and reductive oxidative 17β-hydroxysteroid dehydrogenase (17β-HsB) activity.

Inhibitory action on reactive oxygen species

Seed oils are devoid of the typical phenolic compounds present in olive oil. The lignans (+)-1-acetoxypinoresinol and (+)-pinoresinol are major components of the olive oil phenolics. The fraction (+)-pinoresinol is present in the seeds of *Sesamum indicum* and (+)-1-acetoxypinoresinol is present in the bark of *O. europaea*. If we compare virgin olive oil and refined virgin olive oil we find that the sum of hydroxytyrosol and tyrosol is ten times greater in the former. The effect of olive oil on the enzyme xanthine oxidase, as measured by the production of uric acid, indicates that the minimum amount of uric acid (μM) was produced with the extra virgin olive oil, intermediate with the refined virgin olive oil and the maximum with seed oils. The amount of uric acid produced with seed oils is four times greater in comparison with extra virgin olive oil and two times greater in comparison with refined virgin olive oils. The effect of total phenolic compounds within olive oil on xanthine oxidase activity, which is measured by the production of uric acid (μM), indicates that the greater the total phenol level (μg/ml), the less uric acid is produced. Total phenols over 400 μg/ml reduce uric acid production three- to 15-fold. The inhibition of xanthine oxidase activity by extracts of oils indicated that 100–500 μl methanolic extract of extra virgin olive oil could reduce the production of ROS four- to fivefold, while with refined olive oil or seed oil the production of ROS was at its maximum.

Table 27.1 lists the inhibitory concentration of various phenolic compounds on the production of ROS that attack salicylic acid in an assay of hypoxanthine/xanthine oxidase.

Methanolic extracts of olive oils have significantly greater antioxidant capacities in comparison with seed oils, as measured by the hydroxylation of salicylic acid by HO^-. Antioxidants interfere with the hydroxylation of salicylic acid by donating a proton to HO^-. Inhibition of xanthine oxidase activity will reduce the amount of superoxide generated, and consequently the production of HO^- by Fenton's reaction. Total phenols inhibit directly HO^- attack on

Table 27.1. Inhibitory concentration of phenolic compounds of olive oil on reactive oxygen species that attack salicylic acid in an assay of hypoxanthine/xanthine oxidase (from Owen et al., 2000).

Phenolic compound	Inhibitory concentration (mM)
Hydroxytyrosol	1.34
3,4-Dihydroxybenzoic acid	3.03
Tyrosol	2.51
p-Hydroxybenzoic acid	1.69
Vanillic acid	2.70
Caffeic acid	6.05
Syringic acid	3.19
p-Coumaric acid	2.33
Ferulic acid	1.56
o-Coumaric acid	2.32
Cinnamic acid	2.33
Benzoic acid	2.75
Catechol	2.50
Oleuropein glucoside	7.80
Trolox	12.24
Dimethylsulphoxide	2.30

salicylic acid and indirectly via inhibition of xanthine oxidase activity. Hydroxytyrosol and tyrosol act as antioxidants via proton donation, while lignans of olives have a dual action, i.e. proton donation to HO$^-$ and inhibition of xanthine oxidase. Therefore, the major phenolics of olive oil have significant dietary value and, moreover, they increase the defence against ROS and decrease the activity of xanthine oxidase, which is one of the factors responsible for carcinogenesis. The high concentration of lignans in olive oils has been shown to inhibit skin, breast, colonic and lung cancer. The mechanism of lignin inhibition of carcinogenesis is based on its antiviral and antioxidant activities, and also in part on its anti-oestrogenic effects. Lignans have been shown to inhibit oestradiol-induced human breast carcinoma cells.

Protection of human erythrocytes from oxidative damage

Polyphenols are widely distributed in the vegetable kingdom and are present in high concentrations in the Mediterranean diet, as components of olive oil. However, the benefits of olive oil consumption on human health can be ascribed not only to the elevated oleic acid content but also to the antioxidant properties of its minor components, including polyphenols and – especially– hydroxytyrosol (3,4-dihydroxyphenyl ethanol), which is present in high concentrations in extra virgin olive oil (free or esterified), accounting for 70–80% of total polyphenols. Hydroxytyrosol is an efficient scavenger of peroxyl radicals and is responsible for protection of the oil against the auto-oxidation of unsaturated fatty acids. Furthermore, it was proved to act as a powerful inhibitor of peroxidation of human lipoproteins by *in vitro* studies. Hydroxytyrosol counteracts free radical-induced cytotoxicity in human intestinal epithelial cells in culture.

The red blood cells (RBCs) are exposed to oxidative hazards due to their specific role as oxygen carriers. Therefore, auto-oxidation of haemoglobin (Hb) produces anion superoxide radicals, which are transformed to hydrogen peroxide (H_2O_2). These compounds, when Fe ions are present, form the highly reactive hydroxytyrosol radical. These radicals damage both plasma membrane and cytosolic components, leading to oxidative haemolysis.

A balance exists between the production of ROS and their destruction by the endogenous defence system (enzymes and vitamin E). When no such balance exists (oxidative stress) it leads to chronic oxidative stress, as in the case of hereditary anaemias, which can be ascribed to either an impaired antioxidant defence system or overproduction of ROS (β-thalassaemia). In human RBCs, the molecular target of H_2O_2 is Hb, which is converted to the oxidized forms of Hb and ferryl Hb. Other alterations of RBCs by H_2O_2 include side-chain protein alterations and lipoperoxidation. These oxidative modifications lead to changes in the shape of the RBC and to haemolysis.

Even a small increase in phospholipid hydroperoxides in membranes produces marked changes in the molecular organization of the lipid bi-layer. The well-known oxidative stress-induced reduction in cellular energy charge affects the RBC transport system in a different way. Due to oxidation, stress methionine transport decreases. This amino acid, which is not utilized in protein biosynthesis, is converted to an s-adenosylmethionine (SAMe) donor. This substance is the key intermediate by which methionine is converted to cysteine, which is a precursor of glutathione.

Inhibitory action on reactive nitrogen species

The concentration of phenols in olive oil varies from 100 to 800 mg/kg. This concentration is a function of species, location, climate, stage of maturation and the conditions of processing and storage of olive oil.

Phenolics from virgin olive oil are also powerful scavengers of superoxide anions and hydrogen peroxide, and prevent both the generation of ROS and the aggregation of platelets, which result in disorders of the cardiovascular system. Besides the ROS there are also the 'reactive nitrogen species', such as the radical nitric oxide (NO^-), which is involved in cancer and other diseases when in excess concentration (De la Puerta *et al.*, 2001).

This NO^- toxicity involves a fast reaction with superoxide radical (O_2^-), producing peroxynitrite ($ONOO^-$). This radical induces the peroxidation of lipids, oxidizes methionine and S-H into proteins and causes DNA damage.

Hydroxytyrosol possesses a highly protective action against this radical, leading to the nitration of tyrosine and DNA damage. Nitric oxide is also a messenger molecule in neurotransmission. Oleuropein is able to scavenge nitric oxide, since it reduces the amount of nitrite formed by the reaction between oxygen and nitric oxide, produced from nitroprusside. Furthermore, most olive oil phenolics are able to scavenge $ONOO^-$, except tyrosol, which is less active. Therefore, components of olive oil contribute to protection against oxidative processes and damage by reactive NO^--derived nitrogen species involved in several chronic diseases. Furthermore, oleuropein has proved significantly more toxic to Gram-positive than to Gram-negative bacteria. The o-diphenol system present on its backbone structure is very likely responsible for the antimycoplasmal activity of oleuropein. Oleuropein was also found to be effective against *Mycoplasma fermentans* strains, which are naturally resistant to erythromycin and to tetracyclines.

Antimycoplasmal activity

Among olive oil phenols, oleuropein has been shown to inhibit or delay the rate of growth of mycoplasmas (Furneri *et al.*, 2002), and a range of bacteria and fungi. Furthermore, the major phenolic compounds of green olives have shown various antimicrobial properties, such as the inhibition by oleuropein of *Lactobacillus plantarum*, *Pseudomonas fragi*, *Staphylococcus carnosus*, *Enterococcus faecalis*, *Bacillus cereus*, *Salmonella* spp. and fungi. Oleuropein also has antimicrobial activity against Gram-positive and Gram-negative bacterial strains (*Salmonella* spp., *Vibrio* spp. and *Staphylococcus aureus*).

The antimicrobial action of phenols is related to their ability to denature proteins, generally by causing leakage of cytoplasmic constituents like protein, glutamate or potassium and phosphate from bacteria, possibly due to disruption or damage to the cell membrane. Oleuropein affects a significant leakage of glutamate, potassium and inorganic phosphate from *E. coli*, has no effect on the rate of glycolysis and causes a decrease in the ATP content of cells.

Phenolic extracts from olive mill waste water and their hypocholesterolaemic effects

Dietary phenolic compounds recovered from OMWW have antioxidative effects and are very potent in high cholesterol-fed rats. The concentration of hydroxytyrosol in OMWW is 100- to 500-fold higher than that in olive oil. The overabundance of free radicals in cells causes chain reactions and lipid peroxidation, processes linked with the development of both atherosclerosis and cancer.

THE MEDITERRANEAN DIET

Introduction

The Mediterranean diet is rich in virgin olive oil as a source of fat, and in vegetables, fruits, legumes and other plant foods, and low in saturated and trans-fatty acids and cholesterol. The Mediterranean diet is associated with longevity and low cardiovascular risk (Dominguez and Barbagallo, 2006). A relationship between diet and cardiovascular health was shown scientifically many decades ago. There are many reports in the literature concerning the protective role of the Mediterranean diet in human health (Ferro-Luzzi and Sette, 1989; Ferro-Luzzi and Ghiselli, 1993; Landa et al., 1994; Keys, 1995; Haber, 1997; Mancini and Rubba, 2000; Trichopoulou et al., 2000; Alarcon de la Lastra et al., 2001; Wahrburg et al., 2002). However, by the middle of the 20th century, due to industrialization and dietary changes, the need to establish preventive measures against the ill effects of poor diet became more pressing. Diets rich in saturated fats are associated with cholesterol accumulation and the risk of coronary heart disease. The Mediterranean countries, which in their diet consume considerable amounts of virgin olive oil, are characterized by a low level of heart attacks. This cardiovascular protection is achieved by the effects of virgin olive oil on plasma lipid risk factors and on other factors such as blood pressure and insulin sensitivity. The Mediterranean diet provides additional benefits by acting on other cardiovascular risk factors, including the reduction of blood pressure in individuals with normal or high blood pressure and the improvement of carbohydrate metabolism, not only in healthy people but also in patients with the type 2 diabetes.

Low-density lipoproteins are protected from oxidative damage in individuals consuming much virgin olive oil, in comparison with those consuming polyunsaturated fatty acid-enriched diets. Furthermore, the Mediterranean diet creates a less prothrombotic environment by modification of platelet aggregation and total plasma factor VII.

The Mediterranean diet and ageing

A great number of people in developed countries reach an advanced age, but with concomitant diseases such as Alzheimer's, Parkinson's, vascular dementia, cancer and diabetes. A diet rich in olive oil, high in monounsaturated fats, is associated with a reduced risk from age-related diseases. Furthermore, mitochondrial membranes are very sensitive to free radical attack, due to the presence of a double carbon–carbon bond in the lipid tails of their phospholipids, and unsaturated (oleic) fatty acids decrease cellular oxidative stress. Virgin olive oil strengthens membranes by increasing their resistance to free radical-induced modifications. Hence, virgin olive oil preserves mitochondrial function and its electron transport chain, with both DNA and blood better protected against oxidation.

Minor components of the Mediterranean diet

The Mediterranean diet uses mostly olive oil, which contains monounsaturated fatty acids and several minor components with biological activity, as opposed to seed or other oils. One such component is squalene (see above), which is, in part, responsible for the low incidence of cancer seen in the Mediterranean countries. The quantities of α-tocopherol (vitamin E) and carotenoids are low; however, these constituents exert an antioxidant role in the human body. The sterol acyl coenzyme A, an inhibitor of acyltransferase activity, leads to lower levels of plasma LDL cholesterol. Oleanolic acid also has antioxidant properties. The most important phenolic compounds contained in virgin olive oil, and therefore in the Mediterranean diet, are tyrosol, hydroxytyrosol and oleuropein, which are absorbed by the human digestive system. These components have both antioxidant and chemoprotective activity and a role in endothelial function improvement. Furthermore, these compounds can modify haemostasis, inhibit platelet aggregation and exert antithrombotic activity. Phenolic compounds from the Mediterranean diet delay atherosclerosis.

The Mediterranean diet and cancer prevention

The diet and fatty acid intake are important factors associated with development of cancer – especially gastrointestinal, breast and prostatic. In Mediterranean countries, where virgin olive oil is the main source of fat, the incidence of cancer is lower in comparison with that in northern European countries or the USA. The main mechanisms involved in protection from cancer are the following:

- Prevention of DNA oxidative damage.
- Prevention of changes to cell membranes.
- Prevention of gene expression modulation, which leads to DNA damage.
- Avoidance of an altered expression pattern of new cancer genes (oncogene HER2).

Biological and Integrated Olive Culture

BIOLOGICAL OLIVE CULTURE

Biological (or organic) olive culture is an agricultural management system that reduces significantly the use of chemical fertilizers and phytoprotective materials. Therefore, biological culture means food production that is not harmful to the environment with regard to the use of fertilizer and phytoprotective chemicals.

Organic olive cultivation is often considered to be superior to conventional and other alternative forms of agriculture and to be of greater value to society (Lampkin and Padel, 1994; Conacher and Conacher, 1998; Fabbri and Ganino, 2002; Parra-López and Calatrava-Requena, 2005). The regulations surrounding organic olive cultivation with regard to techniques used cover three categories: obligatory, recommended and forbidden. The two most important aspects are the following:

- The initial planning of the biological olive orchard.
- Provision of pruning, irrigation and pest control.

Biological pest protection achieves ecological stability by maintaining the pathogen population at a level that does not cause significant economic loss. Pest and disease control may be performed in the following ways:

- Selection of an olive variety tolerant to pests and diseases.
- Use of appropriate propagation material and optimum time of planting.
- Maintenance of optimum conditions for olive tree growth.
- Balanced fertilization and use of green composts.
- Destruction of certain species of plants that provide shelter for harmful insects and diseases.
- Soil fumigation and disinfection of propagation material.

With regard to those chemicals used in pest and disease control, for biological olive culture the use of the following techniques only is permitted:

- Use of various traps (pheromone or colour traps).

- Biological ways of dealing with pests and diseases, e.g.: (i) optimizing the performance of *Bacillus thuringiensis* endotoxins; (ii) enhancing beneficial organisms that are parasites of *Dacus* or *Prays oleae*, such as *Opius concolor*, *Eupelmus urozonus*, *Chrysopus*, etc.; (iii) using companion plants and botanical extracts as alternative pest controls.
- Use of selected insects that are parasites of those insects that cause damage to the olive tree or fruit.

The following phytoprotective materials are permitted in biological olive culture:

- Beeswax for protection of large sections of the orchard during pruning.
- Neem (*Azadirachta indica*) extract used as an insecticide.
- Zelatin used as an insecticide.
- Vegetable oils used as insecticide, acaricide and/or fungicide.
- Rotenone (*Derris* spp. and *Lonchocarpus* spp.) and pyrethrin (*Chrysanthemum cinerariaefolium*) used as insecticides.
- Copper in the form of hydroxide or sulphate used as fungicide. The permitted quantity is 0.6 kg/1000 m² of olive orchard.
- Calcium polysulphate used as fungicide, insecticide or acaricide.
- Paraffin oil used as insecticide or acaricide.
- Potassium permanganate ($KMnO_4$) used as fungicide or bactericide.
- Sulphur used as fungicide or acaricide.
- Various traps.

PRODUCTS PERMITTED FOR USE IN BIOLOGICALLY CULTIVATED OLIVE ORCHARDS

Fertilizers

The following materials are permitted for use: the manure of agricultural animals, dried chicken manure, composts from animal manure, animal excretions, peat, perlite, vermiculite, mixtures of vegetative materials, animal by-products (fish, meat or bone flour), wood ash, milled phosphate rocks, K salts such as sylvinite, potassium sulphate, $CaCO_3$ and gypsum.

Phytoprotective materials

The following are permitted for use: Azadirachtin, which is produced from *Azadirachta indica* (the neem tree), beeswax, zelatin, hydrolyzed proteins, lecithin, water extract of *Nicotiana tabacum*, vegetable oils (pine, mentha, etc.), pyrethrins produced from the plant *Chrysanthemum cinerariaefolium*, extracts from *Quassia amara* (a South American shrub) and rotenone from *Derris* spp. and *Lonchocarpus* spp.

Microorganisms

Various microorganisms, such as *Bacillus thuringiensis*, granulosis virus, etc., are used in the control of dangerous and damaging organisms

Substances used in traps

The substances used in biological traps include metaldehyde, pheromones, pyrethrins and ammonium phosphate.

Other materials

Copper (hydroxide, sulphate, etc.), ethylene, soap, paraffin oil, $KMnO_4$, SiO_2, S and calcium polysulphate can also be used in biological olive culture.

Organic olive growing has expanded in recent decades in Italy, Spain and Greece. This can be explained partly by the health-giving properties of virgin olive oil, which results in high prices for the grower. Of course, although organic olive growing results in lower production and/or higher cost, due to its higher quality, people prefer to consume organically produced olive oil. Organic olive growing in Greece started officially in 1993 with a few hectares; by 1998 it represented 10,000 ha and, by 2001, 15,500 ha.

TECHNIQUES OF CULTIVATION IN ORGANIC OLIVE GROWING

Soil cultivation and fertilization

Every 2 years, in autumn, vetch is sown plus a cereal such as barley, for green compost. Also during autumn, compost or manure (ovine or caprine) is added, at 20–25 l/tree, depending on its age and size. Composts containing K and micronutrients, such as B, may also be used at that time.

Cover crops

Cover crops play a vital role in organic olive cultivation (Pardini *et al.*, 2002, 2003). Use of cover crops is necessary in order to control weeds and soil erosion and to improve the quality of the olive oil. Furthermore, cover crops increase the content of the organic matter of soil and improve overall soil fertility, nutrient availability, soil structure and soil microorganisms. Cover crops also reduce the pest and disease problem, reduce nutrient leaching and supply food for livestock. When cover crops include legume species, the N content of the soil is increased. There is a disadvantage, however, in the competition created with olives and a concomitant decrease in olive yields.

Weed control

Winter weeds are controlled by soil cultivation during spring or by cutting (milling cutter). In steep orchards the use of machinery is difficult and it is recommended to use a hedge trimmer and leave the cut weeds on the soil surface.

Irrigation

In flat areas, where water is available, drip irrigation is a method that conserves water and reduces the cost of irrigation.

Pest and disease control

- Olive fly (*Bactrocera oleae*). Various traps can be used, such as ecotrap traps, Delta traps of bait smeared with glue and bottles of different colours and shapes filled with water and bait.
- Mediterranean black scale (*Saissetia oleae*). Appropriate pruning achieves an adequate population reduction in this insect. Furthermore, sprays with Cu or oils are helpful.
- Olive moth (*Prays oleae*). Use of pheromone traps, *Bacillus thuringiensis* at the initiation of flowering, wettable sulphur and control of irrigation are all helpful measures.
- Peacock spot (*Cycloconium oleaginum*). Correct pruning (open centre) and spraying with Cu products are recommended.
- Olive knot (*Pseudomonas savastanoi*). Spraying with Cu immediately after pruning is recommended.

Harvesting

In biological olive cultivation table olives are harvested by hand. The olives of the cultivars destined for olive oil production drop spontaneously onto plastic nets or they are harvested by using shakers.

Olive oil extraction

The olive fruits are immediately transported, on the same day, to the olive oil processing unit. The temperature of the water for oil extraction should be <30°C, the oil thus maintaining all its aromatic substances. The extracted olive oil is stored in stainless steel containers.

APPLICATION OF THE INTEGRATED MANAGEMENT OF OLIVE ORCHARDS

The prevailing socio-economic conditions, the very fast technological progress, the increase in world population, the progress in ecological aspects and the increased demands of consumers for high-quality products have resulted in the development of new agricultural management systems, such as the conventional, organic or biological and integrated olive culture.

Conventional or chemical olive culture, representing a high proportion on a worldwide basis, in its attempts to increase productivity and ease global food shortages, has damaged the ecosystem (Parra-López and Calatrava-Requena, 2006): productivity increases but natural resources and biodiversity are consequently reduced. Without underestimating its role in the 'Green revolution', this form of olive culture may create a series of unpleasant consequences.

Ever-bearing olive culture is a new form of agriculture that tries to achieve adequate productivity but at the same time with minimum cost and with respect for the natural environment. Ever-bearing olive culture may be described as either: (i) ecological, biological or organic olive culture (Cobb *et al.*, 1999); or (ii) integrated or balanced olive culture (Pacini *et al.*, 2003).

The main targets of this type of olive culture are the following:

- Development of methods of cultivation that respect the natural environment and are economically and practically feasible.
- Maintenance of the ever-bearing feature.
- Improvement of product quality.
- Protection of growers' health during the process of production, the production of high-quality olive oil and improved health benefits for olive oil consumers.
- Maintenance of biodiversity.
- Use of alternative means of dealing with pests, diseases and weeds to the use of chemicals.
- Maintenance of soil fertility.

Certain conditions should be fulfilled in order for integrated olive management to be feasible:

- Adequate research of market requirements.
- Development of the necessary technology.

- Development of methods for quality control of the product.
- Practical and theoretical training of growers.
- Selection of certain areas for olive culture on the basis of soil and climatic conditions. Frost- and wind-affected areas and soils with bad drainage should be avoided for olive culture.
- Choice of the best planting system and varieties.
- Soil cultivation and rational irrigation and fertilization. For integrated olive oil production soil fertility, mineral nutrients, organic material in the soil and normal soil microorganisms should all be maintained. Fertilizers should be added only when there is a need, based on soil and leaf analysis. Water availability is a prerequisite for integrated olive culture; however, high soil moisture can promote the development of diseases.
- Choice of the most appropriate pruning system.
- Integrated pest and disease management (use of biological and biotechnological methods). In order to maintain the ecological stability and the natural biodiversity of an olive orchard it is recommended to reduce the use of herbicides and also to maintain 'weed zones'; such zones constitute shelter for certain organisms.
- Care in maintaining the quality of the product during harvesting, transport to the processing units and the correct storage of both oil and table olives. Olives are harvested when the fruit reaches its peak content and quality of olive oil. During the transport of olives, fruit bruising should be avoided, since oil quality quickly deteriorates. After harvesting, olives should be processed very quickly in order to avoid deterioration and to increase oil acidity. Furthermore, the extracted oil should be stored appropriately to avoid oxidation.

29

CHEMICAL AND INTEGRATED WEED MANAGEMENT IN OLIVE ORCHARDS

CHEMICAL WEED MANAGEMENT

Weed control in olive orchards is necessary, since if not correctly managed weeds can compete with olive trees for water, nutrients and sunlight, and can reduce olive tree productivity. The weeds present in olive orchards may be annual, biennial or perennial plants. Such plants complete their life cycle from a few months to 3 years. Weed management is a fundamental agricultural practice, since existing weeds can influence the population of insects, mites, nematodes and diseases. Weed control should start before planting of olive trees, since their control later is not easy. Furthermore, when the weeds become dry there is the danger of fire and destruction of the olive grove (UCIPM, 2001).

Herbicides are valuable when applied to both deciduous and evergreen trees. Herbicides can be classified as either *pre-emergence* or *post-emergence*. Another classification system is *in-contact herbicides* and *herbicides with hormonal action*. The toxicity of hormonal herbicides is closely related to the vegetative activity of the plant where the herbicide has been applied. In order for weeds to be killed, the products of photosynthesis after spraying of a herbicide should be transported to the root system. When perennial weeds are exposed to water stress conditions, an insignificant quantity of herbicide is transported to the root system, causing necrosis of the weed. This is not so critical in annual weeds, since only the parts above ground should die.

Hormonal herbicides are employed to control weeds without damaging the main crop. Some selective herbicides are applied to leaves and some on the soil. After spraying of the foliage the herbicide is transported from leaves to young shoots or from older leaves to roots. This transport system follows the translocation of carbohydrates via the phloem. The herbicide's selectivity is affected by leaf morphology, the degree of herbicide absorption and transport and activation of certain enzymes.

Herbicide absorption

Herbicides are usually absorbed by the roots or leaves. The amount of herbicide absorption depends on the vegetative stage of the weed, since the leaf cuticle is thicker in old leaves in comparison with young ones. The number and size of stomata also affects herbicide entry into the leaf. Weeds differ in both the number of stomata/mm^2 of leaf surface and the location of stomata (on the upper or lower leaf surface).

The efficient use of herbicides depends on their ability to pass through the cuticle and plasmalemma. The entrance of herbicides is via stomata, if they are open during the time of herbicide application. Application of herbicides to green and tender shoots of weeds is frequently as effective as when applied to the leaves. The bark of woody shoots is usually the obstacle to entrance of herbicides. However, in some barks there are openings that allow the herbicide ingress. Lenticels may also constitute a route for herbicidal entrance.

The most dangerous weeds causing losses to olive culture are *Cynodon dactylon*, *Oxalis* and *Rubus fruticosus*. Other weeds present in olive orchards include *Sorghum halepense*, *Convolvulus arvensis*, *Cirsium arvense*, and *Amaranthus* (see Table 29.1).

WEED MANAGEMENT

Weed management varies from orchard to orchard and, of course, upon the weed species present in the orchard, the soil characteristics, the irrigation system and the age of the orchard.

Soil properties

Soil texture and organic matter determine the weed species present in the orchard, the activity and adsorption of herbicides, the number of times soil should be cultivated and when planting should take place. In sandy soils less herbicide is used, and these soils require more cultivation to deter weeds. On the contrary, clay soils are cultivated less frequently and can adsorb a greater quantity of herbicide. Soil characteristics, where herbicides are applied, affect their runoff. Soil texture and organic matter content affect the speed of infiltration and runoff, which are greater in sandy soils (Galindo et al., 2006). Organic matter affects herbicide mobility and adsorption: soil compaction decreases the speed of herbicide infiltration and increases herbicide concentration in the water flow. Stability of soil aggregates affects the rate of infiltration, as does the slope of the orchard, since a greater angle of incline increases runoff.

Table 29.1. Common and scientific names of weeds found regularly in the olive grove.

Common name	Scientific name
Asparagus	*Asparagus officinalis*
Barley	*Hordeum murinum* subsp. *leporinum*
Bermuda grass	*Cynodon dactylon*
Bindweed	*Convolvulus arvensis*
Blackberry	*Rubus* spp.
Bluegrass	*Poa annua*
Brome grass	*Bromus* spp.
Canary grass	*Phalaris canariensis*
Chickweed	*Stellaria media*
Clover	*Trifolium* or *Medicago* spp.
Cocklebur	*Xanthium* spp.
Dandelion	*Taraxacum officinalis*
Dock	*Rumex crispus*
Fescue	*Festuca rubra*
Foxtail	*Setaria* spp.
Goosefoot	*Chenopodium murale*
Johnson grass	*Sorghum halepense*
Mallow (cheeseweed)	*Malva parviflora*
Mustard	*Brassica* spp.
Nettle	*Urtica* spp.
Nightshade	*Solanum* spp.
Oat	*Avena fatua*
Pigweed	*Amaranthus* spp.
Purslane	*Portulaca oleracea*
Star thistle	*Centaurea salstitialis*
Thistle	*Salsola fragus*

Irrigation

Total precipitation, its distribution and the method of irrigation all affect the frequency of cultivation and the choice of herbicides. The amount of water required following herbicide application is 10–25 mm, and large amounts of irrigation water result in herbicide movement into the soil below the tree canopy. Therefore, the efficiency of the herbicide is reduced.

Age of the orchard

We can distinguish three periods during the lifespan of the olive orchard at which application of herbicides is necessary: (i) before planting; (ii) the young olive orchard; and (iii) orchard 4 years old or over.

Before planting
It is ideal to start weed control before planting of the orchard. In that way weed competition during planting will be reduced. The methods available to control weeds before planting include the following:

1. Digging and irrigation to prevent germination of weeds. Afterwards, it is necessary to dig again in order to kill any remaining weeds. This method controls mostly annual weeds.
2. Summer digging, when the soil is dry. Thereby, the rhizomes of perennial weeds are divided into small parts and, with sunshine, they desiccate and die.
3. Soil solarization. The moist soil is covered with a plastic sheet during the summer. The radiant energy trapped in the soil increases soil temperature, killing various pathogens and weed seeds.
4. Use of pre-emergence or post-emergence herbicides before planting. Post-emergence herbicides have some residual effect on the soil and are better applied before planting.

Young olive orchards
In young orchards both chemical and non-chemical methods are available:

1. Hand-weeding around trees and destroying weeds between rows by discing.
2. Use of plastic mulch around young trees.
3. Application of pre-emergence and post-emergence herbicides in one of three ways: (i) a circle around trees; (ii) in zones along the tree row; and (iii) total coverage of the orchard floor with herbicides. More than one herbicide is needed in order to control the various species of weeds. It is worth noting that pre-emergence herbicides do not tackle existing perennial weeds in the orchard.
4. Cover crops also can decrease weed population in the area between tree rows; these should not compete with olive trees. Some of the plants appropriate as cover crops include the autumn-sown cereals, subterranean clover and Bermuda grass.

Olive orchards 4 years old and older
When olive trees are established, weeds are not as competitive as before. However, perennial weeds can still reduce fruit yield. These weeds may be controlled in the following ways:

1. Destruction of weeds between tree rows by agricultural machinery, i.e. by mowing or discing. However, cultivation may damage the roots of olive trees and increase the danger of pathogenic infection.
2. Use of mulches to reduce weed growth.
3. Control of very young weeds between rows by flaming. This method controls the broadleaf weeds by destroying with intense and localized heat.

4. Lastly, pre-emergence and post-emergence herbicides can be used. Sometimes, more than one application of herbicide is required in order to eliminate the problem.

TYPES OF HERBICIDES USED AND MODES OF ACTION

Herbicides are divided into pre-emergence and post-emergence (Elmore, 1994).

Pre-emergence

These herbicides are sprayed on the soil and inhibit germination of weed seeds. The herbicide should enter the soil at a depth of 3–8 cm, through the effects of either rain or irrigation. Pre-emergence herbicides include Diuron (Karmex), simazine (Princep), oryzalin (Surflan), oxyfluorfen (Goal) and napropamide (Devrinol). Pre-emergence herbicides maintain their effectiveness from several weeks to 1 year.

Table 29.2 presents some pre-emergence herbicides used in olive orchards, and their rates of application.

Post-emergence

These herbicides are foliarly applied to either young or perennial weeds, and they can be divided into two categories: (i) *contact herbicides*; and (ii) *herbicides*

Table 29.2. Some pre-emergence herbicides used to control weeds in olive orchards, and their rates of application.

Herbicide	Rate of application (g/1000 m^2)	Time of application	Commercial name	Weeds controlled
Chlorthiamid	600–700	Early winter or early spring	Prefix	Annual cereals
Diuron	170	Winter/early spring	Casoron Karmex	Broadleaf plants Cereals, annual broadleaf plants
Simazine	300–400	Winter	Gesatop	Cereals, annual broadleaf plants

with hormonal activity (Hilton *et al.*, 1969). Contact herbicides destroy the sprayed part of the weed. One representative of this category is Paraquat. Hormonal-like herbicides are sprayed on the leaves and are transported to the rest of plant. Therefore, these are effective in killing perennial weeds, which have rhizomes. The rate of herbicide application depends on various factors, such as age of tree, weed species, irrigation and soil properties.

In olive orchards with weeds a mixture of post-emergence herbicides is often used, e.g. (i) diuron + simazine + aminotriazole; the rate of application of aminotriazole is 350–500 g/1000 m^2; or (ii) dimezon + simazine + Paraquat (gramoxon); the rate of application is 50–100 g/1000 m^2.

The use of herbicides is permitted in trees of 4 years old and above, but should be avoided in gravelly and light soils, since tree damage is possible. For control of existing weeds (cereals or broadleaf) the systemic herbicide Roundup (glyphosate) (UCIPM, 2001) is used, at a rate of 90–180 g/1000 m^2. Furthermore, the herbicide Paraquat (gramoxon) could be used to control broadleaf weeds and the annual cereals at a rate of 50–100 g/1000 m^2.

During spring, when there are perennial weeds such as *Cynodon dactylon*, *Convolvulus*, etc., the dose of Roundup required is 200–480 g/1000 m^2; and for *Cyperus rotundus* the herbicide MSMA (Veliuron) is used at a concentration of 250 g/1000 m^2. For the weed *Rubus* the herbicide Fenoprop (Kuron) is used at 250–700 g/1000 m^2, in early August. In modern chemical weed control the following regimes are common:

1. Use of new herbicides with slow release of the active ingredient.
2. Use of hormone-like herbicides.
3. Study of herbicide behaviour regarding their duration within the soil and their ease of transport in the soil or air.
4. Simazine as a pre-emergence herbicide and glyphosate as a post-emergence one are the two most important herbicides applied in olive orchards in order to control *Amaranthus blitoides* and *Lolium rigidum*. Simazine is a pre-emergence herbicide that inhibits electron transport in photosynthesis; glyphosate is a post-emergence herbicide that inhibits the biosynthesis of aromatic amino acids. Glyphosate, when applied to leaves, inhibits their elongation and causes deterioration of the apical part of the fruit. Simazine, when it was sprayed on *A. blitoides*, achieved complete control for 20 days at rates of 1.2 and 4 kg/ha. Glyphosate as a post-emergence herbicide is very efficient for both weeds.
5. Paraquat is recommended for the control of cereal and broadleaf weeds, as is trifluralin.
6. Ansar 529 HC is recommended for the control of *Sorghum halepense* and *Cyperus rotundus*. The rate of application is 650 g/1000 m^2 diluted in 30–40 l water.

Olive is cultivated mostly in dry and hilly areas and is subjected to intense competition from various weeds for water and nutrients. Due to the presence of weeds, the use of harvesting nets is not easy. Furthermore, the weeds constitute a shelter for pests and olive diseases but, if their height is controlled by cutting, weeds could be useful, since they add organic matter to the soil and help protect it from erosion.

In olive orchards weed control is achieved by various strategies: (i) soil cultivation; (ii) frequent cutting of weeds; (iii) use of post-emergence herbicides; and (iv) biological means of weed control.

Soil cultivation

This is the most common method of controlling weeds. The use of agricultural machinery to divide the rhizomes into small pieces only makes the problem worse, and frequent soil cultivation increases production costs.

Use of grass-cutting machines

This method is efficient for annual weeds but not for woody species. Weeds should be cut early, before producing ripe seeds. The cut weeds remain on the soil surface (mulching), to decompose with time.

One of the systems that can be used is *non-cultivation* of the soil combined with frequent cutting of weeds, in order to reduce competition and the trees' requirements for water and nutrients. The sensitivity of some common weeds of olive groves to various herbicides is presented in Table 29.3.

INTEGRATED WEED MANAGEMENT

Integrated weed control is a combination of chemical, biological, cultural and other means, the application of which reduces the weed population to such a level that is not antagonistic to olive culture. Furthermore, it maintains environmental balance in the agro-ecosystem.

Parameters of integrated weed management

Preventive measures

1. Control of weeds before orchard establishment; fumigation of planting material before establishment in orchard.
2. Cleaning of agricultural machinery before use.
3. Use of decomposed manure for olive orchard fertilization.
4. Control of the purity of cover crop seeds.

Table 29.3. Sensitivity of selected weeds to current herbicides.

Weed	Aminotriazole	Bromacil	Carbetamide	Chloroamide	2,4-D	D-NOC	Dinoseb	Diquat	Paraquat
Agropyron repens	S	T	I-T	T	T	T	T	T	T
Cynodon dactylon	I	T	I-T	T	T	T	T	T	–
Cirsium arvense	S	T	T	S	S	–	–	–	–
Convolvulus spp.	I	T	T	–	S	–	–	S	–
Polygonum spp.	T	S	I-T	–	T	S	S	–	T
Rumex spp.	S	T	I-T	–	T	–	–	–	T
Sorghum halepense	T	T	T	T	T	T	T	T	T
Rubus spp.	S	T	T	–	–	T	T	–	–
Ranunculus spp.	I	T	S	–	–	T	T	–	–

S, sensitive; I, intermediate; T, tolerant.

Cultivation measures

1. Establishment of rotation crops, as a fundamental agricultural practice.
2. Balanced fertilization.
3. Mulching. Mulching of the in row has certain advantages and disadvantages.

- Advantages: (i) increase in organic content of soil; (ii) increase in soil biological activity; and (iii) reduction in both water evaporation from the surface and in soil erosion.
- Disadvantages: (i) increase in the risk of frost damage; (ii) it should be carried out every year; (iii) the mulch may be blown away by strong winds; and (iv) non-composted plant material may spread new weeds or create a fire hazard.

4. Cover crops and permanent sward. Weed control in rows can be achieved by sowing a permanent sward that shades the soil and absorbs the available moisture of the soil surface. Thereby, surface sward reduces weed germination.

Mechanical control

There will be varying effects of various soil cultivation systems on the biological cycle of weeds, to nutrient cycling, to the availability of soil moisture, on the microclimate and on pests and diseases.

Biological control

1. Production of bio-herbicides from various microorganisms.
2. Production of phytotoxins.
3. Sowing of leguminous plants in order to control perennial weeds.
4. Vapour fumigation of the soil at 70°C before planting.
5. Allelopathy.
6. Use of parts remaining from cultivated plants having biological activity against the weed seeds in olive orchards.
7. Crop rotation by selecting crops having an inhibitory effect on weeds.

Future perspectives of integrated weed management

Integrated weed management is the new strategy in the area of plant protection and is compatible with future economical, environmental and social conditions.

30

PESTS AND DISEASES

INTRODUCTION

The olive tree is attacked by various insects, fungi, bacteria and viruses. The most serious attacks include those by *Bactrocera oleae* (*Dacus oleae*), *Prays oleae*, *Saissetia oleae*, *Hemiberlesia lataniae*, *Frankliniella occidentalis*, *Spilocaea oleaginea*, *Verticillium dahliae*, *Mycocentrospora cladosporioides*, *Pseudomonas syringae* pv. *savastanoi* and *Armillaria mellea*. Control of pests utilizes various methods, e.g. cultivation (El-Hakim and Kisk, 1988), biological (El-Khawas, 2000) or IPM (Elmore *et al.*, 2001). We will begin by studying insect pests, followed by fungal, bacterial and finally, viral problems.

INSECTS

Bactrocera oleae (Gmelin) (olive fruit fly, Diptera: Tephritidae)

The olive fruit fly, *Bactrocera oleae* (Gmelin) or *Dacus oleae*, is a very damaging pest of olive fruit in most olive-growing countries. The damage is caused by the *Dacus* larva, which drills tunnels in the fruit pulp resulting in 30% loss of the crop. The adult *Dacus* feeds on nectar, honeydew and other liquid food sources. *Dacus oleae* was first detected in California in 1998.

Morphology
The adult female is about 5 mm long and has a wingspan of approximately 10 mm. The wings are transparent, marked with brown and have a spot at each tip. The thorax is black, with three parallel lines. The abdomen is black, covered with grey pubescence. The inner portion of the scutellum is black, the posterior portion is yellow and odours are emitted by the adults (Economopoulos *et al.*, 1971). The sheath of the ovipositor is black, with the ovipositor itself reddish in colour.

Life cycle

The insect has two to five generations per year in Mediterranean countries. It survives in the pupal form and overwinters several centimetres below the soil surface, the first adults appearing from March to May, depending on temperature and latitude. During summer a period of 6–10 days elapses before mating, and this period is longer with lower temperatures. Starting in June, the female of *Dacus* lays ten to 12 eggs daily, one per olive fruit, and the total number of eggs laid during the lifespan of an insect (1–7 months) is 200–250. The females deposit their eggs beneath the fruit skin with their ovipositor. From each egg a legless larva derives, which feeds on fruit pulp; this results in fruit drop (see Fig. 30.1). The durations of egg, larval and pupal stages are 2–4, 10–14 and 10 days, respectively. Mating occurs near the end of the daylight period. Females, in order to attract males, produce a multi-component pheromone containing 1,7-dioxaspiro(5.5)undecane, which is an attractant for male *Dacus*. Females mate several times during their lifetime. Pheromone is produced also by males and attracts other males but not females.

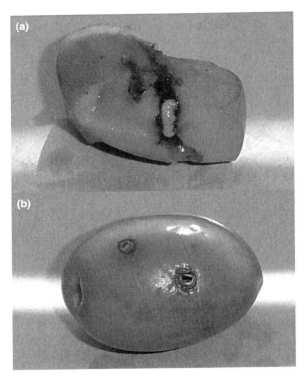

Fig. 30.1. Olive fruits damaged by *Dacus oleae*, olive fruit fly: (a) galleries drilled by *Dacus* pupae, and a pupa; (b) opening created by the exit of the perfect insect (from Therios, 2005b).

Control

Various ways of controlling *Dacus* have been used (Collier and Van Steenway, 2003), including γ-ray male sterilization (Economopoulos, 1972), mass trapping (Haniotakis *et al.*, 1986) and use of chemical repellents of females (Lo Scalzo *et al.*, 1994). The most common method is the use of insecticide in bait or air sprays. Other means, more acceptable environmentally, include the use of radiation to sterilize males, pheromones and biological methods. The required radiation for sterilization of both sexes is 8–12 Krad. Furthermore, synthetic analogues of pheromone have been produced, such as 1,5,7-trioxaspirol(5.5) undecane, which is equally as attractive as the natural compound but does not last in the traps as long.

In order to use synthetic pheromones, small pieces of paper or plywood are dipped in an aqueous (0.1%) solution of deltamethrin for 15 min, then sex pheromone or ammonium bicarbonate added – both *Dacus* attractants. This method provides a low-cost means of control. Other methods include the use of traps for trapping adult insects, biological control and cultivation (early fruit harvesting).

Acetylcholinesterase (ACE) is the molecular target of organophosphate (OP) insecticides, and two mutations conferring different levels of OP insensitivity have been identified in the olive fly *Bactrocera oleae*. Numerous sensitive and two insensitive alleles have been described from the worldwide distribution of the fly. The insensitive alleles probably reach a high level in the Mediterranean region, reaching frequencies of up to 100%. The high frequency of insensitive alleles occurs in areas where OPs have been used extensively.

PHEROMONES   There is little knowledge of the chemistry, enzymology or molecular biology of the biosynthesis of fruit fly pheromones to facilitate species-specific monitoring and control (Mazomenos, 1983). Studies with *Bactrocera oleae* suggest the operation of a single, major biosynthetic pathway. The major component of the pheromone is racemic 1,7-dioxaspiro (5.5) undecane, which is accompanied by low levels (3%) of hydroxyl derivatives.

Adult olive fruit fly populations are monitored using yellow sticky traps baited with sex pheromone and/or ammonium bicarbonate. Sex pheromone attracts male *Dacus* insects, while ammonium bicarbonate attracts mainly female insects. Females need protein and are attracted by ammonia, which is a product of protein decomposition. The yellow colour of the trap attracts both male and female insects. The population of insects trapped is a function of insect population, temperature and humidity of the atmosphere. Another use of this trap is to determine the time of insecticide application, by monitoring the population of *Dacus* trapped.

For control by insecticide, two methods are available: (i) full coverage sprays; and (ii) bait sprays, both methods using OPs, such as dimethoate or fenthion. The number of sprays depends on temperature and humidity, or rainfall during the summer, and involves one to two cover sprays and/or six to seven bait sprays. The number of sprays is smaller (two to three) in hot and dry areas. However, the number of sprays depends on the insect population caught in the traps and whether the fruit is destined for table olives or olive oil production.

Organophosphates are replaced by the insecticide Spinosad in bait sprays, since this is a microbial product, very effective against *Dacus*, but with low toxicity to humans and animals. Spinosad is mixed with the new fruit fly bait known as GF-120, and applied weekly from mid-June up to just before harvesting time.

Another insecticide of low toxicity to vertebrates is a pyrethroid compound (deltamethrin) applied on yellow plywood or paper strips, plus sex pheromone and/or ammonium bicarbonate. *Dacus* is attracted and killed by the lethal dose of pyrethroid.

In recent years the biological control of the olive fruit fly has commenced. In the Mediterranean area *Dacus oleae* is parasitized by the insect *Psyttalia* (or *Opius*) *concolor*. This was introduced to Italy, France, Greece and then California. However, *P. concolor* is ineffective against *Dacus*, since there is no synchronization between the biological stages of the two species, i.e. when female *P. concolor* appears in the spring, no fruit fly larvae are available.

Saissetia oleae (olive black scale, Hemiptera: Coccidae)

Saissetia oleae or black scale is a serious pest in many areas around the world (Katsoyannos, 1992; Van Steenwyk *et al.*, 2002; Kovanci and Kumral, 2004; UCIPM, 2004; Teviotdale *et al.*, 2005). It is native to South Africa and is widespread throughout Mediterranean countries and California. The same scale can also be found in other fruiting trees such as fig, citrus, apple, etc.

Morphology

The colour of the young adult is brown, becoming black later. The shape of mature females is hemispherical, and their size is 2–5 mm long × 1–4 mm wide. The characteristic feature of this pest is the appearance of an 'H' shape on its back. The insect lays eggs under the female scale; their length is 0.3 mm and their colour becomes pink within 2–3 days initially, and reddish orange before hatching. From the eggs after hatching nymphs appear, small in length (0.4 mm) and yellow to brown in colour. These nymphs start to crawl for some days until they find a suitable site for feeding. They are called crawlers and, after a period of feeding lasting 3–8 weeks, their size doubles to 1.0–1.3 mm, with a recognisable dorsal 'H' (second instar nymphs). The next stage is the immature adult stage, which is 2 mm in length and brown; lastly, the pest enters the adult stage. The laying capacity of one female black scale is from 1200 to 4000 eggs.

Life cycle

This species overwinters in the form of a nymph and early in spring becomes adult, starting to lay eggs in May. After hatching crawlers appear in July, feeding on leaves, shoots or sometimes on fruit. The insect has one or two generations depending on weather conditions, pruning and pest control.

Economic damage

Since the insect feeds on tree sugars, these trees become weak, leaf abscission occurs and flower bud differentiation is significantly reduced. Furthermore, young scales produce honey-like substances. This leads to a secondary infection by the fungus *Capnia*, resulting in deterioration of fruit quality and a decrease in photosynthesis.

Control

The insect can be controlled by cultivation practices, insecticidal sprays or by natural enemies of the black scale.

PRUNING In trees left unpruned or lightly pruned and a relatively dense canopy, damage and infestation are severe. Regular pruning exposes the black scale to hot and dry conditions and increases its percentage mortality. However, if the weather is cool and humid this favours black scale, especially when the tree canopy is dense. In unpruned olive trees the shady and moist environment inside the tree canopy protects the scale, which can then survive even very hot summers. Furthermore, irrigation may create favourable conditions for the black scale population to increase.

INSECTICIDES Due to the existence of the scale's covering, pesticides are effective only at the crawler stage during the summer. When the population of black scale is low to medium oil treatments during the dormant period are effective. The most effective pesticides are OPs.

NATURAL ENEMIES There are many natural enemies, such as *Metaphycus helvolus*, *M. bartletic*, *Scutellista cyanea*, *Chrysoperla* spp., *Hippodamia convergens* and *Hyperaspis* sp. From these enemies only *Hyperaspis* spp. can significantly reduce the population of black scale.

Prays oleae (Bernard) (olive moth, Lepidoptera: Yponomeutidae)

This insect is known also as olive kernel borer. It is widespread in the Mediterranean area and in other olive-growing countries, causing severe damage, though a little less harmful than the effects caused by *Dacus*. The moth feeds on olive trees, cultivated and wild, and all olive cultivars are prone to damage if no control measures are taken. Other species where this moth can be found include other genera of the Oleaceae family such as *Ligustrum*, *Jasminum* and *Phillyrea*. *Prays oleae* has three generations: the first on floral buds, the second in the fruit endocarp and the third on the leaves where the insect overwinters.

Life cycle
The insect overwinters as an adult and appears in spring; the females deposit their eggs in the calyx of flowers. The larva, appearing after 9–12 days, feeds on the floral parts – initially on anthers and subsequently on stigma, style and ovary. The larvae later connect the entire inflorescence with a silky thread. The second generation attacks the young fruits close to the stalk and the larva enters the endocarp, where it stays for 80–135 days. After this period the larva leaves the endocarp and exits via the same opening, resulting in fruit drop. The adult insects appear after 8–14 days. The third generation lays eggs on the upper leaf surface. The hatched larvae (7–14 days) enter the leaf by drilling tunnels and digest foliar material. At the end of winter the adults appear and a new cycle begins.

Economic damage
The damage caused by this moth ranks second only to that of *Dacus oleae*, and is serious with high insect populations or low flowering conditions. Damage affects flowers, fruits and leaves and results in fruit drop.

Chemical and biological control
CULTIVATION PRACTICES Unfavourable climatic conditions greatly limit the population of this insect. Pruning and maintenance of open canopies reduce RH, and with RH below 50% or with temperatures above 30°C egg death results. That is why this insect does not appear in dry areas.

BIOLOGICAL CONTROL The insecticidal bacterium *Bacillus thuringiensis* Berliner (subp. Kurstaki) when sprayed is very effective in reducing larval populations. Furthermore, the insect *Chrysoperla carnea* Stephens is a natural enemy of *P. oleae*, since it is predatory to the insect's eggs, larvae and pupae.

CHEMICAL CONTROL The use of insecticides such as dimethoate controls population levels.

PHEROMONES The use of female sex pheromones is useful as a control method (Campion *et al.*, 1979; Kavallieratos *et al.*, 2005).

Palpita unionalis (Hübner) (jasmine moth, Lepidoptera: Pyralidae)

Life cycle
Palpita unionalis is active at night; mating lasts 1–3 h and eggs are laid at twilight. The eggs are laid singly or in series on the lower leaf surface. Newly hatched larvae feed on young, tender leaves, and older ones on the parenchyma of established leaves. The larvae spin several leaves or parts of leaves together with silken threads. The mean total life of the first generation is 45–48 days, while that of the second generation is 38 days (Alexopoulou-Vassilaina and Santorini, 1973; El-Kifl *et al.*, 1974; Amin and Saleh, 1975; Badaawi *et al.*, 1976; Shehata *et al.*, 2003). The larvae drill tunnels having a crescent-like shape, and can also infect inflorescences and young fruits causing fruit drop, since they feed on the seed embryo.

Control
BIOLOGICAL CONTROL *Bacillus thuringiensis* has toxic effects on moth larvae when they feed on it. Use of the commercial product marketed under the name Dipel 2X at a concentration 0.4 g/l achieves 100% eradication of the larvae within 48 h.

CULTIVATION PRACTICES Sucker eradication reduces the problem, since the insect prefers to lay its eggs on the suckers. The larval stage lasts 16 days at 17–23°C and 15 days at 21.6–25.5°C for the first and second generations, respectively. The number of eggs laid per female ranges from 630 to 658 and from 425 to 496 for the first and second generations, respectively.

CHEMICAL CONTROL Oil spraying is effective against light to moderate infestations, when used in conjunction with pruning to open the orchard canopy. Oil is not sprayed between 20 August and harvest for olives intended for green ripe processing, since that causes fruit spotting. It is preferable to spray at night or early in the morning if the temperature is greater than 32°C during the day.

When the infestation is severe, the insecticide Carbaryl (Sevin 80S) is included in the spray solution. This spray is, however, very destructive to most natural enemies of *Palpita*. Another insecticide that might be incorporated is Methidathion (Supracide 25WP).

Frankliniella occidentalis (Pergande) (western flower thrips, Thysanoptera: Thripidae)

This insect is also known as western flower thrips. Many plant species are hosts of this insect.

Morphology and life cycle

The adults of *F. occidentalis* are very small insects (1 mm). Their colour ranges from white to yellow, with small brown spots on the abdomen. The thorax is orange in colour and the abdomen is brown. Adults overwinter in weeds and in other places in the orchard. During spring they lay eggs and deposit them in buds, shoots and flowers. After hatching, nymphs feed on shoots, fruits or leaves. The nymphs, after attaining full size and development, are ready to pupate; they drop to the ground and pupate in protected places. The maximum adult population occurs in June and, as the weeds dry out, the adults move to various crops such as olive trees. This insect has five to six generations per year. The damage is due to the fact that thrips feed on leaves and shoots but, because they absorb sap from the fruits, the affected fruit shows scars and is rendered unsuitable for processing. This problem is serious with green olives intended for processing.

Control

Various insecticides are recommended for thrips, and cultivation techniques are very effective. One such technique is to disc the area close to the olive grove that is covered with weeds, to avoid an increase in thrips population and subsequent migration to olive orchards. Managing the vegetation in and around olive orchards is important in reducing the potential for damage from western flower thrips. Avoid discing orchard cover crops while trees are in bloom. Spray treatments are applied at full bloom.

Parlatoria oleae (Colvée) (olive scale, Homoptera: Diaspididae)

Parlatoria oleae is also known as olive scale. This is widespread in the Mediterranean countries, China, the Middle East, India, Turkey and the USA (California, Arizona). This is a serious pest and requires systematic pesticide treatment every year to reduce economic losses. *Parlatoria* can be found in about 200 plant species.

Morphology and life cycle

The adult is 2 mm in length with a greyish oval covering. The young female's body under the scale cover is reddish to deep purple; the covering of the male is elongate, white and flat. The insect has two generations per year and overwinters as an immature mated female; the overwintering forms begin to lay eggs in early May. The emergence of the first-generation crawlers occurs in late June to early July, when the young fruits are attacked and deformed. The second generation in August causes semi-circular dark purple–black scales on the green fruits; this damage is severe in table olives. Infestation may also occur on branches and leaves, which results in a significant decrease in productivity.

Economic damage and control

This insect can cause leaf abscission, drying of twigs and reduction in photosynthesis; however, the most serious damage is that on the fruits (Foda, 1973). These become disfigured, with purple spotting on green olives, which then become unsuitable for processing (see Fig. 30.2). *Parlatoria* can be attacked by two parasites, i.e. *Aphytis maculicornis* MASI and *Coccophagoides utilis* Doutt. If chemical treatments are necessary two sprays are required, the first in late May–June for the first generation and July–August for the second generation. Application should be carried out when crawlers appear.

Hemiberlesia lataniae (Signoret) (latania scale, Hemiptera: Diaspididae)

Morphology

Latania scale is a scale similar in size to the adult olive scale. The difference is in the waxy shell covering, which is more conical, with a small black spot to one side of the centre. Also, the body of the female is yellow while it is reddish purple on the olive scale. The insect produces many generations per year.

Life cycle

The insect overwinters in the form of nymph (second instar). Early in spring it matures and the female insect lays eggs in batches of 15–20, located beneath the scale covering. After hatching the crawlers start to feed, appearing in May,

Fig. 30.2. Olive fruit damage in the form of spot due to *Parlatoria oleae*, olive scale (from Therios, 2005b).

July and September. Latania, except on olive trees, feeds on avocado, acacia, euonymus, kiwi fruit, rose and other species.

Economic damage
The scales infest leaves, bark and fruits, producing a dark purple spot of very clear outline. Infested fruits, especially for table olive processing, become worthless.

Chemical and biological control
For biological control the species *Aphytis melinus* and other parasites are used. However, where chemical treatment is necessary oil sprays or oil sprays plus Sevin 80S or Supracide 25WP are used. Spraying is performed in late May and in late July to August. The best results are obtained when scale crawlers start moving onto the fruit. Treatment between these two sprays is not necessary. Furthermore, a postharvest treatment (October/November) is also effective. Spraying is effective, especially when used in conjunction with pruning to open the orchard canopy.

Aspidiotus nerii (oleander scale (Bouche) or ivy scale, Hemiptera: Diaspididae)

Many plants are hosts to this insect such as camellia, cherry, grapefruit, lemon, magnolia, orange, rose and olive, where infestations may damage the fruit.

Morphology and life cycle
It resembles greedy scale, the only difference being that the covering of its scale is less convex. It overwinters in the form of adult females, which start to lay eggs in April. The insect has two generations, in April and in July/August.

Economic damage and control
During heavy infestations this scale infests olive fruit, producing green spots on purple fruit and certain deformations. The infested fruit becomes worthless. Biological control agents may decrease the scale population.

Phloeotribus scarabaeoides (Bernard) (olive beetle, Coleoptera: Scolytidae)

This is known also as olive beetle and is present in all Mediterranean areas. The adult is 2.0–2.5 mm long.

Life cycle
This insect overwinters in the form of either larva, nymph or adult in shoot tunnels. The female drills galleries in the branches and trunk. In these

channels mating takes place and the female lays eggs. Within 1–2 weeks larvae appear, which open small channels under the bark. In another 3–4 weeks the larva is transformed into an adult, which leaves the branch. The insect has four generations, the adults appearing in March/April, April–July, August/September and August–November. The infested trees have low vigour and low productivity.

Control

- Eradication of the pruning material by burying, before the insect lays its eggs.
- Burning the suckers, which can host a large number of larvae.
- Keeping the trees healthy by appropriate fertilization, irrigation and pruning.
- Application of organophosphate insecticides is recommended when the infection is severe.

Oxyenus maxwelli (K.) (olive mite)

This mite is widespread in Mediterranean and other countries.

Morphology and life cycle
It is very small in size, with four legs, with body colour yellowish to orange. It overwinters in the form of adult in the tree bark. The females lay eggs in early spring through to summer. During periods of high temperature and low humidity its population is reduced.

Economic damage and control
It infests mainly young olive leaves, these leaves developing a silver colour and longitudinal curl. The mite also infests inflorescences, causing pistil abortion. Some of the symptoms due to infestation include inflorescence abscission and necrosis and drop of buds. In cases of severe infestation chemical control is recommended.

Zeuzera pyrina L. (leopard moth, Lepidoptera: Cossidae)

Damage due to this moth has been reported in species from over 30 botanical families. In olives the insect is present mostly in Mediterranean countries, and also Syria. The insect can damage all olive varieties: resistant varieties have not yet been reported.

Life cycle

Adults lay their eggs during summer and, after hatching, the larvae attack lateral buds and drill sub-cortical channels. The larvae exit the branch by creating a hole and subsequently enter a new branch. Their presence is obvious by the sawdust released from the opening (frass). The biological cycle lasts 1 year, with larvae being fully developed in late winter and overwintering in the form of chrysalids. The adults occur in May to June.

Economic damage and control

The tunnels opened cause serious damage and young trees may dry out. The secondary branches also die and tree vigour is reduced. For control we immerse a piece of cotton wool in a liquid producing toxic vapours and with that we plug the opening of the tunnel. Furthermore, in young trees spraying with systemic insecticides may solve the problem.

FUNGAL DISEASES

Spilocaea (or *Cycloconium*) *oleaginea* (olive peacock spot)

This disease is known as olive peacock spot, peacock spot and/or bird's eye or cycloconium, and is caused by the fungus *Spilocaea* (or *Cyclonium*) *oleaginea* (Cast) Hughes. The disease is common in many areas of the world.

Symptoms

The fungus attacks leaves, fruit and fruit-bearing stems, the symptoms being observed most often on the upper leaf surfaces of leaves located in the lower part of the tree canopy (Graniti, 1962; Proietti *et al.*, 2002a). Peacock spot causes cutin degradation, appearing on leaves in the form of sooty blotches having a diameter of 2–6 mm (Sparavano and Graniti, 1978). The colour of these spots changes to muddy green and later to black, with a yellow halo, giving the appearance of the 'eye' spot on the peacock's tail feathers (see Fig. 30.3). One leaf may have more than one spot, which causes the infected leaves to abscise. Leaf abscission reduces productivity and enhances biennial bearing. New spots appear in late winter to early spring. Outbreaks are sporadic and the disease may need many years before it causes economic damage. In the margins of the spots spores develop, high temperatures hindering spore germination. Therefore, high temperatures inactivate the disease. Cultivar and pruning intensity also affect the attacks of *Spilocaea oleaginea* (Tombesi and Ruffolo, 2006).

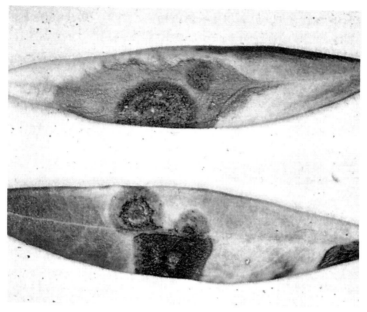

Fig. 30.3. Olive leaf infection by *Spilocaea oleaginea* (peacock spot). The infection appears to be in the form of spots, resembling the 'eye' of a peacock's tail feather (from Therios, 2005b).

Life cycle and control

The fungus survives on the infected leaves that do not abscise. During autumn the margins of the spots grow and produce conidia. Rainfall or watering transports the conidia to the lower part of the canopy, where they infect the lower leaves. The optimum temperature for conidial germination is 21°C, while temperatures greater than 30°C hinder germination. Infection during winter requires more time for development of symptoms in comparison with spring infection. A Cu-based fungicide applied in late autumn, before the rainy season, provides adequate control; sometimes, a second spray is applied in January, but its effectiveness is questionable. Furthermore, later spraying does not protect against *Spilocaea oleaginea*. It is important to apply the Cu fungicide yearly.

Complete coverage is necessary, requiring a power sprayer with high pressure to cover the leaves on both sides, especially in the inner parts of the tree. Both old and young leaves are sensitive and, furthermore, most varieties are susceptible. If rainfall is frequent, both winter and spring infections may occur. Peacock spot is not affected significantly by the level of tree nutrition. However, high N and low Ca concentrations render the tree more susceptible to infection. The forms of Cu that can be used to protect olive trees from peacock spot include copper sulphate tribasic, copper oxide, copper hydroxide and copper oxychloride. Infection is worst in areas of low elevation, under conditions of heavy dew, fog or high humidity, under a dense tree canopy and following excessive sprinkler irrigation.

The guidelines for effective spraying programmes include the following:

- Spraying is effective if carried out twice a year, in late autumn and in early spring.
- Spraying should be carried out before any rain and, should it rain within 2 h following application, a further spraying is necessary.
- Avoid spraying during the flowering period, since Cu may negatively affect budding and fruit set.
- When the level of infection is high (> 10% of leaves infected) the recommended combinations are: (i) a combination of Stroby WG (10–15 g/100 l) + Cuprofix (500 g/100 l); or (ii) Stroby WG (10–15 g/100 l) + Kocide 2000 (190 g/100 l), both combinations being mixed prior to spraying. The volume of spray solution per 1000 m^2 orchard is 100–150 l.

Verticillium dahliae (olive verticillium wilt, Hypocreales: Incertae sedis)

The pathogen of this disease is the fungus *Verticillium dahliae* Kleb. This is a serious disease causing significant damages in olive culture and killing both young and mature trees. It is widespread in many areas of the world (Thanassoulopoulos *et al.*, 1979; Bellahcene *et al.*, 2000, 2005).

Symptoms
The leaves on some branches suddenly collapse and die early in the growing season, the symptoms becoming more acute as the season progresses (Cirulli, 1981; Blanco-López *et al.*, 1984). The vascular tissues exhibit little or no discoloration and the dead leaves do not abscise.

Life cycle
The fungus *V. dahliae* is very common in many areas and in many agricultural soils. It survives in both soil and infected roots from year to year (Schnathorst, 1981; Tjamos, 1993). The form surviving in the soil is the microsclerotia, which are formed inside the infected plants and, as tissues decay, microsclerotia are liberated into the soil. Microsclerotia remain inactive for some time, and their germination starts when the roots commence growth. The fungus enters and grows into the vessels of the current year's growth, resulting in wilt and death of the plants. *Verticillium* infection occurs mostly during late winter and spring, when the soil is cool and having adequate moisture. Rising temperatures decrease levels of *Verticillium* infection. Another

wave of infection occurs in early autumn, when the soil cools again. High summer temperatures kill the fungus. The fungus has various hosts, such as cotton, tomato, pepper, melon, stone fruits and pistachio.

Management and control
The most important way of protecting trees from *Verticillium* wilt is the avoidance of the fungus. Therefore, soils that have previously been planted for a number of years with crops very susceptible to *Verticillium*, i.e. aubergine, etc., should not be given over to olive. Furthermore, *Verticillium* spores and inocula can be reduced in number prior to planting by soil fumigation or disinfection by solar energy. Other methods of control include cultivation of grass cover crops or by use of a combination of the above methods. When we plant olives in an area where previously susceptible perennial plants have been grown, it is necessary to remove most of the existing root system. With regard to rootstocks not resistant to *Verticillium* wilt most were reported to be so, apart from *Olea oblonga* and the cultivar 'Ascolano'.

For solar disinfection the soil is covered with a transparent plastic film throughout the summer, when increased temperatures kill the fungus. For soil fumigation we withhold water during the summer; the drier the soil the more efficient is the penetration of methyl bromide. Methyl bromide is injected to 45–75 cm depth and is covered with a gas-proof cover; the greater the dose of methyl bromide the greater the depth of penetration. The cover remains in place for two weeks, and afterwards it is removed and the soil is aerated for 1 month before planting. Methyl bromide may be applied from late summer to early autumn. It is generally accepted that soil fumigation is effective in nursery soils but is not appropriate in existing olive groves.

Armillaria mellea (armillaria root rot, oak root fungus, Agaricales: Marasmiaceae)

Armillaria root rot is not a serious disease of olive trees, although the fungus can survive for decades in soils on dead roots. The pathogen involved is *Armillaria mellea* (Vahl.) Quel.

Symptoms
Infected trees become weak and exhibit low vigour. This symptom appears initially on one side of the tree only, but over several years it gradually involves the whole tree. Discoloration of the bark and outer wood of the roots and tree crown is obvious. Therefore, the colour of infected roots with *A. mellea* is white to yellowish, fan-shaped mycelia appearing between the bark and wood. Another symptom is the appearance of brown to black rhizomorphs and mushrooms at the crown of *Armillaria*-infected trees.

Life cycle and control

Armillaria mellea is very common in soils where oak trees previously grew. The fungus lives for years in the dead roots of trees, providing the soil is not dry. As a root grows it comes in contact with infected roots and the disease is spread thus through the orchard, from tree to tree. The infected roots represent inocula retained in the soil profile. To reduce the danger from *A. mellea* root rot we avoid planting olive groves where forest or oak trees were recently grown. Another technique in infected soils is to reduce irrigation so that the crown and the upper root system grow in a dry soil. If there are any infected trees these should be excavated and the area should be fumigated in order to delay the spread of disease. With regard to rootstocks, no resistant examples exist, and damaged trees cannot be treated.

Many cultivation practices are applicable in controlling this fungus. Initially, the soil from around the base of the tree is removed to a depth of 20–30 cm, creating a small depression. In this way the trunk is exposed and the upper roots and crown become dry. Furthermore, during winter the collection of rainwater in the depression is avoided by appropriate drainage; the depression remains open for a number of years.

Another technique is fumigation, which is very efficient in shallow soils; in deep or heavy clay soils soil fumigation is not efficient. It is necessary before fumigation to remove all infected trees and a significant part of their root system. Sometimes it is necessary to remove trees adjacent to those having symptoms, since it is likely that these trees will also be infected. The infected material should be burned. Methyl bromide is injected to a depth of 45–75 cm and the soil is covered with a gas-proof material that remains for 2 weeks, after which it is aerated for 1 month before planting.

An alternative method is to dry the soil by reducing irrigation during the summer or by using cover crops such as safflower or sudan grass.

Mycocentrospora (*Cercospora*) *cladosporioides* (Hyphales: Deuteromyocotina)

Symptoms

Leaves become slightly chlorotic. The lower side of the leaves is discolored, with the conidia of the fungus appearing as a black dust, but no spots are evident; such leaves may abscise. Furthermore, spots appear on the fruits and such olives are characterized by lack of uniform ripening.

The disease is common in humid areas. This fungus has a similar life cycle to that of *Spilocaea oleaginea*. Leaf spotting is more difficult to control than in peacock spot and requires more careful treatment. A workable solution to the problem is preventive spraying of olive trees following harvest (before winter rains) and a second application in spring, if the weather is wet or rainy. Defoliation will result in less shoot growth, decreased fruit set and poor flower bud formation the following year. High and frequent rainfall increases *Cercospora* infection. When the winter is dry problems are minimized.

BACTERIAL DISEASES

Pseudomonas syringae pv. *savastanoi* (Smith Stevens) (olive knot, Psedomonadales: Pseudomonadaea)

This pathogen creates rough galls or swellings about 1.2–5.0 cm in diameter on twigs, branches, trunks, roots and leaves; galls may also be formed on trunks or limbs following wounding. Olive knot reduces productivity, since it destroys twigs and branches (Iacobellis, 2001).

Symptoms

This disease was first described in olive by the Greek philosopher Theophrastus (4th century BC). Due to heavy infestation the affected limbs are dwarfed, defoliated or killed and the whole plant may be stunted. The pathogen is also responsible for similar diseases of other genera of the Oleaceae family such as *Forsythia*, *Jasminum*, *Phillyrea* and *Oleander* (*Nerium oleander*). Due to infection, a decrease in production and fruit size is recorded. Furthermore, the green olives from infected limbs obtain off-flavours or rancidity in taste. Other strains of *P. savastanoi* include *P. savastanoi* pv. *nerii* and *P. savastanoi* pv. *fraxini*.

The main symptom is the appearance of rough galls appearing on twigs and small branches in areas where wounds, leaf scars, pruning cuts and mechanical injuries exist. These knots, especially at harvest, are very large and hinder water or assimilate transport, leading to defoliation (Sisto *et al*., 2004). The root system is not affected by the fungus; furthermore, stomata and lenticels are not routes of entry for olive knot bacteria.

The optimum temperature for *Pseudomonas* infection is 21–24°C, the maximum 32°C and the minimum 5–10°C. Rainfall encourages infection, the most susceptible period for infection being October–June. The knots develop when the tree is actively growing during spring or early summer.

Life cycle and control

In the knots the bacteria survive and are spread easily by water over the whole year. Various factors affect *P. savastanoi* inoculation (Penyalver *et al*., 2006): the most appropriate temperatures for infection are the lows in autumn or spring. The occurrence of infection requires openings provided by leaf scars, pruning wounds or cracking due to freezing. No tolerant cultivars are known.

For *Pseudomomas* control it is necessary to make preventive fungicide sprays to reduce bacterial entry through the scars. Furthermore, pruning should be conducted carefully during the dry season, in order to remove galls from twigs.

After pruning disinfection follows, with a Cu-based solution; the required number of sprays for control is three. The spraying schedule is one in the autumn, one before the winter rains begin and another in spring, following leaf abscission. No specific bactericides are available, so infected trees are treated with Cu compounds or by application of preventive techniques that reduce the pathogen population (Lavermicocca *et al.*, 2003). Copper sulphate applications are required in cases such as wounding due to frost, hail, pruning, harvesting or disease, e.g. peacock spot, *Verticillium* wilt or *Cercospora* infection).

Agronomic practices devised for limitation or eradication of the disease include the following:

- Avoid harvesting during rainy days and avoid inflicting wounds on the trees.
- Use of pathogen-free stock (internally and externally).
- Use of stock with a phytosanitary certification protocol.
- Use of rapid detection methods for the pathogen.
- Use of resistant cultivars (Hassani *et al.*, 2003; Iannotta *et al.*, 2006).
- Evaluation of the available olive germplasm to find cultivars resistant or tolerant to olive knot disease; this is the most important means of disease management.
- Selection of cultivars tolerant/resistant to freezing.

VIRAL DISEASES

A number of viruses infect olive trees in various areas (Faggioli *et al.*, 2005). Such viruses include cherry leaf roll (Savino and Gallitelli, 1981a), olive latent ring spot (Savino *et al.*, 1983), cucumber mosaic virus (Savino and Gallitelli, 1981b), an elongated virus (Faggioli and Barba, 1995) and two necroviruses (Felix and Clara, 2002). Modern techniques are used for virus detection (Grieco *et al.*, 2000; Bertolini *et al.*, 2001), and various sanitation methods for virus infection have been proposed (Bottalico *et al.*, 2004).

31

BIOTECHNOLOGICAL ASPECTS OF OLIVE CULTURE

INTRODUCTION

Olive culture was spread to all the various countries of the Mediterranean region accompanying human migration, but in modern times certain changes must be considered in order to face several issues: (i) cultivar identification; (ii) regeneration; (iii) maturation; (iv) abiotic stress tolerance; (v) plant size and architecture; (vi) resistance to pests and diseases; (vii) water and salt tolerance; (viii) parthenocarpy; and (ix) self-compatibility. Rugini *et al.* (2006) have provided an overview of olive biotechnology.

IDENTIFICATION AND IMPROVEMENT OF CULTIVARS

The current trend is to use a small number of cultivars with excellent characteristics through selection and biotechnology (Guerin *et al.*, 2002) for the production of table olives and olive oil. The current trend is the production of mono-cultivar olive oil with excellent characteristics. In Spain, olives have been managed as self-compatible trees and no other cultivars are included as pollinators (Cuevas *et al.*, 2001). This means that molecular markers should be used to distinguish cultivars precisely and not base selection purely on their morphological characteristics (Hatzopoulos *et al.*, 2002).

MICROPROPAGATION

Some very important olive cultivars such as 'Kalamon', 'Frantoio' and 'Picholine' are difficult to propagate *in vitro* (Zuccherelli and Zuccherelli, 2002). Attempts have been conducted to root such cultivars, and many factors affecting rooting were performed, such as C:N ratio, B concentration and use of putrescine, which promotes the activity of total peroxidases. Basal treatments of explants with H_2O_2 promote early and improved rooting. Furthermore, improved production of difficult-to-root cultivars can be achieved by basal etiolation of shoots during rooting and the use of polyamines and H_2O_2 (Rugini et al., 1997). The issue of low rooting capacity of valuable olive cultivars could be circumvented by the use of transgenic plants with the *rolB* gene of *Agrobacterium rhizogenes*, which renders plants more sensitive to auxin.

ABIOTIC STRESS TOLERANCE

Cold tolerance and ozone stress

Climatic changes create problems in olive culture due to frost damage. Most olive cultivars can resist temperatures no lower than $-12°C$ after acclimation due to freezing-tolerance mechanisms (Thomashow, 1999). However, in certain olive-growing areas lower temperatures are common. Cold-hardiness is a very important trait for olive improvement (Bartolozzi and Fontanazza, 1999). Improved cold tolerance could involve one or more of the following:

- The superoxide dismutase gene (*SOD*), which is responsible for repair of ozone-damaged cells (Van Camp et al., 1993).
- The *CBF1* gene, which increases tolerance to cold (Jaglo-Ottosen et al., 1998).
- The involvement of the cryoprotective proteins COR/LEA/dehydrin (Late Embryogenesis Abundant proteins). Among the genes that are the most highly induced during cold acclimation are: (i) the 'classical' *COR* (cold-regulated) genes, alternatively designated *KIN* (cold-induced); (ii) *RD* (responsive to dehydration); (iii) *LTI* (low temperature-induced); and (iv) *ERD* (early responsive to dehydration) (Tomashow, 1998). The proteins encoded by these genes are extremely hydrophilic and are either novel or members of the dehydrins (Close, 1997).

Water stress

The protein osmotin is involved in water stress. Xiong et al. (2002) reported on cell signalling following cold, drought and salt stress. The problems of drought and salinization could be solved by using drought- and salt-resistant cultivars, which are also appropriate for high-density plantings.

Salt tolerance

The genes *SOS1*, *SOS2* and *SOS3* are postulated to encode regulatory components for salt tolerance. Do such genes exist in olives? Evidence indicates a critical role for K in salt tolerance (Zhu et al., 1998). Furthermore, high salinity affects cellular and molecular responses (Hasegawa et al., 2000).

SIZE OF TREE

The new trend of dense plantings in olive orchards requires trees of both reduced size and altered shape. A number of genes exist in olive for controlling size, and a number of cultivars should be studied for their possible use as olive rootstocks, in order to produce trees adapted to mechanical harvesting and/or pruning.

RIPENING OF FRUIT

Many olive cultivars are late in ripening, or their ripening is not synchronous, creating problems at harvesting time. The introduction of genes that block ethylene synthesis is very important with regard to timing of ripening.

RESISTANCE TO PESTS AND DISEASES

The need for more healthy products for human consumption, containing no traces of phytoprotective chemicals in either table olives or olive oil, triggers a quest for genes tolerant to diseases and pests. Two very serious problems are *Verticillium* (verticillium wilt) and *Spilocaea oleaginea* (peacock spot). Several ways of improving the olive's defence system are outlined below.

- Through antifungal genes, genes for hydrolytic enzymes (glucanase, chitinase) (Broglie *et al.*, 1991) and genes coding for the inhibition of polygalacturonase.
- Resistance to bacteria has been obtained by the introduction of genes coding for bacteriocidal polypeptides such as thionin. Of prime importance is the need to find genes conferring resistance to *Bactrocera oleae*.
- Resistance to *Prays oleae*, through the creation of plants resistant to *P. oleae* by genetic transformation of certain cultivars with toxin protein genes from *Bacillus thuringiensis*.
- *Cycloconium*-resistant plants have thicker leaves and could be selected from tetraploid meristems.
- *Verticillium*-resistant plants obtained by somaclonal variation from a zygotic embryo callus.
- Resistance to viruses and phytoplasmas. Many viruses and phytoplasmas have been isolated in olive; therefore, the use of healthy plants for new planting is recommended, and thermotherapy of meristem tip culture is a useful strategy.

PARTHENOCARPY

Some self-sterile cultivars and others produce a significant proportion of parthenocarpic fruits. Therefore, their fruit set and productivity is reduced since parthenocarpic fruits are very small in size. Introduction of genes that induce parthenocarpy allow the normal development of parthenocarpic fruit.

QUALITY OF OLIVE OIL

Olive oil quality is determined by its oleic acid and polyphenol contents, both substances playing an important role in human health. Therefore, by suppression of oleate desaturase it would be possible to increase the percentage of oleic acid (C18:1) by almost threefold. The same measures could be employed to increase the polyphenol content of olive oil.

LOW CHILLING REQUIREMENTS

It is important to find new varieties with low chilling requirements appropriate for southern areas. This will help expand the area of olive cultivation.

SELF-COMPATIBILITY

Self-compatible (self-fruitful) cultivars are important in high-density plantings and for mono-cultivar olive oil production.

REFERENCES

Abeles, F.B. (1973) *Ethylene in Plant Biology*. Academic Press, New York and London, 302 pp.
Adakalic, M., Barranco, D., Leò, L. and De la Rosa, R. (2004) Influence of harvesting date on the germination and emergence of seeds of five olive cultivars. In: *5th International Symposium on Olive Growing*, 27 September–2 October 2004, Izmir, Turkey, Abstract 220.
Adiri, N. (1975) Isolation of protoplasts from olive leaves. MSc thesis from the Hebrew University of Jerusalem.
Agati, G., Pinelli, P., Ebner, S.C., Romani, A., Cartelat, A. and Cerovic, Z.G. (2005) Nondestructive evaluation of anthocyanins in olive (*Olea europaea*) fruits by in situ chlorophyll fluorescence spectroscopy. *Journal of Agricultural Food Chemistry* 53, 1354–1363.
Agrawal, S.B. and Agrawal, M. (2000) *Environmental Pollution and Plant Responses*. Lewis Publishers, Boca Raton, Florida.
Akilioglu, M. (1991) The use of plant growth regulators and the control of alternate bearing in olive. *Olea* 21, 2.
Al-Absi, K., Qrunfleh, M. and Abu-Sharar, T. (2003) Mechanism of salt tolerance of two olive (*Olea europaea* L.) cultivars as related to electrolyte concentration and toxicity. *Acta Horticulturae* 618, 281–290.
Alarcon de la Lastra, C., Barranco, M.D., Motilva, V. and Harrerias, J.M. (2001) Mediterranean diet and health: biological importance of olive oil. *Current Pharmaceutical Design* 7(10), 933–950.
Alberdi, M. and Corcuera, L.J. (1991) Cold acclimation in plants. *Phytochemistry* 30, 3177–3184.
Alcalá, A.R. and Barranco, D. (1992) Prediction of flowering time in olive for the Córdoba Olive Collection. *HortScience* 27, 1205–1207.
Alexopoulou-Vassilaina, P. and Santorini, A.P. (1973) Some data on the biology of *Palpita unionalis* Kb. (Lepidoptera: Pyralidae), under laboratory conditions. *Annals of the Benaki Phytological Institute (N.S.)* 10, 320–326.
Al-Jalil, H.F., Al-Omari, K.K. and Abu-Ashour, J. (1999) Comparative suitability for mechanical harvesting of two olive cultivars. *Agricultural Mechanization in Asia, Africa and Latin America (AMA)* 30(1), 38–40.
Allouche, N., Fki, I. and Sayadi, S. (2004) Toward a high yield recovery of antioxidants and purified hydroxytyrosol from olive mill wastewaters. *Journal of Agriculture and Food Chemistry* 52, 267–273.

Alohé, J.D., Corpas, F.J., Rodriguez-Garcia, M.I. and del Rio, L.A. (1998) Identification and immunolocalization of superoxide dismutase isoenzymes of olive pollen. *Physiologia Plantarum* 104, 772–776.

Alves, M.R., Cunha, S.C., Amaral, J.S., Pereira, J.A. and Oliveira, M.B. (2005) Classification of PDO olive oils on the basis of their sterol composition by multivariate analysis. *Analytica Chimica Acta* 549 (1–2), 166–178.

Amane, M., Lumaret, R., Hany, V., Ouazzani, N., Debain, C., Vivier, G. and Deguilloux, M.F. (1999) Chloroplast–DNA variation in cultivated and wild olive (*Olea europaea* L.). *Theoretical and Applied Genetics* 99, 133–139.

Amin, A.H. and Saleh, M.R.A. (1975) Seasonal activity of the olive leaf moths, *Palpita unionalis* (Hübner) (Lepidoptera: Pyralidae), in Kharga-Oasis, New Valley, Egypt. *Annals of the Agricultural Science, Faculty of Agriculture, Ain Shams University, Egypt* 20(1), 35–41.

Amiot, M.J., Fleuriet, A. and Macheix, J.J. (1986) Importance and evolution of phenolic compounds in olive during growth and maturation. *Journal of Agricultural Food Chemistry* 34, 823–826.

Amiot, M.J., Fleuriet, A. and Macheix, J.J. (1989) Accumulation of oleuropein derivatives during olive maturation. *Phytochemistry* 28, 67–70.

Amiri, M.E. (2004) Study of mass propagation of olive (*Olea europaea* L.) by tissue culture. In: *5th International Symposium on Olive Growing*, 27 September–2 October 2004, Izmir, Turkey, Abstract 223.

Andrews, M. (1986) The partitioning of nitrate assimilation between the root and shoot of higher plants. *Plant, Cell and Environment* 9, 511–519.

Andrikopoulos, N.K. (1989) The tocopherol content of Greek olive oils. *Journal of the Science of Food and Agriculture* 46, 503–509.

Androulakis, I.L. and Loupassaki, M.H. (1990) Studies on the self-fertility of some olive cultivars in the area of Crete. *Acta Horticulturae* 286, 159–162.

Angelakis, A.N., Bontoux, L. and Lazarova, V. (2002) Main challenges for water recycling and re-use in EU countries. In: *Proceedings of 'Water Recycling in the Mediterranean Region'*, 26–29 September 2002, Heraklio, Crete.

Angelopoulos, K., Dichio, B. and Xiloyannis, C. (1996) Inhibition of photosynthesis in olive trees (*Olea europaea* L.) during water stress and rewatering. *Journal of Exploratory Botany* 47, 1093–1100.

Angiolillo, A., Mencuccini, M. and Baldoni, L. (1999) Olive genetic diversity assessed using amplified fragment length polymorphisms. *Theoretical and Applied Genetics* 98, 411–421.

Antognossi, E., Cartechini, A. and Preziosi, P. (1975) Indagine sulla individuazione dei migliori impollinatory per olive da mensa della cultivar Ascolana Tenera. *Proceedings of the 2nd Seminar in International Oleiculture*, 6–7 October, 1975, Cordoba, Spain.

Antognozzi, E. and Catalano, F. (1985) Risposta varietale dell' olivo ai Danni da freddo. *Annali della Facoltà di Agraria di Perugia*, XXXIX, 185–198.

Antognozzi, E., Cartechini, A., Tombesi, A. and Proietti, P. (1990a) Transmission and efficiency of vibrations on 'Moraiolo' olive harvesting. *Acta Horticulturae* 286, 413–416.

Antognozzi, E., Cartechini, A., Tombesi, A. and Proietti, P. (1990b) Effect of cultivar and vibration characteristics on mechanical harvesting of olives. *Acta Horticulturae* 286, 417–420.

Antognozzi, E., Famiani, F., Proietti, P., Pannelli, G. and Alfei, B. (1994) Frost resistance of some olive cultivars during the winter. *Acta Horticulturae* 356, 152–155.

Antognozzi, E., Pilli, M., Proietti, P. and Romani, F. (1990c) Analysis of some factors affecting frost resistance in olive trees. In: *XXIII International Horticultural Congress*, Florence, Italy. Abstracts of contributed papers (2. Poster), p. 4289.

Antonopoulou, C., Dimassi, K., Chatzissavvidis, C., Therios, I. and Papadakis, I. (2006) Effect of BA and GA3 on the micropropagation and vitrification of olive (*Olea europaea* L.) explants (cv. 'Chondrolia Chalkidikis'). In: *Proceedings of the 2nd International Seminar Olivebioteq*, 5–10 November 2006, Marsala-Mazara del Vallo, Italy, 1, pp. 477–480.

Aparicio, R., Morales, M.T. and Alonso, V. (1997) Authentication of European virgin olive oils by their chemical compounds, sensory attributes and consumers' attitudes. *Journal of Agriculture and Food Chemistry* 45, 1076–1083.

Aragüés, R., Puy, A., Royo, A. and Espada, J.L. (2005) Three-year field response of young olive trees (*Olea europaea* L., cv. 'Arbequina') to soil salinity: trunk growth and leaf ion accumulation. *Plant and Soil* 271, 265–273.

Aviram, M. and Elias, K. (1993) Dietary olive oil reduces low-density lipoprotein uptake by macrophages and decreases the susceptibility of lipoprotein to undergo lipid peroxidation. *Annals of Nutrition and Metabolism* 37, 75–84.

Ayers, B.S. and Westcot, D.W. (1985) *Water Quality for Agriculture*. Irrigation and drainage paper 29, FAO, Rome.

Ayerza, R. and Coates, W. (2004) Supplemental pollination increasing olive (*Olea europaea*) yields in hot, arid environments. *Exploratory Agriculture* 40, 480–491.

Ayerza, R. and Sibbett, G.S. (2001) Thermal adaptability of olive (*Olea europaea* L.) to the Arid Chaco of Argentina. *Agriculture, Ecosystems and Environment* 84, 277–285.

Aziz, N.H., Farag, S.E., Mousa, L.A. and Abo-Zaid, M.A. (1998) Comparative antibacterial and antifungal effects of some phenolic compounds. *Microbios* 93, 43–54.

Bacon, M.A. (2004) Water use efficiency in plant biology. In: Bacon, M.A. (ed.) *Water Use Efficiency in Plant Biology*. Blackwell Publishing/CRC Press, Boca Raton, Florida, 327 pp.

Badaawi, A., Awadallah, A.M. and Foda, S.M. (1976) On the biology of the olive leaf moth *Palpita unionalis* Hb. (Lepidoptera: Pyralidae). *Zeitschift ang Entomologisches* 81(1), 103–110.

Baggio, G., Pagnam, A., Muraca, M., Martini, S., Opportuno, A., Bonanome, A., Bottista Ambrosio, G., Ferrari, S., Guarini, P., Piccolo, D., Monzato, E., Corrocher, R. and Crepaldi, G. (1988) Olive oil-enriched diet: effect on serum lipoprotein levels and biliary cholesterol saturation. *American Journal of Clinical Nutrition* 47, 960–964.

Balatsouras, G. (1990) Edible olive cultivars, chemical composition of fruit, harvesting, transportation, processing, sorting and packaging, styles of black olives, deterioration, quality standards, chemical analysis, nutritional and biological value of the end product, pp. 291–330.

Balatsouras, G. (1995) *Table Olives: Varieties, Chemical Composition of Fruit, Commercial Types, Quality Characteristics, Packaging and Marketing*. 2nd edn, Athens, 438 pp. [in Greek]. Publisher unknown.

Baldoni, L., Guerriero, C., Sossey-Aloui, K., Abbott, A.G., Angiolillo, A. and Lumaret, R. (2002) Phylogenetic relationships among *Olea* species based on nucleotide variation at a non-coding chloroplast DNA region. *Plant Biology* 4, 346–351.

Bandelj, D., Jakse, J. and Javornik, B. (2002) DNA fingerprinting of olive varieties by microsatellite markers. *Food Technology and Biotechnology* 40(3), 185–190.

Bandelj, D., Jakse, J. and Javornik, B. (2004) Assessment of genetic variability of olive varieties by microsatellite and AFLP markers. *Euphytica* 136(1), 93–102.

Bari, A., Martin, A., Boulouba, B., Gonzalez-Andujar, J.L., Barranco, D., Ayad, G. and Padulosi, S. (2003) Use of fractals and moments to describe olive cultivars. *Journal of Agricultural Science* 141, 63–71.

Barranco, D., Cimato, A., Fiorino, P., Rallo, L., Touzani, A., Castañeda, C., Sefarani, F. and Trujillo, I. (eds) (2000) *Catálogo Mundial de las Variedades de Olivo*, p. 360.

Barranco, D., Ruiz, N. and Gómez-del-Campo, M. (2005) Frost tolerance of eight olive cultivars. *HortScience* 40(3), 558–560.

Bartoli, R., Fernandez-Banares, F., Navarro, E., Costellà, E., Mañé, J., Alvarez, M., Pastor, C., Cabré, E. and Gassull, M.A. (2000) Effect of olive oil on early and late events of colon carcinogenesis in rats: modulation of arachidonic acid metabolism and local prostaglandin E (2) synthesis. *Gut* 46, 191–199.

Bartolini, G., Mazuelos, C. and Troncoso, A. (1991) Influence of Na_2SO_4 and NaCl salts on survival, growth and mineral composition of young olive plants in inert sand culture. *Advances in Horticultural Science* 5, 73–76.

Bartolozzi, F. and Fontanazza, G. (1995) Preliminary results on olive germplasm for frost hardiness. *Olea* 23, 27.

Bartolozzi, F. and Fontanazza, G. (1999) Assessment of frost tolerance in olive (*Olea europaea* L.). *Scientia Horticulturae* 81, 309–319.

Bartolozzi, F., Rocchi, P., Camerini, F. and Fontanaza, G. (1999) Changes of biochemical parameters in olive (*Olea europaea* L.) leaves during an entire vegetative season, and their correlation with frost resistance. *Acta Horticulturae* 474, 435–440.

Bazakos, C. (2007) Means to detect adulteration in olive oil. Ms thesis, Mediterranean Agronomic Institute of Chania, Crete.

Beede, R.H. and Goldhamer, D.A. (1994) Olive irrigation management. In: Ferguson, L., Sibbett, G.S. and Martin, G.C. (eds) *Olive Production Manual*. University of California, Division of Agriculture and Natural Resources, Berkeley, California, Publication 3353, pp. 61–68.

Belaj, A., Cipriani, G., Testolin, R., Rallo, L. and Trujillo, I. (2004) Characterization and identification of the main Spanish and Italian olive cultivars by Simple-sequence-repeat markers. *HortScience* 39, 1557–1561.

Belaj, A., Trujillo, I., Rossa, R., Rallo, L. and Gimenez, M.J. (2001) Polymorphism and discrimination capacity of randomly amplified polymorphic markers in an olive germplasm bank. *Journal of the American Society of Horticultural Science* 126, 64–71.

Bellahcene, M., Fortas, Z., Fernandez, D. and Nicole, M. (2005) Vegetative compatibility of *Verticillium dahliae* isolated from olive tress (*Olea europaea* L.) in Algeria. *African Journal of Biotechnology* 4(9), 963–967.

Bellahcene, M., Fortas, Z., Geiger, J.P., Matallah, A. and Henni, D. (2000) Verticillium wilt in olive in Algeria: geographical distribution and extent of the disease. *Olivae* 82, 41–43.

Beltran-Heredia, A.J., Torregrosa, A.J., Garcia-Araya, J.F., Dominguez-Vargas, J.R. and Tierno, J.C. (2001) Degradation of olive mill wastewater by the combination of Fenton's reagent and ozonation with an aerobic biological treatment. *Water Science and Technology* 44, 103–108.

Ben Ruina, B., Taamallah, H. and Ammar, E. (1999) Vegetation water used as fertilizer on young olive plants. *Acta Horticulturae* 474, 353–355.

Benavente-Garcia, O., Castillo, J., Lorente, J., Ortuño, A. and Del Rio, J.A. (2000) Antioxidant activity of phenolics extracted from *Olea europaea* L. leaves. *Food Chemistry* 68, 457–462.

Benitez, M.L., Pedrajas, V.M., del Campillo, M.C. and Torrent, J. (2002) Iron chlorosis in olive in relation to soil properties. *Nutrient Cycling in Agroecosystems* 62, 47–52.

Benlloch, M., Arboleda, F., Barrano, D. and Fernández-Escobar, R. (1991) Response of young olive trees to sodium and boron excess in irrigation water. *HortScience* 26, 867–870.

Benlloch, M., Marín, L. and Fernández-Escobar, R. (1994) Salt tolerance of various olive varieties. *Acta Horticulturae* 356, 215–217.

Ben-Shalow, N., Kahn, V., Harel, E. and Mayer, A.M. (1977) Olive catechol oxidase changes during fruit development. *Journal of Science, Food and Agriculture* 28, 545–550.

Ben-Tal, Y. (1987) Improving ethephon's effect on olive fruit abscission by glycerine. *HortScience* 22, 869–871.

Ben-Tal, Y. (1992) Quantification of ethephon requirements for abscission in olive fruits. *Plant Growth Regulation* 11, 397–403.

Ben-Tal, Y. and Lavee, S. (1976a) Increasing the effectiveness of ethephon for olive harvesting. *HortScience* 11, 489–490.

Ben-Tal, Y. and Lavee, S. (1976b) Ethylene influence on leaf and fruit detachment in 'Manzanillo' olive trees. *Scientia Horticulturae* 4, 337–340.

Ben-Tal, Y. and Lavee, S. (1984) Girdling olive trees, a partial solution to biennial bearing. The influence of consecutive mechanical girdling on flowering and yield. *Rivista Ortoflorofrutticultura Italiana* 68, 441–452.

Ben-Tal, Y. and Wodner, M. (1997) Chemical loosening of olive pedicels for mechanical harvesting. *Acta Horticulturae* 356, 297–301.

Ben-Tal, Y., Lavee, S. and Klein, I. (1979) The role of the source of ethylene in the development of an abscission zone in olive pedicels. In: Geissbuhler, H. (ed.) *Advances in Pesticide Science: Proceedings of the 4th International Congress of Pesticide Chemistry*, 24–28 July 1978, Zurich, Switzerland, 2, 347–350.

Berenguer, M.J., Grattam, S.R., Connell, J.H., Polito, V.S. and Vossen, P.M. (2004) Irrigation management to optimize olive oil production and quality. *Acta Horticulturae* 664, 79–85.

Berenguer, M.J., Vossen, P.M., Grattan, S.R., Connell, J.H. and Polito, V.S. (2006) Tree irrigation levels for optimum chemical and sensory properties of olive oil. *HortScience* 4, 427–432.

Bertolini, E., Olmos, A., Martinez, M.C., Gorris, M.T. and Cambra, M. (2001) Single-step multiplex RT-PCR for simultaneous and colorimetric detection of six RNA viruses in olive trees. *Journal of Virological Methods* 96, 33–41.

Besnard, G. and Bervillé, A. (2002) On chloroplast DNA variations in the olive (*Olea europaea* L.) complex: comparison of RFLP and PCR polymorphisms. *Theoretical and Applied Genetics* 104(6–7), 1157–1163.

Besnard, G., Breton, C., Baradat, P., Khadari, B. and Bervillé, A. (2001) Cultivar identification in olive based on RAPD markers. *Journal of the American Society of Horticultural Science* 126, 668–675.

Besnard, G., Khadari, B., Villemur, P. and Berville, A. (2000) Cytoplasmic male sterility in the olive (*Olea europaea* L.). *Theoretical and Applied Genetics* 100, 1018–1024.

Bianchi, G. (2003) Lipids and phenols in table olives. *European Journal of Lipid Science and Technology* 105, 229–242.

Bieleski, R.L. (1982) Sugar alcohols. In: Loewus, F. and Tanner, W. (eds) *Encyclopedia of Plant Physiology. New Series VI3A, Plant Carbohydrates. 1. Intercellular Carbohydrates.* Springer-Verlag, New York, pp. 158–192.

Bisingnano, G., Tomaino, A., Lo Cascio, R., Crisafi, G., Uccella, N. and Saija, A. (1999) On the *in vitro* antimicrobial activity of oleuropein and hydroxytyrosol. *Journal of Pharmacy and Pharmacology* 51, 971–974.

Blanco-López, M.A., Jiménez-Díaz, R.M. and Caballero, J.M. (1984) Symptomatology, incidence and distribution of *Verticillium* wilt of olive trees in Andalusia. *Phytopathologia Mediterranea* 23, 1–8.

Bogani, P., Cavalieri, D., Petruccelli, R., Polsinelli, L. and Roselli, G. (1994) Identification of olive tree cultivars by using random amplified polymorphic DNA. *Acta Horticulturae* 356, 98–101.

Bonachela, S., Orgaz, F., Villalobos, F.J. and Fereres, E. (2001) Soil evaporation from drip-irrigated olive orchards. *Irrigation Science* 20(2), 65–71.

Bongi, G. (1986) Oleuropein: an *Olea europaea* secoiridoid active on growth regulation. *Acta Horticulturae* 179, 245–249.

Bongi, G. and Loreto, F. (1989) Gas-exchange properties of salt-stressed olive (*Olea europaea* L.) leaves. *Plant Physiology* 90, 1408–1416.

Bongi, G., Mancuccini, M. and Fontanazza, G. (1987) Photosynthesis of olive leaves: effect of light, flux density, leaf age, temperature, peltates and H_2O vapour pressure deficit on gas exchange. *Journal of the American Society of Horticultural Science* 112, 143–148.

Bottalico, G., Saponari, M., Campanale, A., Mondelli, G., Gallucci, C., Serino, E., Savino, V. and Martelli, G.P. (2004) Sanitation of virus-infected olive trees. *Journal of Plant Pathology* 86(4), 311.

Bouat, A. (1961) Variabileté de l'alimentation minérale chez l'olivier. *Informations Olèicoles Internationales (N.S.)* 16, 19–31.

Bouaziz, A. (1990) Behavior of some olive varieties irrigated with brackish water and grown intensively in the central part of Tunisia. *Acta Horticulturae* 286, 247–250.

Bouma, J. (1997) Precision agriculture: introduction to the spatial and temporal variability of environmental quality. In: Lake, J.V., Bock, G.L. and Goode, J.A. (eds) *Precision Agriculture: Spatial and Temporal Variability of Environmental Quality.* Ciba Foundation 210, Symposium Wiley, Wageningen, Netherlands, pp. 5–17.

Bravo, L. (1999) Polyphenols: chemistry, dietary sources, metabolism and nutritional significance. *Nutritional Reviews* 56, 317–333.

Breton, C., Medail, F., Pinatel, C. and Berville, A. (2006) From olive tree to Oleaster: origin and domestication of *Olea europaea* L. in the Mediterranean basin. *Cahiers Agriculture* 15(4), 329–336.

Briante, R., Patumi, M., Limongelli, S., Febbraio, F., Vaccaro, C., Di Salle, A., La Cara, F. and Nucci, R. (2002a) Changes in phenolic and enzymatic activities during fruit ripening in two Italian cultivars of *Olea europaea* L. *Plant Science* 162(5), 791–798.

Briante, R., Patumi, M., Terenziani, S., Bismuto, E., Febbraio, F., and Nucci, R. (2002b) *Olea europaea* L. leaf extract and derivatives: antioxidant properties. *Journal of Agriculture and Food Chemistry* 50(17), 4934–4940.

Briccoli Bati, C., Nuzzo, V. and Godino, G. (2000) Preliminary agronomic evaluation of two cultivars of trees obtained from micropropagation methods. *Acta Horticulturae* 586, 867–870.

Briccoli Bati, C., Basta, P., Tocci, C. and Turco, C. (1994) Influence of irrigation with saline water upon young olive plants. *Olivae* 53, 35–38 [in Italian].

Broglie, K., Chet, I., Holliday, M., Cressman, R., Biddle, P., Knowlton, S., Mauvaic, C.J. and Broglie, R. (1991) Transgenic plants with enhanced resistance to the fungal pathogen *Rhizoctonia solani*. *Science* 254, 1194–1197.

Brooks, R.M. (1948) Seasonal incidence of perfect and staminate olive flowers. *Proceedings of the American Society of Horticultural Science* 52, 213–218.

Bucci, R., Magri, D., Magri, A.L., Marini, D. and Marini, F. (2002) Chemical authentication of extra virgin olive oil varieties by supervised chemometric procedures. *Journal of Agriculture and Food Chemistry* 50, 413–418.

Burke, M.J., Gusta, L.V., Quamme, H.A., Weiser, C.J. and Li, P.H. (1976) Freezing and injury in plants. *Annual Review of Plant Physiology* 27, 507–528.

Caballero, J.M. and del Rio, C. (1990) Rootstock influence on productivity parameters of two olive cultivars. *Abstracts of the 23rd International Horticultural Congress*, Florence, Italy.

Cadogan, G. (1980) *Palaces of Minoan Crete*. Corrected edition. Methuen, London.

Cain, J.C. (1972) Hedgerow orchard design for most efficient interception of solar radiation. Effects of tree size, shape, spacing and row direction. *New York State Agricultural Experimental Station Search Agriculture* 2, 1–14.

Campbell, W.H. (1988) Nitrate reductase and its role in nitrate assimilation in plants. *Physiologia Plantarum* 74, 214–219.

Campion, D.G., McVeigh, L.J. and Polyrakis, J. (1979) Laboratory and field studies of the female sex pheromone of the olive moth, *Prays oleae*. *Experientia* 35(9), 1146–1147.

Cañas, L.A. and Benbadis, A. (1988) Plant regeneration from cotyledon fragments of the olive tree (*Olea europaea* L.). *Plant Science* 54, 65–74.

Cañas, L.A., Wyssmann, A.M. and Benbadis, M.C. (1987) Isolation, culture and division of olive (*Olea europaea* L.) protoplasts. *Plant Cell Reproduction* 6, 369–371.

Capasso, R., Evidente, A., Avolio, S. and Solla, F. (1999) A highly convenient synthesis of hydroxytyrosol and its recovery from agricultural waste waters. *Journal of Agriculture and Food Chemistry* 47, 1745–1748.

Capasso, R., Evidente, A., Schivo, L., Orru, G., Marcialis, M.A. and Cristinzio, G. (1995) Antibacterial polyphenols from olive oil mill waste waters. *Journal of Applied Bacteriology* 79, 393–398.

Carriero, F., Fontanazza, G., Cellini, F. and Giorro, (2002) Identification of simple sequence repeats (SSRs) in olive (*Olea europaea* L.). *Theoretical and Applied Genetics* 104, 301–307.

Castro, J., Fernández, A., Aguilera, P., Orgaz, F., Garcia, J.A. and Jimenez, B. (2006) Oil quality and response to irrigation in traditional olive orchards. In: *Proceedings of the 2nd International Seminar Olivebioteq*, 5–10 November 2006, Marsala-Mazara del Vallo, Italy, 2, pp. 157–160.

Cegarra, J., Paredes, C., Roing, A., Bernal, M.P. and Garcias, D. (1996) Use of olive mill wastewater compost for crop production. *International Biodeterioration and Biodegradation* 38(3–4), 193–203.

Centritto, M., Lucas, M.E. and Jarvis, P.G. (2002) Gas exchange, biomass, whole-plant water-use efficiency and water uptake of peach seedlings in response to elevated [CO_2] and water availability. *Tree Physiology* 22, 699–706.

Centritto, M., Wahbi, S., Serraj, R. and Chaves, M.M. (2005) Effects of partial rootzone drying (PRD) on adult olive trees (*Olea europaea*) in field conditions under arid climate. II. Photosynthetic responses. *Agriculture, Ecosystems and Environonment* 106, 303–311.

Centritto, M. (1998) Tree responses to rising global CO_2 concentration and temperature: observation and mechanisms. In: Chinese Academy of Forestry (ed.) *Forest Towards the 21st century: Forest Science and Technology under the Global Strategy of Sustainable Development.* China Agricultural Scientech Press, Beijing, pp. 318–329.

Centritto, M., Loreto, F. and Chartzoulakis, K. (2003) The use of low $[CO_2]$ to estimate diffusional and non-diffusional limitations of photosynthetic capacity of salt-stressed olive saplings. *Plant and Cell Environment* 26, 585–594.

Chaari-Rkhis, A., Trigui, A. and Drira, A. (1999) Micropropagation of Tunisian cultivars olive trees: preliminary results. *Acta Horticulturae* 474, 79–81.

Charlet, M. (1965) Observation sur le comportement au froid de certaines varietes et de portegreffes d'oliviers en France. *Information Oleicole International* 31, 13–39.

Chartzoulakis, K., Loupassaki, M., Bertaki, M. and Androulakis, I. (2002) Effects of NaCl salinity on growth, ion content and CO_2 assimilation rate of six olive cultivars. *Scientia Horticulturae* 96, 235–247.

Chartzoulakis, K., Patakas, A. and Bosabalidis, A.M. (1999) Changes in water relations, photosynthesis and leaf anatomy induced by intermittent drought in two olive cultivars. *Environmental and Experimental Botany* 42, 113–120.

Chartzoulakis, K., Psarras, G., Vemmos, S. and Loupassaki, M. (2004) Effects of salinity and potassium supplement on photosynthesis, water relations and Na, Cl, K and carbohydrate concentration of two olive cultivars. *Agricultural Research* 27(1), 75–84.

Chartzoulakis, K.S. (2005) Salinity and olive: growth, salt tolerance, photosynthesis and yield. *Agricultural Water Management* 78, 108–121.

Chatzissavvidis, C. (2002) Study of boron toxicity in olive plants. PhD thesis, School of Agriculture, Aristotle University, Thessaloniki, Greece, p. 379.

Chatzissavvidis, C. and Therios, I. (2003) The effect of different B concentrations on the nutrient concentrations of one olive (*Olea europaea* L.) cultivar and two olive rootstocks. In: Stefanoudaki, E. (ed.) *Proceedings of the International Symposium on the Olive Tree and the Environment*, 1–3 October 2003, Chania, Greece, pp. 214–220.

Chatzissavvidis, C., Therios, I. and Antonopoulou, C. (2007) Effect of nitrogen source on olives growing in soils with high boron content. *Australian Journal of Experimental Agriculture* 47, 1491–1497.

Chatzissavvidis, C.A., Therios, I.N. and Antonopoulou, C. (2004) Seasonal variation of nutrient concentration in two olive (*Olea europaea* L.) cultivars irrigated with boron water. *Journal of Horticultural Science and Biotechnology* 79, 683–688.

Chatzissavvidis, C.A., Therios, I.N. and Molassiotis, A.N. (2005) Seasonal variation of nutritional status of olive plant as affected by boron concentration in nutrient solution. *Journal of Plant Nutrition* 28, 309–321.

Chatzistathis, Th., Therios, I., Patakas, A. and Gianakoula, A. (2006) The influence of manganese nutrition on the photosynthetic rate, transpiration, stomatal conductance and chlorophyll fluorescence of two olive cultivars. In: *Proceedings of the 2nd International Seminar Olivebioteq*, 5–10 November 2006, Marsala-Mazara del Vallo, Italy, 1, pp. 485–488.

Cherif, S., Rahal, N., Haouala, M., Hizaoui, B., Dargauth, F., Gueddliche, M., Kallel, Z., Balansard, G. and Boukef, K. (1996) A clinical trial of a titrated *Olea* extract in the treatment of essential arterial hypertension. *Journal de Pharmacie Belgique* 51, 69–71.

Christakis, G., Fordyce, M.K. and Kurtz, C.S. (1982) *The Biological and Medical Aspects of Olive Oil*. International Olive Oil Council, Madrid.

Chuine, I., Cour, P. and Rousseau, D.D. (1998) Fitting models predicting dates of flowering of temperate zone trees using simulated annealing. *Plant Cell Environment* 21, 455–466.

Cimato, A., Cantini, C. and Sillari, B. (1990) A method of pruning for the recovery of olive productivity. *Acta Horticulturae* 286, 251–254.

Cirulli, M. (1981) Attuali cognizioni sulla Verticilliosi dell'olivo. *Informatore Fitopatologico* 31, 101–105 [in Italian].

Clodoveo, M.L., Delcuratolo, D., Gomes, T. and Colleli, G. (2007) Effect of different temperatures and storage atmospheres on 'Coratina' olive oil quality. *Food Chemistry* 102, 571–576.

Close, T.J. (1997) Dehydrins: a commonality in the response of plants to dehydration and low temperature. *Physiologia Plantarum* 100, 291–296.

Cobb, D., Feber, R., Hopkins, A., Stockdale, L., O'Riordan, T., Clements, B., Firbank, L., Goulding, K., Jarvi, S.S. and Macdonald, D. (1999) Integrating the environmental and economic consequences of converting to organic agriculture: evidence from a case study. *Land Use Policy* 16(4), 207–221.

Cohen, L.A. and Wydner, E.I. (1990) Do dietary monounsaturated fatty acids play a protective role in carcinogenesis and cardiovascular disease? *Medical Hypotheses* 31, 83–89.

Collier, T.R. and Van Steenway, K.R.A. (2003) Prospects for integrated control of olive fruit fly are promising in California. *California Agriculture* 57(1), 28–31.

Conacher, J. and Conacher, A. (1998) Organic farming and the environment, with particular reference to Australia: a review. *Biological Agriculture and Horticulture* 16(2), 145–171.

Conde, C., Silva, P., Agasse, A., Lemoine, R., Delrot, S., Tavares, R. and Gerós, H. (2007) Utilization and transport of mannitol in *Olea europaea* and implications for salt stress tolerance. *Plant and Cell Physiology* 48(1), 42–53.

Coni, E., Di Benedetto, R., Di Pasquale, M., Masella, R., Modesti, D., Mattei, R. and Carlini, E.A. (2000) Protective effect of oleuropein, an olive oil biophenol, on low density lipoprotein oxidizability in rabbits. *Lipids* 35, 45–54.

Connell, J.H. and Catlin, P.B. (1994) The olive tree and fruit. Root physiology and rootstock characteristics. In: *Olive Production Manual*, University of California, Division of Agriculture and Natural Resources, Berkeley, California, Publication 3353, pp. 43–50.

Connor, D.J. (2005) Adaption of olive (*Olea europaea* L.) to water-limited environments. *Australian Journal of Agricultural Research* 56, 1181–1189.

Connor, D.J. (2006) Towards optimal designs for hedgerow olive orchards. *Australian Journal of Agricultural Research* 57(10), 1067–1072.

Costagli, G., Gucci, R. and Rapoport, H.F. (2003) Growth and development of fruits of olive 'Frantoio' under irrigated and rainfed conditions. *Journal of Horticultural Science and Biotechnology* 78(1), 119–124.

Cramer, G.R., Lauchli, A. and Polito, V.S. (1985) Displacement of Ca^{2+} by Na^+ from the plasmalemma of root cells. *Plant Physiology* 79, 207–211.

Cresti, M., Ciampolini, F., Tattini, M. and Cimato, A. (1994) Effect of salinity on productivity and oil quality of olive (*Olea europaea* L.) plants. *Advances in Horticultural Science* 8, 211–214.

Cresti, M., Linskens, H.F., Mulcahy, D.L., Bush, S., Di Stilio, V., Xu, M.Y., Vignani, R. and Cimato, A. (1996) Preliminary communication about the identification of DNA in leaves and in olive oil of *Olea europaea*. *Advances in Horticultural Science* 10, 105–107.

Cuevas, J. and Polito, V.S. (2004) The role of staminate flowers in the breeding system of *Olea europaea* (Oleaceae): an andromonoecious, wind-pollinated taxon. *Annals of Botany* 93(5), 547–553.

Cuevas, J. and Rallo, L. (1990) Response to cross-pollination in olive trees with different levels of flowering. *Acta Horticulturae* 286, 179–182.

Cuevas, J., Diaz-Hermoso, A.J., Galián, D., Hucso, J.J., Pinillos, V., Pricto, M., Sola, D. and Polito, V.S. (2001) Response to cross pollination and choice of pollinators for the olive cultivars (*Olea europaea* L.) 'Manzanilla de Sevilla', 'Hojiblanca' and 'Picual'. *Olivae* 85, 26–32.

Cuevas, J., Pinney, K. and Polito, V.S. (1999) Flower differentiation pistil development and pistil abortion in olive *Olea europaea* L. 'Manzanillo'. *Acta Horticulturae* 474, 293–296.

Cuevas, J., Rallo, L. and Rapoport, H.F. (1994a) Crop load effects on floral quality in olive. *Scientia Horticulturae* 59, 123–130.

Cuevas, J., Rallo, L. and Rapoport, H.F. (1994b) Staining procedure for the observation of olive pollen tube behavior. *Acta Horticulturae* 356, 264–267.

Cuevas, J., Rallo, L. and Rapoport, H.F. (1995) Relationships among reproductive processes and fruitlets abscission in Arbequina olive. *Advances in Horticultural Science* 9, 92–96.

D'Andria, R., Morelli, G., Patumi, M. and Fontanazza, G. (2002) Irrigation regime affects yield and oil quality of olive trees. *Acta Horticulturae* 586, 273–276.

D'Angeli, S. and Altamura, M.M. (2007) Osmotin induces cold protection in olive trees by affecting programmed cell death and cytoskeleton organization. *Planta* 225, 1147–1163.

D'Angeli, S., Malho, R. and Altamura, M.M. (2003) Low temperature sensing in olive tree: calcium signaling and cold acclimation. *Plant Science* 165, 1303–1313.

Dag, A., Avidan, B., Birger, R. and Lavee, S. (2006) High-density olive orchards in Israel. In: *Proceedings of the 2nd International Seminar Olivebioteq*, 5–10 November 2006, Marsala-Mazara del Vallo, Italy, 2, 31–35.

Damtoft, S., Franzyk, H. and Jensen, S.R. (1993) Biosynthesis of secoiridoid glucosides in Oleaceae. *Phytochemistry* 34, 1291–1299.

Darral, N.M. and Wareing, P.F. (1981) The effect of nitrogen nutrition on cytokinin activity and free amino acids in *Betula pendula* Roth and *Acer pseudoplatanus* L. *Journal of Experimental Botany* 32, 369–379.

De la Puerta, R., Martinez Dominguez, M.E., Ruiz-Gutierrez, V., Flavill, J.A., Robin, J. and Hoult, S. (2001) Effects of virgin olive oil phenolics on scavenging of reactive nitrogen species and upon nitrergic neurotransmission. *Life Sciences* 69, 1213–1222.

de la Rosa, R., James, C.M. and Tobutt, K.R. (2002) Isolation and characterization of polymorphic microsatellites in olive (*Olea europaea* L.) and their transferability to other genera in the *Oleaceae*. *Molecular Ecology Notes* 2, 265–267.

de la Rosa, R., James, C.M. and Tobutt, K.R. (2004) Using microsatellites for paternity testing in olive progenies. *HortScience* 39(2), 351–354.

De Melo-Abreu, J.P., Barranco, D., Cordeiro, A.M., Tous, J., Rogado, B.M. and Villalobos, F.J. (2004) Modeling olive flowering date using chilling for dormancy release and thermal time. *Agricultural Forest Meteorology* 125, 117–127.

Delgado, A., Benlloch, M. and Fernández-Escobar, R. (1994) Mobilization of boron in olive trees during flowering and fruit development. *HortScience* 29, 616–618.

Demetriades, S.D., Gavalas, N.A. and Holevas, K.D. (1960a) Boron deficiency in olive orchards of the island of Lesbos. *Chronicles of the Benaki Institute of Phytopathology* 3, 123–134 [in Greek].

Demetriades, S.D., Holevas, K.D. and Gavalas, N.A. (1960b) A non-parasitic disease of olive probably caused by abnormal Ca/Mg ratio within the plant. *Chronicles of the Benaki Institute of Phytopathology* 3(N.S.), 130–138.

Demiral, M.A. (2005) Comparative response of two olive (*Olea europaea* L.) cultivars to salinity. *Turkish Journal of Agriculture and Forestry* 29, 267–274.

Denney, J.O. and Martin, G.C. (1994) Ethephon tissue penetration and harvest effectiveness in olive as a function of solution pH, application time and BA or NAA addition. *Journal of the American Society of Horticultural Science* 119, 1185–1192.

Denney, J.O., McEachern, G.R. and Griffiths, J.F. (1985) Modeling the thermal adaptability of the olive (*Olea europaea* L.) in Texas. *Agricultural and Forest Metereology* 35, 309–327.

Di Giacomo, G., Brandani, V. and Del Re, G. (1991) *Evaporation of olive mill vegetation waters. Desalination* 81, 249–259.

Di Giovacchino, L. and Preziuso, S. (2006) Utilization of olive mill by-products. In: *Proceedings of the 2nd International Seminar Olivebioteq*, 5–10 November 2006, Marsala-Mazara del Vallo, Italy, 1, pp. 379–389.

Di Giovacchino, L., Marsilio, V., Costantini, N. and Di Serio, G. (2005) Use of olive mill wastewater (OMW) as fertilizer of the agricultural soil: effects on crop production and soil characteristics. In: *Proceedings of the 3rd European Bioremediation Conference*, Chania, Greece, Paper P165.

Di Giovacchino, L., Solinas, M. and Miccoli, M. (1994) Effect of extraction systems on the quality of virgin olive oil. *Journal of the American Oil Chemists' Society* 71, 1189–1194.

Dias, A.B., Peca, J.O., Pinheiro, A.C., Costa, S., Almeida, A., Santos, L., de Souza, D.R. and Lopes, J. (1999) Effect of tree size and variety on olive harvesting with an impact shaker. *Acta Horticulturae* 474, 219–222.

Diaz, A., De la Rosa, R., Martin, A. and Rallo, P. (2006a) Development, characterization and inheritance of new microsatellites in olive (*Olea europaea* L.) and evaluation of their usefulness in cultivar identification and genetic relationship studies. *Tree Genetics & Genomes* 2, 165–175.

Diaz, A., Martín, A., Rallo, P., Barranco, D. and De La Rosa, R. (2006b) Self-incompatibility of 'Arbequina' and 'Picual' olive assessed by SSR markers. *Journal of the American Society of Horticultural Science* 131(2), 250–255.

Diaz-Espejo, A., Walcroft, A., Fernández, J.E., Hafidi, B., Palomo, M.J. and Girón, I.F. (2006) Modeling photosynthesis in olive leaves under drought conditions. *Tree Physiology* 26, 1445–1456.

Dichio, B., Xiloyannis, C., Angelopoulos, K., Nuzzo, V., Bufo, S.A. and Celano, G. (2003) Drought induced variations of water relations parameters in *Olea europaea*. *Plant and Soil* 257, 381–389.

Dichio, B., Xiloyannis, C., Celano, G. and Angelopoulos, K. (1994) Response of olive trees subjected to various levels of water stress. *Acta Horticulturae* 356, 211–214.

Dimassi, K., Therios, I. and Passalis, A. (1999) Genotypic effect on leaf mineral levels of 17 olive cultivars grown in Greece. *Acta Horticulturae* 141, 345–348.

Division of Agricultural Science, University of California (1975) *Home Pickling of Olives*. Division of Agricultural Science, University of California, Berkeley, California, Leaflet 2758, 12.

Dominguez, L.J. and Barbagallo, M. (2006) One hundred years of ... health! Mediterranean diet and longevity: role of olive oil. In: *Proceedings of the 2nd International Seminar Olivebioteq*, 5–10 November 2006, Marsala-Mazara del Vallo, Italy, Seminars and invited lectures, pp. 245–254.

Donaire, J.P., Belver, A., Rodriguez-Garcia, M.I. and Megias, L. (1984) Lipid biosynthesis, oxidative enzyme activities and cellular changes in growing olive fruit. *Rensta Espanola de Fisiologia* 40(2), 191–203.

Drossopoulos, J.B. and Niavis, C.A. (1988) Seasonal changes of the metabolites in the leaves, bark and xylem tissues of olive tree. II. Carbohydrates. *Annals of Botany* 62, 321–327.

Dry, P.R. and Loveys, B.R. (2000) Partial drying of the rootzone of grape. I. Transient changes in shoot growth and gas exchange. *Vitis* 39, 3–7.

Duran Grande, M. and Izquierdo Tamayo, A. (1964) Study on histological structure of the *Olea europaea* L. fruit. I. Cv. Zorzalena (O.E. Argentata). *Grasas Aceites* 15, 72–85.

Economopoulos, A.P. (1972) Sexual competitiveness of γ-ray sterilized males of *Dacus oleae*. Mating frequency of artificially reared and wild females. *Environmental Entomology* 1, 490–497.

Economopoulos, A.P., Giannakakis, A., Tzanakakis, M.E. and Voyadjoglou, A.V. (1971) Reproductive behavior and physiology of the olive fruit fly. 1. Anatomy of the adult rectum and odours emitted by adults. *Annals of the Entomological Society of America* 64, 1112–1116.

El-Hakim, A.M. and Kisk, S.A. (1988) Cultural methods for the control of olive pests. *Bulletin of the Faculty of Agriculture, Cairo University* 39(1), 345–351.

El-Khawas M.A. (2000) Integrated control of insect pests on olive trees in Egypt with emphasis on biological control. PhD thesis, Faculty of Science, Cairo University, Egypt, 247 pp.

El-Kifl, A.H., Abdel-Salam, A.L. and Rahhal, A.M.M. (1974) Biological studies on the olive leaf moth, *Palpita unionalis* Hb. (Lepidoptera: Pyralidae). *Bulletin de la Société d'Enterologie d'Egypte* 58, 337–344.

Elmore, C.L. (1994) Weed management in olives. In: *Olive Production Manual*. University of California, Division of Agriculture and Natural Resources, Berkeley, California, Publication 3353, 91–95.

Elmore, C.L., Cudney, D.W. and Donaldson, D.R. (2001) *UC IPM Pest Management Guidelines*. Olive UC ANR Publication 3452, University of California, Berkeley, California.

Epstein, E., Norlyn, J.D., Rush, D.W., Kingshury, R.W., Kelly, D.B., Cunningham, G.A. and Wrona, A.F. (1980) Saline culture of crops. A genetic approach. *Science* 210, 399–404.

Eriş, A. and Barut, E. (1991) Growth regulators used for decreasing the severity of alternate bearing in olive. *Olea* 21, 11.

European and Mediterranean Plant Protection Organization (EPPO) (2006) *Pathogen-tested Olive Trees and Rootstocks*, OEEP/EPPO Bulletin 36, Paris, pp. 77–83.

Evans, A.J. (1903) The palace of Knossos. Provisional Report of the Excavations for the year 1903, pp. 1–124.

Fabbri, A. (2006) Olive propagation: new challenges and scientific research. In: *Proceedings of the 2nd International Seminar Olivebioteq*, 5–10 November 2006, Marsala-Mazara del Vallo, Italy, Seminars and invited lectures, pp. 411–421.

Fabbri, A. and Ganino, T. (2002) Organic olive growing in Italy. *Advances in Horticultural Science* 16(3–4), 204–217.

Fabbri, A. Bartolini G., Lambardi, M. and Kailis, S.G. (2004) *Olive Propagation Manual*. CSIRO Publishing, Canberra.

Fabbri, A., Hormaza, J.I. and Polito, V.S. (1995) Random amplified polymorphic DNA analysis of olive (*Olea europaea* L.) cultivars. *Journal of the American Society of Horticultural Science* 120, 538–542.

Faci, J.M., Berenguer, M.J., Espada, J.L. and Garcia, S. (2002) Effect of variable irrigation supply in olive (*Olea europaea* cv. 'Arbequina') in Araga (Spain). I. Fruit and oil production. *Acta Horticulturae* 586, 342–344.

Fadzilla, N.M., Robert, P., Finch, R.P. and Burdon, R.H. (1997) Salinity, oxidative stress and antioxidant response in shoot cultures of rice. *Journal of Experimental Botany* 48, 325–351.

Faggioli, F. and Barba, M. (1995) An elongated virus isolated from olive *Olea europaea* L. *Acta Horticulturae* 386, 593–599.

Faggioli, F., Ferretti, L., Albanese, G., Sciarroni, R., Pasquini, G., Lumia, V. and Barba, M. (2005) Distribution of olive tree viruses in Italy as revealed by one step RT-PCR. *Journal of Plant Pathology* 87(1), 49–55.

FAOSTAT (2003) *FAO Primary Crops Statistical Database*. Food and Agriculture Organization, Rome.

Farinelli, D., Boco, M. and Tombesi, A. (2002) Intensity and growth period of the fruit components of olive varieties. *Acta Horticulturae* 568, 607–610.

Farinelli, D., Boco, M. and Tombesi, A. (2006) Results of four years of observations on self-sterility behaviour of several olive cultivars and significance of cross-pollination. *Proceedings of the 2nd International Seminar Olivebioteq*, 5–10 November 2006, Marsala-Mazara del Vallo, Italy, 1, pp. 275–282.

Fehri, B., Aiache, J.M., Memmi, A., Korbi, S., Yacoubi, M.T., Mrad, S. and Lamaison, J.L. (1994) Hypotension, hypoglycemia and hypouricemia recorded after repeated administration of aqueous leaf extract of *Olea europaea* L. *Journal de Pharmacie Belgique* 49, 101–108.

Felix, M.R.F. and Clara, M.I.E. (2002) Two necrovirus with properties of *Olive latent virus 1* and *Tobacco necrosis* virus from olive in Portugal. *Acta Horticu*lturae 586, 725–728.

Fereres, E. and Castel, J.R. (1981) *Drip Irrigation Management*. Division of Agricultural Sciences, University of California, Berkeley, California, Publication leaflet 21259.

Fereres, E., Pruit, W.D., Beutel, J.A., Henderson, D.W., Holzaptel, E., Schulbach, H. and Uriu, K. (1981) In: Fereres, E. (ed.) *Evapotranspiration and Irrigation Scheduling*. Drip Irrigation Management, Division of Agricultural Sciences, University of California, Berkeley, California, Publication leaflet 21259, pp. 8–13.

Ferguson, L. (2006) The table olive industry in California. In: *Proceedings of the 2nd International Seminar Olivebioteq*, 5–10 November 2006, Marsala-Mazara del Vallo, Italy, Seminars and invited lectures, pp. 199–204.

Ferguson, L., Krueger, W.H., Reyes, H. and Methney, P. (2002) Effect of mechanical pruning on California Black Ripe (*Olea europaea*) cv. 'Manzanillo' table olive yield. *Acta Horticulturae* 586, 281–284.

Ferguson, L., Sibbett, G.S. and Martin, G.C. (1994) *Olive Production Manual*. University of California, Division of Agriculture and Natural Resources, Oakland, California, Publication 3353, 156 pp.

Fernandes Serrano, J.M., Serrano, M. and Amaral, E. (2002) Effect of different hormone treatments on rooting of *Olea europaea* cv. 'Galega vulgar' cuttings. *Acta Horticulturae* 586, 875–877.

Fernández, J.E. (2006) Irrigation management in olive. In: *Proceedings of the 2nd International Seminar Olivebioteq*, 5–10 November 2006, Marsala-Mazara del Vallo, Italy, Seminars and invited lectures, pp. 295–305.

Fernández, J.E. and Moreno, F. (1999) Water use by the olive tree. *Journal of Crop Production* 2, 101–162.

Fernández, J.E., Diaz-Espejo, A., Infante, J.M., Durán, P., Palomo, M.J., Chamorro, V., Girón, I.F. and Villagarcia, L. (2006) Water relations and gas exchange in olive trees under regulated deficit irrigation and partial rootzone drying. *Plant and Soil* 284, 271–287.

Fernández, J.E., Moreno, F. and Martin-Aranda, J. (1990) Study of root dynamics of olive trees under drip irrigation and dry farming. *Acta Horticulturae* 286, 263–266.

Fernández, J.E., Moreno, F., Girón, I.F. and Blazquez, O.M. (1997) Stomatal control of water use in olive tree leaves. *Plant Soil* 190, 179–192.

Fernández, J.E., Palomo, M.J., Diaz-Espejo, A. and Giron, I.F. (1999) Calibrating the compensation heat-pulse technique for measuring sap flow in olive. *Acta Horticulturae* 474, 455–457.

Fernández, J.E., Palomo, M.J., Diaz-Espejo, A., and Girón, I.F. (2003) Influence of partial soil wetting on water relation parameters of the olive tree. *Agronomie* 23, 545–552.

Fernández, J.E., Palomo, M.J., Diaz-Espejo, A., Clothier, B.E., Green, S.R., Girón, I.F. and Moreno, F. (2001) Heat-pulse measurements of sap flow in olives for automating irrigation: tests, root flow and diagnostics of water stress. *Agricultural Water Management* 51, 99–123.

Fernández-Bolanos, J., Rodriguez, R., Guillen, R., Jimenez, A. and Heredia, A. (1995) Activity of cell wall-associated enzymes in ripening olive fruit. *Physiologia Plantarum* 93, 651–658.

Fernández-Escobar, R. and Marín, L. (1999) Nitrogen fertilization in olive orchards. *Acta Horticulturae* 474, 333–335.

Fernández-Escobar, R., Barranco, D. and Benlloch, M. (1993) Overcoming iron chlorosis in olive and peach trees using a low-pressure trunk-injection method. *HortScience* 28, 192–194.

Fernández-Escobar, R., Benlloch, M., Herrera, E. and Garcia-Novelo, J.M. (2004a) Effect of traditional and slow-release N fertilizers on growth of olive nursery plants and N losses by leaching. *Scientia Horticulturae* 101, 39–49.

Fernández-Escobar, R., Benlloch, M., Navarro, C. and Martin, G.C. (1992) The time of floral induction in the olive. *Journal of the American Society of Horticultural Science* 117, 304–307.

Fernández-Escobar, R., Garcia Barragán, T. and Benlloch, M. (1994) Estado nutritive de las plantaciones de olivar en la provincial de Granada. *ITEA* 90, 39–49.

Fernández-Escobar, R., Garcia-Novelo, J.M., Sánchez-Zamora, M.A., Uceda, M., Beltrán, G. and Aguilera, M.P. (2002) Efeto del abonado nitrogenado en la producciòn y la calidad del aceite de oliva. Direcciòn General de Investigaciòn y Formaciòn Agraria y Resquera (ed.) *Jounadas de Investigaciòn y Transferencia de Technologia al Sector Oleicola*, Cordoba (Spain), pp. 299–302.

Fernández-Escobar, R., Moreno, R. and Garcia-Creus, M. (1999) Seasonal changes of mineral nutrients in olive leaves during the alternate-bearing cycle. *Scientia Horticulturae* 82, 25–45.

Fernández-Escobar, R., Moreno, R. and Sanchez-Zamora, M.A. (2004b) Nitrogen dynamics in the olive bearing shoot. *HortScience* 39(6), 1406–1411.

Ferrante, A., Hunter, D.A. and Reid, M.S. (2004) Towards a molecular strategy for improving harvesting of olives (*Olea europaea* L.). *Postharvest Biology and Technology* 31, 111–117.

Ferrara, E., Lorusso, G. and Lampanelli, F. (1991) A study of floral biology and the technological features of seven olive cultivars of different origins. *Acta Horticulturae* 474, 279–283.

Ferro-Luzzi, A. and Ghiselli, A. (1993) Protective aspects of the Mediterranean diet. In: Zappia, V., Salvatore, M. and Della Ragione, F. (eds) *Advances in Nutrition and Cancer*. Plenum Press, New York, pp. 137–144.

Ferro-Luzzi, A. and Sette, S. (1989) The Mediterranean diet: an attempt to define its present and past composition. *European Journal of Clinical Nutrition* 43(2), 13–29.

Fiorino, P. and Mancuso, S. (2000) Differential thermal analysis, supercooling and cell viability in organs of *Olea europaea* at subzero temperatures. *Advances in Horticultural Science* 14, 23–27.

Flora, L.L. and Madore, M.A. (1993) Stachyose and mannitol transport in olive (*Olea europaea* L.). *Planta* 189, 484–490.

Foda, S.M.A. (1973) Studies on *Margaronia* (*Glyphodes*) *unionalis* and its control. MSc thesis, Faculty of Agriculture, Ain Shams University, Egypt.

Fodale, A.S., Mulè, R., Briccoli Bati, C. and Tagliavini, M. (2006) Tolerance to brackish water of *in vitro* selected olive seedlings. In: *Proceedings of the 2nd International Seminar Olivebioteq*, 5–10 November 2006, Marsala-Mazara del Vallo, Italy, 1, 393–396.

Fontanazza, G. and Cappelletti, M. (1993) Evolución sistemas de cultivo del olivo: de los olivares intensivos mecanizados a las plantaciones densas. *Olivae* 48(10), 28–36.

Fontanazza, G., Bartolozzi, F. and Vergati, G. (1998) Olivo 'FS–17'. *Rivista di Frutticoltura e di Ortofloricoltura* 60(7/8), 61.

Fortes, C., Forastiere, F., Anatra, F. and Schmid, G. (1995) Re: consumption of olive oil and specific food groups in relation to breast cancer risk in Greece. *Journal of the National Cancer Institute* 87, 1020–1021.

Fouad, M.M., Fayek, M.A., Selin, H.H. and El-Sayed, M.E. (1990) Rooting of eight olive cultivars under mist. *Acta Horticulturae* 286, 57–60.

Foyer, C. and Galtier, N. (1996) Source–sink interaction and communication in leaves. In: Zamski, E. and Schaffer, A.A. (eds) *Photoassimilate Distribution in Plants and Crops: Source–Sink Relationships*. Marcel Dekker Inc., New York, 905 pp.

Frakulli, F. and Voyiatzis, D.G. (1999) The growth retardants paclobutrazol and triapenthenol affect water relations and increase drought resistance of olive (*Olea europaea* L.). *Acta Horticulturae* 474, 427–429.

Frenguelli, G., Bricchi, E., Romano, B., Mincigrucci, G. and Spieksma, F.T.M. (1989) A predictive study of the beginning of the pollen season for *Gramineae* and *Olea europaea* L. *Aerobiologia* 5, 64–70.

Furneri, P.R., Marino, A., Saija, A., Uccella, N. and Bisignano, G. (2002) *In vitro* antimycoplasmal activity of oleuropein. *International Journal of Antimicrobial Agents* 20, 293–296.

Gaddi, A.V., Bove, M., Cicero, A., Nascetti, S. and Covas, M.I. (2006) Role of olive oil in the health: results from the Eurolive study. In: *Proceedings of the 2nd International Seminar Olivebioteq*, 5–10 November 2006, Marsala-Mazara del Vallo, Italy, Special seminars and invited lectures, pp. 273–280.

Galán, C., Garcia-Mozo, H., Cariñanos, P., Alcázar, P. and Dominguez-Vilches, E. (2001) The role of temperature in the onset of the *Olea europaea* L. pollen season in southwestern Spain. *International Journal of Biometeorology* 45, 8–12.

Galán, C., García-Mozo, H., Vásquez, L., Ruiz, L., Guardia, C.D. and Trigo, M.M. (2005) Heat requirement for the onset of the *Olea europaea* L. pollen season in several sites in Andalusia and the effect of the expected future climate change. *International Journal of Biometeorology* 49, 184–188.

Galindo, P., de Luna, E., Castro, J., Polo, M.J. and Navarro, C. (2006) Herbicides used in olive orchards: behavior in superficial runoff flow. In: *Proceedings of the 2nd International Seminar Olivebioteq*, 5–10 November 2006, Marsala-Mazara del Vallo, Italy, 2, pp. 69–75.

Galli, C. and Visioli, F. (1999) Antioxidant and other activities of phenolics in olives/olive oil, typical components of the Mediterranean diet. *Lipids* 34, 523–526.

Ganino, T., Bartolini, G. and Fabbri, A. (2006) The classification of olive germplasm – a review. *Journal of Horticultural Science and Biotechnology* 81(3), 319–334.

Garcia, J.L., Avidan, N., Troncoso, A., Sarmiento, R. and Lavee, S. (2000) Possible juvenile-related proteins in olive tree tissues. *Scientia Horticulturae* 85(4), 271–284.

Garcia, J.L., Liñán, J., Sarmiento, R. and Troncoso, A. (1999) Effect of different N forms and concentrations on olive seedlings growth. *Acta Horticulturae* 474, 323–327.

Garcia, J.L., Troncoso, J., Sarmiento, R. and Troncoso, A. (2002) Influence of carbon and concentration on the *in vitro* development of olive zygotic embryos and explants raised from them. *Plant Cell, Tissue and Organ Culture* 69(1), 95–100.

Garcia, J.M., Seller, S. and Perez-Camino, M.C. (1996) Influence of fruit ripening on olive oil quality. *Journal of Agriculture and Food Chemistry* 44, 3516–3520.

Garcia-Fèrriz, L., Ghorbel, R., Ybarra, M., Mari, A., Belay, A. and Trujillo, I. (2000) Micropropagation from adult olive trees. *Acta Horticulturae* 586, 879–882.

Garcia-Ortiz, A., Fernandez, A., Pastor, M. and Humanes, J. (1997) In: Barranco, D., Fernandez-Escobar, R. and Rallo, L. (eds) *El Cultivo del Olivo*. 2nd edn, Mundi-Prensa, Madrid, pp. 307–343.

Garrido-Fernández, A., Lòpez Lòpez, A. and Garia, P. (2006) Table olives in Spain. In: *Proceedings of the 2nd International Seminar Olivebioteq*, 5–10 November 2006, Marsala-Mazara del Vallo, Italy, Seminars and invited lectures, pp. 205–212.

Gaspar, T., Kevers, C. and Hausman, J.F. (1997) Indissociable chief factors in the inductive phase of adventitious rooting. In: Altman, A and Waisel, Y. (eds) *Biology of Root Formation and Development*. Plenum Press, New York, pp. 55–63.

Gaspar, T., Kevers, C., Hausman, J.F. and Ripetti, V. (1994) Peroxidase activity and endogenous free auxin during adventitious root formation. In: Lumsden, P.J., Nicholas, J.R. and Davis, W.J. (eds) *Physiology, Growth and Development of Plants in Culture*. Kluwer Academic Publishers, Dordrecht, Netherlands, p. 430.

Gavalas, N.A. (1978) *Inorganic Nutrition and Fertilization of Olive*. Benaki Phytopathological Institute, Kifissia, Athens.

Gerasopoulos, D., Metzidakis, I. and Naoufel, E. (1999) Ethephon sprays affect harvest parameters of 'Mastoides' olives. *Acta Horticulturae* 474, 223–226.

Giametta, G. (1988) Olive mechanical pruning tests. In: *2nd International Meeting on Mediterranean Tree Crops*, 2–4 November 1988, Chania, Crete, pp. 207–215.

Giametta, G. (2001) Innovazione nella meccanizzazione della raccolta delle olive. *Olivo e Olio* 10, 35–38.

Giametta, G. and Zimbalatti, G. (1993) Possibilities of mechanical pruning in traditional olive groves. In: *2nd International Symposium on Olive Growing*, 6–10 September 1993, Jerusalem, Israel, pp. 311–314.

Giametta, G. and Zimbalatti, G. (1997) Mechanical pruning in new olive groves. *Journal of Agricultural Engineering and Research* 68, 15–20.

Giorgelli, F., Lorenzini, G., Minnocci, A., Panicucci, A. and Vitagliano, C. (1994) Effects of long-term SO_2 pollution on olive trees. *Acta Horticulturae* 356, 185–188.

Giorgio, V., Gallotta, A., Camposeo, S., Roncasaglia, R. and Dradi, G. (2006) Advances in improving micropropagation of olive (*Olea europaea* var. *sativa* L.): preliminary results on 18 olive varieties belonging to Italian and Spanish germplasm. In: *Proceedings of the 2nd International Seminar Olivebioteq*, 5–10 November 2006, Marsala-Mazara del Vallo, Italy, 1, pp. 441–444.

Giorio, P. and d'Andria, R. (2002) Sap flow estimated by compensation heat-pulse velocity technique in olive trees under two irrigation regimes in Southern Italy. *Acta Horticulturae* 586, 401–403.

Gioulekas, D., Chatzigeorgiou, G., Lykogiannis, S., Papakosta, D., Mpalafoutis, C. and Spieksma, F.T.M. (1991) *Olea europaea* 3-year pollen record in the area of Thessaloniki, Greece, and its sensitizing significance. *Aerobiologia* 7, 57–61.

Godwin, I.D., Aitken, E.A.B. and Smith, L.W. (1997) Application of Inter Simple Sequence Repeat (ISSR) markers to plant genetics. *Electrophoresis* 18, 1524–1528.

Goldhamer, D., Dunai, J. and Ferguson, L. (1994) Irrigation requirements of olive trees and responses to sustained deficit irrigation. *Acta Horticulturae* 356, 172–176.

Goldhamer, D.A. and Fereres, E. (2001) Irrigation scheduling protocols using continuously recorded trunk diameter measurements. *Irrigation Science* 20(3), 115–125.

Gomez-Rico, A., Desamparados Salvator, M., La Greca, M. and Fregapane, G. (2006) Phenolic and volatile compounds of extra virgin olive oil (*Olea europaea* L. cv. 'Cornicabra') with regard to fruit ripening and irrigation management. *Journal of Agriculture and Food Chemistry* 54, 7130–7136.

Goor, A. (1962) The place of the olive in the Holy Land and its history through the ages. *Economic Botany* 20, 223.

Graniti, A. (1962) Osservazioni su *Spilocaea oleagina* (Cast.) Hugh.: sulla localizzazione del micelio nelle foglie di olivo. *Phytophathologia Mediterranea* 1, 157–165.

Grattan, S.R., Berenguer, M.J., Connell, J.H., Polito, V.S. and Vossen, P.M. (2006) Olive oil production as influenced by different quantities of applied water. *Agricultural Water Management* 85, 133–140.

Greenway, H. and Munns, R. (1980) Mechanisms of salt stress in non-halophytes. *Annual Review of Plant Physiology* 31, 149–190.

Grieco, F., Alkowni, R., Saponari, M., Savino, V. and Martelli, G.P. (2000) Molecular detection of olive viruses. *Bulletin EPPO/OEPP Bulletin* 29, 127–133.

Griggs, W.H., Hartmann, H.T., Bradley, M.V., Iwakiri, B.T. and Whisler, J.E. (1975) Olive pollination in California. *California Agriculture Experimental Station Bulletin* 869, 49.

Gucci, R. and Cantini, C. (2004) *Pruning and Training Systems for Modern Olive Growing.* CSIRO Publishing, Clayton South, Victoria, Australia, pp. 35, 144.

Gucci, R. and Tattini, M. (1997) Salinity tolerance in olive. *Horticultural Review* 21, 177–214.

Gucci, R., Lombardini, L. and Tattini, M. (1997a) Analysis of leaf water relations in leaves of two olive (*Olea europaea*) cultivars differing in tolerance to salinity. *Tree Physiology* 17(1), 13–21.

Gucci, R., Massai, R., Casano, S. and Costagli, G. (1997b) The effect of leaf age on CO_2 assimilation and stomatal conductance of field-grown olive trees. *Acta Horticulturae* 474, 289–292.

Guerin, J., Mekurra, G., Collins, G., Jones, G., Burr, M., Wirtensohn, M. and Sedgley, M. (2002) Olive cultivar improvement through selection and biotechnology. *Advances in Horticultural Science* 16(3–4), 198–203.

Gutiérrez, F., Perdiguero, S., Gutiérrez, R. and Olias, J.M. (1992) Evaluation of the bitter taste in virgin olive oil. *Journal of the American Oil Chemists' Society* 69, 394–395.

Guy, C.L., Huber, J.L. and Huber, S.C. (1992) Sucrose phosphate synthase and sucrose accumulation at low temperature. *Plant Physiology* 100, 502–508.

Haber, B. (1997) The Mediterranean diet: a view from history. *American Journal of Clinical Nutrition* 66(4), 1053S–1057S.

Hackett, W.P. 1985. Juvenility, maturation and rejuvenation in woody plants. *Horticultural Reviews* 7, 109–156.

Hackett, W.P. and Hartmann, H.T. (1963) Morphological development of olive as related to low temperature requirement for inflorescence formation. *Botanical Gazette* 124, 383–387.

Hackett, W.P. and Hartmann, H.T. (1964) Inflorescence formation in olive as influenced by low temperature, photoperiod and leaf area. *Botanical Gazette* 125, 65–72.

Hackett, W.P. and Hartmann, H.T. (1967) The influence of temperature on floral initiation in the olive. *Physiologia Plantarum* 20, 430–436.

Hagidimitriou, M. and Pontikis, C.A. (2005) Seasonal changes in CO_2 assimilation in leaves of five major Greek olive cultivars. *Scientia Horticulturae* 104(1), 11–24.

Hagidimitriou, M., Katsiotis, A., Menexes, G., Pontikis, C. and Loukas, M. (2005) Genetic diversity of major Greek olive cultivars using molecular (AFLPs and RAPDs) markers and morphological traits. *Journal of the American Society of Horticultural Science* 130(2), 211–217.

Hagin, J., Olsen, S.R. and Shaviv, A. (1990) Review of interaction of ammonium nitrate and potassium nutrition of crops. *Journal of Plant Nutrition* 13, 1211–1226.

Hanbury, D. (1954) On the febrifuge properties of the olive (*Olea europaea* L.). *Pharmaceutical Journal of Provincial Transactions* 1854, 353–354.

Hangeveld, H.G. (2000) *Projections for Canada's Climate Future*. CCD 00–01, Special Edition (http://www.MSc-smc.ec.gc.ca/apac/climate/ccsci-e.cfm).

Haniotakis, G.E., Kozyrakis, K. and Bonatsos, C. (1986) Control of the olive fruit fly, *Dacus oleae* Gmel. (Diptera: Tephritidae) by mass trapping: pilot-scale feasibility study. *Journal of Applied Entomology* 101, 343–352.

Hare, D.D. and Cress, W.A. (1997) Metabolic implication of stress-induced proline accumulation in plants. *Plant Growth Regulation* 21, 79–102.

Hare, P.D., Cress, W.A. and van Staden, J. (1997) The involvement of cytokinins in plant responses to environmental stress. *Plant Growth Regulation* 23, 79–103.

Hare, P.D., Cress, W.A. and van Staden, J. (1998) Dissecting the roles of osmolyte accumulation during stress. *Plant Cell and Environment* 21, 535–553.

Hare, P.D., Cress, W.A. and Van Staden, J. (1999) Proline synthesis and degradation: a model system for elucidating stress-related signal transduction. *Journal of Experimental Botany* 50(333), 413–434.

Hartmann, H., Schnathorst, W.C. and Whisler, W.C. (1971) 'Oblonga', a clonal olive rootstock resistant to *Verticillium* wilts. *California Agriculture* 25, 12–25.

Hartmann, H.T. (1949) Growth of the olive fruit. *Proceedings of the American Society of Horticultural Science* 54, 86–49.

Hartmann, H.T. (1950) The effect of girdling on flower type, fruit set and yields in the olive. *Proceedings of the American Society of Horticultural Science* 56, 217–226.

Hartmann, H.T. (1953) Effect of winter chilling on fruit fullness and vegetative growth in the olive. *Proceedings of the American Society of Horticultural Science* 62, 184–186.

Hartmann, H.T. and Panetsos, C. (1962) Effect of soil moisture deficiency during floral development on fruitfulness in olives. *Proceedings of the American Society of Horticultural Science* 78, 209–217.

Hartmann, H.T. and Porlingis, I. (1957) Effect of different amounts of winter chilling on fruitfulness of several olive varieties. *Botanical Gazette* 119, 102–104.

Hartmann, H.T. and Whisler, J.E. (1970) Some rootstock and interstock influences in the olive (*Olea europaea* L.) cv. 'Sevillano'. *Journal of the American Society of Horticultural Science* 95, 562–565.

Hartmann, H.T. and Whisler, J.E. (1975) Flower production in olive as influenced by various chilling temperature regimes. *Journal of the American Society of Horticultural Science* 100, 670–674.

Hartmann, H.T., Fadl, M.S. and Hackett, W.P. (1967) Initiation of flowering and changes in endogenous inhibitors and promoters in olive buds as a result of chilling. *Physiologia Plantarum* 20, 746–759.

Hartmann, H.T., Kester, D.E., Geneve, R.L. and Davis Jr, F.T. (2001) *Plant Propagation. Principles and Practices.* 7th edn., Culinary and Hospitality Industry Publications Services, Weimar, Texas, 896 pp.

Hartmann, H., Schnathorst, W.C. and Whirler, W.C. (1971) 'Oblonga', a clonal olive rootstock resistant to *Verticillium* wilts. *California Agriculture* 25, 12–25.

Hartmann, H.T., Tombesi, A. and Whisler, J. (1970) Promotion of ethylene evolution and fruit abscission in the olive by 2-chloroethylphosphonic acid and cycloheximide. *Journal of the American Society of Horticultural Science* 95, 635–640.

Hartmann, H.Y. (1951) Time of floral differentiation of the olive in California. *Botanical Gazette* 112, 323–327.

Hasegawa, P.M., Bressan, S.A., Zhu, J.K. and Bohnert, H.J. (2000) Plant cellular and molecular responses to high salinity. *Annual Review of Plant Physiology/Plant and Molecular Biology* 51, 463–499.

Hassan, M.M. and Seif, S.A. (1990) Response of seven olive cultivars to water logging. *Gartenbauwissenschaft* 55, 223.

Hassani, D., Buonaurio, R. and Tombesi, A. (2003) Response of some olive cultivars, hybrid and open-pollinated seedlings to *Pseudomonas savastanoi* pv. *pavastanoi*. In: Iacobellis, N.S. et al. (eds) *Pseudomomas* syringae *and Related Pathogens*. Kluwer Academic Publishers, Dordrecht, Netherlands, pp. 489–494.

Hatzopoulos, P., Banilas, G., Giannoulia, K., Gazis, F., Nikoloudakis, N., Milioni, D. and Haralampidis, K. (2002) Breeding, molecular markers and molecular biology of the olive tree. *European Journal of Lipid Science and Technology* 104, 574–586.

Herenguer, M.J., Vossen, P.M., Grattan, S.R., Connell, J.H. and Polito, V.S. (2006) Tree irrigation levels for optimum chemical and sensory properties of olive oil. *HortScience* 41, 427–432.

Hetherlington, A.M. and Woodward, F.I. (2003) The role of stomata in sensing and driving environmental change. *Nature* 424, 901–908.

Hilton, J.L., Scharen, A.L., St. John, J.B., Moreland, D.E. and Norris, K.H. (1969) Modes of action of pyridazinone herbicides. *Weed Science* 17, 541–547.

Hirayama, T., Ohto, C., Mizoguchi, T. and Shinozaki, K. (1995) A gene encoding a phosphatidylinositol-specific phospholipase C is induced by dehydration and salt stress in *Arabidopsis thaliana*. *Proceedings of the Natural Academy of Sciences of the United States of America* 92, 3903–3907.

Ho, C.T., Lee, C.Y. and Huang, M.T. (eds) (1992) *Phenolic Compounds in Foods and their Effects on Health. Volume 1: Analysis, Occurrence and Chemistry.* American Chemistry Society, Washington, DC.

Hood, S. (1971) *The Minoans. Crete in the Bronze Age.* Thames and Hudson, London.

Howe, G.R., Hirohata T., Hislop, T.G., Iscivich, J.M., Yuan, J-M., Katsouyanni, K., Lubin, F., Marubini, E., Modan, B., Rohan, T., Toniolo, P. and Shunzhang, Y. (1990) Dietary factors and risk of breast cancer: combined analysis of 12 case-control studies. *Journal of the National Cancer Institute* 82, 561–569.

Iacobellis, N.S. (2001) Olive knot. In: Maloy, O.C. and Murray, T.D. (eds) *Encyclopedia of Plant Pathology*. Vol. 2, John Wiley and Sons, New York, pp. 713–715.

Iannotta, N. and Perri, E. (2006) L'esperienza della Spagna nell'olivicoltura superintensiva. *L'Informatore Agrario* 1, 59–63.

Iannotta, N., Noce, M.E., Scalercio, S. and Vizzarri V. (2006) Behavior of olive cultivars to catch the knot disease caused by *Pseudomonas savastanoi* pv. *savastanoi* (Smith). In: *XIII Congresso Nazionale SIPaV*, Foggia, Italy, 12–15 September 2006.

Instanbouli, A. and Neville, P. (1979) Étude de la 'dormance' des semences d'olivier (*Olea europaea* L.). III. Influence des enveloppes sur la germination. *Annales de Sciences Naturelles Botaniques Paris*, 13 Sér. 1, 151–165.

Intrieri, M.C., Muleo, R. and Buiatt,i M. (2007) Chloroplast DNA polymorphisms as molecular markers to identify cultivars of *Olea europaea* L. *Journal of Horticultural Science and Biotechnology* 82(1), 109–113.

IPCC (2001) *Climate Change 2001: Impacts, Adaptation and Vulnerability*. Cambridge University Press, Cambridge, UK.

Israilides, C.J., Vlyssides, A.G., Mourafeti, V.N. and Karvouni, G. (1997) Olive oil wastewater treatment with the use of an electrolysis system. *Bioresource Technology* 61, 163–170.

Jackson, J.E. (1980) Light interception and utilization by orchard systems. *Horticultural Reviews* 2, 208–267.

Jaglo-Ottosen, K.R., Gilmour, G.S.J., Zarka, D.G., Schabenberger, O. and Thomashow, M.F. (1998) Arabidopsis CBF1 overexpression induces Cor genes and enhances freezing tolerance. *Science* 280, 104–106.

Janzen, H.H. and Chang, C. (1987) Cation nutrition of barley as influenced by soil solution composition in a saline soil. *Canadian Journal of Soil Science* 67, 619–629.

Jiménez, A., Rodriguez, R., Fernández-Caro, I., Guillèn, R., Fernández-Bolanos, J. and Heredia, A. (2001) Olive fruit cell wall: degradation of cellulosic and hemicellulosic polysaccharides during ripening. *Journal of Agriculture and Food Chemistry* 49, 2008–2013.

Jones, H. (2004) What is water use efficiency? In: Bacon, M.A. (ed.) *Water Use Efficiency in Plant Biology*. Blackwell Publishing/CRC Press, Boca Raton, Florida, 27–41 pp.

Jordao, P.V., Marcelo, M.E. and Centeno, M.S.L. (1999) Effect of cultivar on leaf-mineral composition of olive tree. *Acta Horticulturae* 474, 349–352.

Juven, B. and Henis, Y. (1970) Studies on antimicrobial activity of olive phenolic compounds. *Journal of Applied Bacteriology* 33, 721–732.

Juven, B., Henis, Y. and Jacoby, B. (1972) Studies on the mechanism of the antimicrobial action of oleuropein. *Journal of Applied Bacteriology* 35, 559–567.

Juven, B., Samish, Z. and Henis, Y. (1968) Identification of oleuropein as a natural inhibitor of lactic acid fermentation. *Israeli Journal of Agricultural Research* 18, 137–138.

Kailis, S.G. and Considine, J.A. (2002) The olive *Olea europaea* L. in Australia: 2000 onwards. *Advances in Horticultural Science* 16(3–4), 299–306.

Kapdan, I.K. and Kargi, F. (2006) Bio-hydrogen production from waste materials. *Enzyme and Microbial Technology* 38, 569–582.

Karabourniotis, G., Kotsabassidis, D. and Manetas, Y. (1995) Trichome density and its protective potential against ultraviolet-B radiation damage to leaf development. *Canadian Journal of Botany* 73, 376–383.

Karabourniotis, G., Papadopoulos, K., Papamarkou, M. and Manetas, Y. (1992) Ultraviolet-B radiation absorbing capacity of leaf hairs. *Physiologia Plantarum* 86, 414–418.

Katsoyannos, P. (1992) *Olive Pests and Their Control in the Near East*. FAO Plant Production and Protection Paper, 179 pp.

Kavallieratos, N.G., Athanassiou, C.G., Balotis, G.N., Tatsi, G.T. and Mazomenos, B.E. (2005) Factors affecting male *Prays olaeae* (Lepidoptera: Yponomeutidae) captures in pheromone-baited traps in olive orchards. *Journal of Economic Entomology* 98(5), 1499–1505.

Kavi Kishor, P.B.K., Hong, Z., Miao, G.H., Hu, C.A.A. and Verma, D.P.S. (1995) Overexpression of Δ1-pyrroline-5-carboxylate synthetase increases proline production and confers osmotolerance in transgenic plants. *Plant Physiology* 108, 1387–1394.

Keys, A. (1995) Mediterranean diet and public health: personal reflections. *American Journal of Clinical Nutrition* 61(6), 1321S–1323S.

Khabou, W and Trigui, A. (1999) Optimization of hardwood-cutting as a method of olive tree multiplication. *Acta Horticulturae* 474, 55–57.

Khayyal, M.T., El-Ghazaly, M.A., Abdallah, D.M., Nassar, N.N., Okpanyi, S.N. and Kreuter, M.H. (2002) Blood pressure lowering effect of an olive leaf extract (*Olea europaea*) in L-NAME induced hypertension in rats. *Arzneimittelforschung* 52, 797–802.

Kirschbaum, M.U.F. (2004) Direct and indirect climate-change effects on photosynthesis and transpiration. *Plant Biology* 6, 242–253.

Klein, I., Ben-Tal, Y. and David, I. (1992) *Olive Irrigation with Saline Water*. Volcani Center Report, Bet-Dagan, Israel [in Hebrew].

Klein, I., Ben-Tal, Y., Lavee, S., De Malach, Y. and David, I. (1994) Saline irrigation of cv. 'Manzanillo' and 'Uovo di Piccione' trees. *Acta Horticulturae* 356, 176–180.

Klein, I., Epstein, E., Lavee, S. and Ben-Tal, Y. (1978) Environmental factors affecting ethephon in olive (*Olea europaea* L.). *Scientia Horticulturae* 9, 21–30.

Kovanci, B. and Kumral, N.A. (2004) Insect pests in olive groves of Bursa (Turkey). In: *5th International Symposium on Olive Growing*, 27 September–2 October 2004, Izmir, Turkey.

Krueger, W.H., Heath, Z. and Deleonardis, D. (2004) *Patch Budding: a Convenient Method for Top-working. Olives.* University of California, Berkeley, California, Publication 8115.

Kyritsakis, A. (1998) *Olive Oil*. 2nd edn, Food and Nutrition Press Inc., Trumbull, Connecticut.

Kyritsakis, A. (2007) *Olive Oil, Conventional and Biological. Table Olives and Olive Paste.* P. Sindika, Thessaloniki, Greece, pp. 131–178 [in Greek].

La Porta, N., Zacchini, M., Bartolini, S., Viti, R. and Roselli, G. (1994) The frost hardiness of some clones of olive cv. 'Leccino'. *Journal of Horticultural Science* 69(3), 433–435.

La Vecchia, C., Negri, E., Franceschi, S., Decarli, A., Giacosa, A. and Lipworth, L. (1995) Olive oil, other dietary fats, and the risk of breast cancer. *Cancer Causes and Control* 6, 545–550.

Lagarda, A. and Martin, G.C. (1983) 'Manzanillo' olive seed dormancy as influenced by exogenous hormone application and endogenous abscisic acid concentration. *HortScience* 18, 869–871.

Lambardi, M. and Rugini, E. (2003) Micropropagation of olive (*Olea europaea* L.). In: Jain, S.M. and Ishii, K. (eds) *Micropropagation of Woody Trees and Fruits.* Kluwer Academic Publishers, Netherlands, pp. 621–646.

Lambardi, M., Benelli, C., De Carlo, A., Fabbri, A., Grassi, S. and Lynch, P.T. (2002) Medium and long-term *in vitro* conservation of olive germplasm (*Olea europaea* L.). *Acta Horticulturae* 586, 109–112.

Lambardi, M., Caccavale, A., Rugini, E. and Caricato, G. (1999) Histological observations on somatic embryos of olive (*Olea europaea* L.). *Acta Horticulturae* 474, 67–70.

Lambardi, M., Rinaldi, L.M.R., Menabeni, D. and Cimato, A. (1994) Ethylene effect on *in vitro* olive seed germination (*Olea europaea* L.). *Acta Horticulturae* 356, 54–57.

Lambert, S.J. (1996) Intense extratropical northern hemisphere winter cyclone events, 1899–1991. *Journal of Geophysical Research* 101, 219–221, 325.

Lampkin, N. and Padel, S. (1994) Conversion to organic farming: an overview. In: Lampkin, N. and Padel, S. (eds) *The Economy of Organic Farming: an International Perspective.* CAB International, Wallingford, UK, pp. 295–311.

Landa, M.C., Frago, N. and Tres, A. (1994) Diet and risk of breast cancer in Spain. *European Journal of Cancer Prevention* 3, 313–320.

Larbi, A., Ayadi, M., Mabrouk, M., Kharroubi, M., Kammoun, N. and Msallem, M. (2006) Agronomic characteristics of some olive varieties cultivated under high-density planting conditions. In: *Proceedings of the 2nd International Seminar Olivebioteq*, 5–10 November 2006, Marsala-Mazara del Vallo, Italy, 2, 135–138.

Larson, K.D., Graetz, D.A. and Schaffer, B. (1991) Flood-induced chemical transformations in calcareous agricultural soils of South Florida. *Soil Science* 152, 33.

Lavee, S. (1989) Involvement of plant growth regulators and endogenous growth substances in the control of alternate bearing. *Acta Horticulturae* 239, 311–322.

Lavee, S. and Avidan, N. (1982) The involvement of phenolic substances in controlling alternate bearing of the olive (*Olea europaea* L.). In: *XXIth International Horticultural Congress*, 29 August–4 September 1982, Hamburg, Germany, Abstract 1370.

Lavee, S. and Avidan, N. (1994) Protein content and composition of leaves and shoot bark in relation to alternate bearing of olive trees (*Olea europaea* L.). *Acta Horticulturae* 356, 143–147.

Lavee, S. and Harshemesh, H. (1990) Climatic effect on flower induction in semi-juvenile olive plants (*Olea europaea*). *Olea* 17, 89.

Lavee, S. and Haskal, A. (1993) Partial fruiting regulation of olive trees (*Olea europaea* L.) with paclobutrazol and gibberellic acid in the orchard. *Advances in Horticultural Science* 7, 83–86.

Lavee, S. and Martin, G.C. (1981) *In vitro* studies of ethephon-induced abscission in olive. II. The relation between ethylene evolution and abscission of various organs. *Journal of the American Society of Horticultural Science* 106(1), 19–26.

Lavee, S. and Schachtel, J. (1999) Interaction of cultivar rootstock and water availability on olive tree performance and fruit production. *Acta Horticulturae* 474, 399–401.

Lavee, S. and Spiegel-Roy, P. (1967) The effect of time of application of two growth substances on the thinning of olive fruits. *Journal of the American Society of Horticultural Science* 91, 180–186.

Lavee, S. and Wodner, M. (1991) Factors affecting the nature of oil accumulation in fruit of olive (*Olea europaea* L.) cultivars. *Journal of Horticultural Science* 66, 583–591.

Lavee, S. and Wodner, M. (2004) The effect of yield, harvest time and fruit size on the oil content of irrigated olive trees (*Olea europaea*), cvs. 'Barnea' and 'Manzanillo'. *Scientia Horticulturae* 99, 267–277.

Lavee, S., Harshemesh, H. and Avidan, N. (1986) Endogenous control of alternate bearing: possible involvement of phenolic acids. *Olea* 17, 61–66.

Lavee, S., Haskal, A. and Ben-Tal, Y. (1983) Girdling olive trees, a partial solution to biennial bearing. I. Methods, timing and direct tree response. *Journal of Horticultural Science* 58, 209–218.

Lavee, S., Rallo, L., Rapoport, H.F. and Troncoso, A. (1996) The floral biology of the olive: effect of flower number, type and distribution on fruit set. *Scientia Horticulturae* 66, 149–158.

Lavee, S., Rallo, L., Rapoport, H.F. and Troncoso, A. (1999) The floral biology of the olive: II. The effect of inflorescence load and distribution per shoot on fruit set and load. *Scientia Horticulturae* 82, 181–192.

Lavee, S., Taryan, J., Levin, J. and Haskal, A. (2002) The significance of cross-pollination for various olive cultivars under irrigated intensive growing conditions. *Olivae* 91, 25–36.

Lavermicocca, P., Valerio, F., Lonigro, S.L., Lazzaroni, S., Evidente, A. and Visconti, A. (2003) Control of olive knot disease with a bacteriocin. In: Iacobellis, N.S. *et al.* (eds) *Pseudomomas* syringae *and Related Pathogens*, Kluwer Academic Publishers, Dordrecht, Netherlands, pp. 451–457.

Lee-Huang, S., Zhang, L., Huang, P.L., Chang, Y.T. and Huang, P.L. (2003) Anti-HIV activity of olive leaf extract (OLE) and modulation of host cell gene expression by HIV-1 infection and OLE treatment. *Biochemical and Biophysical Research Communications* 307, 1029–1037.

Leifert, C., Pryce, S., Lumsden, P.J. and Waites, W.M. (1992) Effect of medium acidity on growth and rooting of different plant species growing *in vitro*. *Plant Cell, Tissue and Organ Culture* 30, 171–179.

Leon, L. and Downey, G. (2006) Preliminary studies by visible and near-infrared reflectance spectroscopy of juvenile and adult olive (*Olea europaea* L.) leaves. *Journal of the Science of Food and Agriculture* 86(6), 999–1004.

Lerutour, B. and Guedon, D. (1992) Antioxidative activities of *Olea europaea* leaves and related phenolic compounds. *Phytochemistry* 31(4), 1173–1178.

Leva, A., Cantos, M., Liñan, J., Troncoso, J., García, M. and Troncoso, A. (2006) Morphological aspects of the *in vitro* formation of cv. 'Manzanillo' olive somatic embryos and plant obtaining. In: *Proceedings of the 2nd International Seminar Olivebioteq*, 5–10 November 2006, Marsala-Mazara del Vallo, Italy, 1, pp. 445–448

Leva, A., Muleo, R. and Petruccelli, R. (1995) Long-term somatic embryogenesis with immature olive cotyledons. *Journal of Horticultural Science* 70(3), 417–421.

Leva, A.R., Petruccelli, R. and Bartolini, G. (1994) Mannitol *in vitro* culture of *Olea europaea* L. (cv. 'Maurino'). *Acta Horticulturae* 356, 43–46.

Leva, A.R., Petruccelli, R., Montagni, G. and Muleo, R. (2002) Field performance of micropropagated olive plants (cv. 'Maurino'): morphological and molecular features. *Acta Horticulturae* 586, 891–894.

Levin, A.G. and Lavee, S. (2005) The influence of girdling on flower type, number, inflorescence density, fruit set and yields in three different olive cultivars ('Barnea', 'Picual' and 'Souri'). *Australian Journal of Agricultural Research* 56(8), 827–831.

Levitt, J. (1980) *Responses of Plants to Environmental Stresses. Vol. I. Chilling, Freezing and High Temperature Stresses*. Academic Press, New York.

Li, C., Weiss, D. and Goldschmidt, E.E. (2003) Girdling affects carbohydrate-related genes expression in leaves, bark and roots of alternate-bearing citrus trees. *Annals of Botany* 92(1), 137–143.

Lipworth, L., Martinez, M.E., Angell, J., Hsieh, C.-C. and Trichopoulos, D. (1997) Olive oil and human cancer: an assessment of the evidence. *Preventive Medicine* 26, 181–190.

Liu, J. and Zhu, J.-K. (1997) Proline accumulation and salt-induced gene expression in a salt-hypersensitive mutant of *Arabidopsis*. *Plant Physiology* 114, 591–596.

Lo Scalzo, R., Scarpati, M.L., Verzegnassi, B. and Vita, G. (1994) *Olea europaea* chemicals repellent to *Dacus oleae* female. *Journal of Chemical Ecology* 20, 1813–1823.

Locy, R.D., Chang, C.-C., Nielsen, B.L. and Singh, N.K. (1996) Photosynthesis in salt-adapted heterotrophic tobacco cells and regenerated plants. *Plant Physiology* 110, 2–321–328.

Lombardo, N., Alessandrino, M., Godino, G. and Madeo, A. (2006) Comparative observations regarding the floral biology of 150 Italian olive (*Olea europaea* L.) cultivars. *Advances in Horticultural Science* 20(4), 247–255.

López-Granados, F., Jurado-Expósito, M., Alamo, S. and Garcia-Torres, L. (2004) Leaf nutrient spatial variability and site-specific fertilization maps within olive (*Olea europaea* L.) orchards. *European Journal of Agronomy* 21, 209–222.

Loreto, F. and Bongi, G. (1987) Control of photosynthesis under salt stress in the olive. In: Prodi, F., Rossi, F. and Christopheri, G. (eds) *International Conference on Agrometeorology*, Fondazione Cesena Agricoltura, Cesena, Italy.

Loreto, F., Centritto, M. and Chartzoulakis, K. (2003) Photosynthetic limitations in olive cultivars with different sensitivity to salt stress. *Plant and Cell Environment* 26, 595–601.

Loukas, M. and Krimbas, C.B. (1983) History of olive cultivars based on their genetic distances. *Journal of Horticultural Science* 58, 121–127.

Loveys, B.R., Stoll, M. and Davies, W.J. (2004) Physiological approaches to enhance water use efficiency in agriculture: exploiting plant signaling in novel irrigation practice. In: Bacon, M.A. (ed.) *Water Use Efficiency in Plant Biology*. Blackwell Publishing/CRC Press, Boca Raton, Florida, 327 pp.

Luna, G., Morales, M.T. and Aparicio, R. (2006) Characterization of 39 varietal virgin olive oils by their volatile compositions. *Food Chemistry* 98(2), 243–252.

Mafra, I., Lanza, B., Reis, A., Marsilio, V., Campestre, C., De Angelis, M. and Coimbra, M.A. (2001) Effect of ripening on texture, microstructure and cell wall polysaccharide composition of olive fruit (*Olea europaea*). *Physiologia Plantarum* 111, 439–447.

Mahan, J.R., McMichel, B.L. and Wanjura, D.F. (1995) Methods for reducing the adverse effects of temperature stress on plants: a review. *Environmental and Exploratory Botany* 35(3), 251–258.

Malik, N.S.A. and Bradford, J.M. (2005) Is chilling a prerequisite for flowering and fruiting in 'Arbequina' olives? *International Journal of Fruit Science* 5, 29–39.

Malik, N.S.A. and Bradford, J.M. (2006a) Regulation of flowering in 'Arbequina' olives under non-chilling conditions: the effect of high daytime temperatures on blooming. *Journal of Food and Agricultural Environment* 4(2), 283–286.

Malik, N.S.A. and Bradford, J.M. (2006b) Changes in oleuropein levels during differentiation and development of floral buds in 'Arbequina' olives. *Scientia Horticulturae* 110, 274–278.

Malik, N.S.A. and Bradford, J.M. (2006c) Flowering and fruiting in 'Arbequina' olives in subtropical climates where olives normally remain vegetative. *International Journal of Fruit Science* 5(4), 47–56.

Mancini, M. and Rubba, P. (2000) The Mediterranean diet in Italy. In: Simopoulos, A. and Visioli, F. (eds) *Mediterranean Diets*. World Review of Nutrition and Diet, Karger Press, Basel, Switzerland, 87, 114–126.

Mancuso, S. (2000) Electrical resistance changes during exposure to low temperature measure chilling and freezing tolerance in olive tree (*Olea europaea* L.) plants. *Plant Cell Environment* 23, 291–299.

Mancuso, S. and Azzarello, E. (2002) Heat tolerance in olive. *Advances in Horticultural Science* 16(3–4), 125–130.

Mancuso, S. and Rinaldelli, E. (1996) Response of young mycorrhizal and non-mycorrhizal plants of olive tree (*Olea europaea* L.) to saline conditions. II. Dymamics of electrical impedance parameters of shoots and leaves. *Advances in Horticultural Science* 10, 135–145.

Manna, C., Galletti, P., Cucciolla, V., Montedoro, G. and Zappia, V. (1999) Olive oil hydroxytyrosol protects human erythrocytes against oxidative damage. *Journal of Nutritional Biochemistry* 10, 159–165.

Mannina, L., Patumi, M., Proietti, N. and Segre, A.L. (2001) PDO (Protected Designation of Origin): geographical characterization of Tuscan extra virgin olive oils using high-field H-1 NMR spectroscopy. *International Journal of Food Science* 13(1), 53–63.

Mannino, P. and Pannelli, G. (1990) Fully mechanized harvesting of olive fruit. Technical and agronomic preliminary evaluations. *Acta Horticulturae* 286(12), 437–440.

Maranto, J. and Krueger, W.H. (1994) Olive fruit thinning. In: Ferguson, L., Sibbett, G.S. and Martin, G.C. (eds) *Olive Production Manual*. University of California, Division of Agriculture and Natural Resources, Berkeley, California, Publication 3353, pp. 87–89.

Marín, L. and Fernández-Escobar, R. (1997) Optimization of nitrogen fertilization in olive orchards. In: Val, J., Montanes, L. and Monge, E. (eds) *Proceedings of the Third International Symposium on Mineral Nutrition of Deciduous Fruit Trees*, Zaragoza, Spain, pp. 411–414.

Marín, L., Benlloch, M. and Fernández-Escobar, R. (1995) Screening of olive cultivars for salt tolerance. *Scientia Horticulturae* 64, 113–116.

Marmiroli, N., Peano, C. and Maestri, E. (2003) Advanced PCR techniques in identifying food components. In: Lees, M. (ed.) *Food Authenticity and Traceability*. Woodhead Publishing Ltd, Cambridge, UK and CRC Press, Boca Raton, Florida, pp. 3–33.

Marschner, H. (1997) *Mineral Nutrition of Higher Plants*. 2nd edn. Academic Press, London.

Marsilio, V. (2006) The use of LAB starters during table olive fermentation. In: *2nd International Seminar Olivebiotech*, 5–10 November 2006, Marsala-Mazara del Vallo, Italy, Special seminars and invited lectures, pp. 221–233.

Marsilio, V. (2006) The use of LAB starters during table olive fermentation. In: *Proceedings of the 2nd International Seminar Olivebioteq*, 5–10 November 2006, Marsala-Mazara del Vallo, Italy, Seminars and invited lectures, pp. 221–233.

Marsilio, V., Seghetti, L., Iannucci, E., Russi, F., Lanza, B. and Felicioni, M. (2005) Use of lactic acid bacteria starter culture during green olive (*Olea europaea* L. cv. 'Ascolana tenera') processing. *Journal of Science and Food Agriculture* 85(7), 1084–1090.

Martin, G.C. (1994a) Botany of the olive. In: Ferguson, L., Sibbett, G.S. and Martin, G.C. (eds) *Olive Production Manual*. University of California, Division of Agriculture and Natural Resources, Berkeley, California, Publication 3353, pp. 19–21.

Martin, G.C. (1994b) Mechanical olive harvest: use of fruit loosening agents. *Acta Horticulturae* 356, 284–291.

Martin, G.C., Denney, J.O., Ketchie, D.O., Osgood, J.W., Connel, J.H., Sibbett, G.S., Kammereck, R., Krueger, W.H. and Nour, G. (1993) Freeze damage and cold hardiness in olive: findings from the 1990 freeze. *California Agriculture* 47(1), 1–12.

Martin, G.C., Ferguson, L. and Polito, V.S. (1994a) Flowering, pollination, fruiting, alternate bearing and abscission. In: Ferguson, L., Sibbett, G.S. and Martin, G.C. (eds) *Olive Production Manual*. University of California, Division of Agriculture and Natural Resources, Berkeley, California, Publication 3353, pp. 19–21.

Martin, G.C., Klonski, K. and Ferguson, L. (1994b) The olive harvest. In: Ferguson, L., Sibbett, G.S. and Martin, G.C. (eds) *Olive Production Manual*. University of California, Division of Agriculture and Natural Resources, Berkeley, California, Publication 3353, 117–128.

Martin, G.C., Lavee, S. and Sibbett, G.C. (1981) Chemical loosening agents to assist mechanical harvesting in olive. *Journal of the American Society of Horticultural Science* 106(3), 325–330.

Martinez, D., Arroyo-Garcôa, R. and Revilla, M.A. (1999) Cryopreservation of *in vitro* grown shoot-tips of *Olea europaea* L., var. 'Arbequina'. *Cryo-Letters* 20(1), 29–36.

Martin-Lopes, P., Lima-Brito, J., Gomes, S., Meirinhos, J., Santos, L. and Guedes-Pinto, H. (2007) RAPD and ISSR molecular markers in *Olea europaea* L.: genetic variability and molecular cultivar identification. *Genetic Resources and Crop Evolution* 54(1), 117–128.

Martin-Moreno, J.M., Willet, t P., Gorgojo, L., Bunegas, J.R., Rodriguez-Artalejo, F., Fernandez-Rondriguez, J.C., Maisonneuve, P. and Boyle, P. (1994) Dietary fat, olive oil intake and breast cancer risk. *International Journal of Cancer* 58, 774–780.

Martins, P.C., Cordeiro, A.M. and Rapoport, H.F. (2006) Flower quality in orchards of olive, *Olea europaea* L., cv. 'Morisca'. *Advances in Horticultural Science* 20(4), 262–266.

Mazomenos, B.A. (1983) Biosynthesis of a sex pheromone of the olive fruit fly *Dacus oleae* Gmel. PhD dissertation, Rijksuniversiteit, Gent, Netherlands, 137 pp.

Mazzuca, S., Spadafora, A. and Innocenti, A.M. (2006) Cell and tissue localization of beta-glucosidase during the ripening of olive fruit (*Olea europaea*) by *in situ* activity assay. *Plant Science* 171(6), 726–733.

Meilan, R. (1997) Floral induction in woody angiosperms. *New Forests* 14(3), 179–202.

Mencuccini, M. (1991) Protoplast culture isolated from different tissues of olive (*Olea europaea* L.) cultivars. *Physiologia Plantarum* 82, 14.

Mencuccini, M. (2003) Effect of medium darkening on *in vitro* rooting capability and rooting seasonality of olive (*Olea europaea* L.) cultivars. *Scientia Horticulturae* 97(2), 129–139.

Mencuccini, M. and Rugini, E. (1993) *In vitro* regeneration from olive cultivar tissues. *Plant Cell, Tissue and Organ Culture* 32, 283–288.

Menendez, L., Vellon, R., Colomer, R. and Lupu, R. (2005) Oleic acid, the main monounsaturated fatty acid of olive oil, suppresses Her-2/neu (erbB-2) expression and synergistically enhances the growth inhibitory effects of trastuzumab (Herceptin TM) in breast cancer cells with Her-2/neu oncogene amplification. *Annals of Oncology* 16(3), 359–371.

Mensik, R.P. and Katan, M.B. (1987) Effects of monounsaturated fatty acids versus complex carbohydrates on HDL in healthy men and women. *Lancet* 338, 122–125.

Metzidakis, I.T. (1999) Field studies for mechanical harvesting by using chemicals for the loosening of olive pedicel on cv. 'Koroneiki'. *Acta Horticulturae* 474, 197–202.

Metzidakis, I.T. and Koubouris, G.C. (2006) Olive cultivation and industry in Greece. In: *Proceedings of the 2nd International Seminar Olivebioteq*, 5–10 November 2006, Marsala-Mazara del Vallo, Italy, Seminars and invited lectures, pp. 133–140.

Michelakis, N. and Barbopoulou, E.A. (2002) Stem places and supporting frames suitability for 'Dendrometers' installation on olive trees. *Acta Horticulturae* 586, 423–427.

Michelakis, N.Z.C., Vouyoucalou, E. and Clapaki, G. (1994) Soil moisture, depletion, evapotranspiration and crop co-efficients for olive trees cv. 'Kalamon', for different levels of soil water potential and methods of irrigation. *Acta Horticulturae* 356, 162–167.

Minguez-Mosquera, M.I. and Garrido-Fernández, J. (1989) Chlorophyll and carotenoid presence in olive fruit (*Olea europaea*). *Journal of Agricultural Food Chemistry* 37(1), 1–7.

Minguez-Mosquera, M.I. and Garrido-Fernández, J. (1989) Chlorophyll and carotenoid presence in olive fruit (*Olea europaea*). *Journal of Agriculture and Food Chemistry* 37(1), 1–7.

Minnocci, A., Panicucci, A. and Vitagliano, C. (1995) Gas exchange and morphological stomatal parameters in olive plants exposed to ozone. In: Lorenzini, E. and Soldatini, G.F. (eds) *Responses of Plants to Air Pollution: Biological and Economic Aspects*. Pacini, Pisa, Italy, pp. 77–81.

Minnocci, A., Panicucci, A., Sebastiani, L., Lorenzini, G. and Vitagliano, C. (1999) Physiological and morphological responses of olive plants to ozone exposure during a growing season. *Tree Physiology* 19, 391–397.

Mitrakos, K. and Diamantoglou, S. (1984) Endosperm dormancy breakage in olive seeds. *Physiologia Plantarum* 62, 8–10.

Mitrakos, K., Alexaki, A. and Papadimitriou, P. (1992) Dependence of olive morphogenesis on callus origin and age. *Journal of Plant Physiology* 139, 269–273.

Montemurro, C., Simeone, R., Pasqualone, A., Ferrara, A. and Bianco, A. (2005) Genetic relationships and cultivar identification among 112 olive accessions using AFLP and SSR markers. *Journal of Horticultural Science and Biotechnology* 80, 105–110.

Morales-Sillero, A., Fernández, J.E. and Troncoso, A. (2006) Table olives and oil quality can be affected by fertigation. In: *Proceedings of the 2nd International Seminar Olivebioteq*, 5–10 November 2006, Marsala-Mazara del Vallo, Italy, 2, pp. 173–176.

Moreno, F., Conejero, W., Martín-Palomo, M.J., Girón, I.F. and Torrecillas, A. (2006) Maximum daily trunk shrinkage reference values for irrigation scheduling in olive trees. *Agricultural Water Management* 84, 290–294.

Moreno, J.M. (2005) *Evaluación Preliminary de los Impactos en España por Efecto del Cambio Climatic*. Ministerio de Medio Ambiente, Spain.

Morgan, J.M. (1984) Osmoregulation and water in higher plants. *Annual Review of Plant Physiology* 35, 299–319.

Moriana, A. and Fereres, E. (2002) Plant indicators for scheduling irrigation of young olive trees. *Irrigation Science* 24, 77–84.

Morini, S., Loreti, F. and Sciutti, R. (1990) Effect of light quality on rooting of 'Leccino' olive cuttings. *Acta Horticulturae* 286, 73–76.

Motilva, M.J., Tovar, M.J., Alegre, S. and Girona, J. (2002) Evaluation of oil accumulation and polyphenol content in fruits of olive tree (*Olea europaea* L.) related to different irrigation strategies. *Acta Horticulturae* 586, 345–348.

Motilva, M.J., Tovar, M.J., Romero, M.P., Alegre, S. and Girona, J. (2000) Influence of regulated deficit irrigation strategies applied to olive trees ('Arbequina' cultivar) on oil yield and oil composition during the fruit ripening period. *Journal of Science and Food Agriculture* 80, 2037–2043.

Munns, R. (1993) Physiological processes limiting plant growth in saline soils: some dogmas and hypotheses. *Plant and Cell Environment* 16, 15–24.

Murray, M. (1999) Mechanisms and significance of inhibitory interactions involving cytochrome P450 enzymes (review). *International Journal of Molecular Medicine* 3(3), 227–238.

Murray, M.B., Cape, J.N. and Fowler, D. (1989) Quantification of frost damage in plant tissues by rates of electrolyte leakage. *New Phytology* 113, 307–311.

Muzzalupo, I., Pellegrino, M. and Perri, E. (2007) Detection of DNA in virgin olive oils extracted from destoned fruits. *European Food Research and Technology* 224(4), 469–475.

Nathawat, N.S., Kuhad, M.S., Goswami, C.L., Patel, A.L. and Kumar, R. (2005) Nitrogen-metabolizing enzymes: effect of nitrogen sources and saline irrigation. *Journal of Plant Nutrition* 28, 1089–1101.

Navarro, C., Fernández-Escobar, R. and Benlloch, M. (1990) Flower bud induction in 'Manzanillo' olive. *Acta Horticulturae* 286, 195–198.

Nergiz, C. and Engez, Y. (2000) Compositional variation of olive fruit during ripening. *Food Chemistry* 69, 55–59.

Newmark, H.L. (1999) Squalene, olive oil and cancer risk. *Annals of the New York Academy of Sciences* 889, 193–203.

Nilsson, M.P. (1972) *The Mycenaean Origin of Greek Mythology. A New Introduction and Bibliography by E. Vermeule.* University of California, Berkeley, California, p. 258.
Norman, J.M. and Welles, J.M. (1983) Radiative transfer in an array of canopies. *Agronomic Journal* 75, 481–488.
Oaks, A. and Hirel, B. (1986) Nitrogen metabolism in roots. *Annual Review of Plant Physiology* 36, 345–365.
Oddo, E., Saiano, F., Alonzo, G. and Bellini, E. (2002) An investigation of the seasonal pattern of mannitol content in deciduous and evergreen species of the Oleaceae growing in Northern Sicily. *Annals of Botany* 90, 239–243.
Ohbogge, J.B. and Jaworski, J.G. (1997) Regulation of fatty acid synthesis. *Annual Review of Plant Biology* 48, 109–136.
Olsen, S.R. and Sommers, L.E. (1982) Phosphorus. In: Page, A.L., Miller, R.H. and Keeney, D.R. (eds) *Methods of Soil Analysis. Part 2: Chemical and Microbiological Properties.* 2nd edn, Agronomy 9, ASA, SSSA, Madison, Wisconsin, pp. 403–430.
Orgaz, F. and Pastor, M. (2005) Fertirrigación del olivo. Programación de riegos. In: Cadabia, C. (ed.) *Fertirrigación Cultivos Horticolas, Frutales y Ornamentales.* Ediciones Mundi-Prensa, Madrid, pp. 496–533.
Orgaz, F., Testi, L., Villalobos, F.J. and Fereres, E. (2006) Water requirements of olive orchards. II: determination of crop coefficients for irrigation scheduling. *Irrigation Science* 24, 77–84.
Orinos, T. and Mitrakos, K. (1991) Rhizogenesis and somatic embryogenesis in calli from wild olive (*Olea europaea* var. *sylvestris* (Miller) Lehr) mature zygotic embryos. *Plant Cell, Tissue and Organ Culture* 27, 183–187.
Osborne, C.P., Chulne, I., Viner, D. and Woodward, F.L. (2000) Olive phenology as a sensitive indicator of future climatic warming in the Mediterranean. *Plant Cell Environment* 23, 701–710.
Ouazzani, N., Lumaret, R., Villemur, P. and di Giusto, F. (1993) Leaf allozyme variation in cultivated and wild olive trees (*Olea europaea* L.). *Journal of Heredity* 84, 34–42.
Owen, R.W., Giacosa, A., Hull, W.E., Haubner, R., Spiegelhalder, B. and Bartsch, H. (2000) The antioxidant/anticancer potential of phenolic compounds isolated from olive oil. *European Journal of Cancer* 36, 1235–1247.
Ozkaya, M.T. and Celik, M. (1999) The effects of various treatments on endogenous carbohydrate content of cuttings in easy-to-root and hard-to-root olive cultivars. *Acta Horticulturae* 474, 51–53.
Ozkaya, M.T., Celik, M. and Algan, G. (1997) Anatomy of adventitious root formation in stem cuttings of the easy-to-root ('Gemlik') and hard-to-root ('Domat') olive cultivars. In: *1st Balkan Botanical Congress*, 19–22 September 1997, Thessaloniki, Greece.
Pacini, C., Wossink, A., Giesen, G., Vazzana, C. and Huirne, R. (2003) Evaluation of sustainability of organic, integrated and conventional farming systems: a farm and field-scale analysis. *Agriculture, Ecosystems and Environment* 95(1), 273–288.
Pallioti, A., Famiani, F., Proietti, P., Boco, M. and Guelfi, P. (1999) Effects of training system on tree growth, yield and oil characteristics in different olive cultivars. *Acta Horticulturae* 474, 189–192.
Palliotti, A. and Bongi, G. (1996) Freezing injury in the olive leaf and effects of mefluidide treatment. *Journal of Horticultural and Scientific Biotechnology* 71(1), 57–63.
Palta, J.P. and Weiss, L.S. (1993) Ice formation and freezing injury: an overview on the survival mechanisms and molecular aspects of injury and cold acclimation in

herbaceous plants. In: Li, P.H. and Christersson, L. (eds) *Advances in Cold Hardiness*. CRC Press, Boca Raton, Florida, pp. 143–176.

Palya, F.T. (1993) *Gene Expression under Low Temperature Stress*. Harwood, New York, pp. 103–130.

Papadopoulos, I. (2006) *Irrigation with Treated Wastewater and Efficient Fertilizer Use*. Agricultural Research Institute of Cyprus, Report 94, 10 pp.

Pardini, A., Faiello, C., Longhi, F. and Tallarico, R. (2003) A study of agronomic parameters of cover crop species utilized in grape and olive organic production. *Advances in Horticultural Science* 17(2), 67–71.

Pardini, A., Faiello, C., Longhi, F., Mancuso, S., Elliot, J. and Snowball, R. (2002) Cover crop species and their management in vineyards and olive groves. *Advances in Horticultural Science* 16(3–4), 225–234.

Parra-López, C. and Calatrava-Requena, J. (2005) Factors related to the adoption of organic farming in Spanish olive orchards. *Spanish Journal of Agricultural Research* 3, 5–16.

Parra-López, C. and Calatrava-Requena, J. (2006) Comparison of farming techniques actually implemented and their rationality in organic and conventional olive groves in Andalusia, Spain. *Biological Agriculture and Horticulture* 24(1), 35–59.

Parry, M. (1992) The potential effects of climate changes on agriculture and land use. *Advances in Ecological Research* 22, 63–91.

Patumi, M., d'Andria, R., Fontanazza, G., Morelli, G., Giorio, P. and Sorrentino, G. (1999) Yield and oil quality of intensively trained trees of three cultivars of olive (*Olea europaea* L.) under different irrigation regimes. *Journal of Horticultural Science and Biotechnology* 74, 729–737.

Patumi, M., d'Andria, R., Marsilio, V., Fontanazza, G., Morelli, G. and Lanza, B. (2002) Olive and olive oil quality after intensive monocone olive growing (*Olea europaea* L. cv. 'Kalamata') in different irrigation regimes. *Food Chemistry* 77, 27–34.

Pavel, E.W. and Fereres, E. (1998) Low soil temperatures induce water deficits in olive (*Olea europaea*) trees. *Physiologia Plantarum* 104, 525–532.

Peltier, J.P., Marigo, D. and Marigo, G. (1997) Involvement of malate and mannitol in the diurnal regulation of the water status in members of Oleaceae. *Trees* 12, 27–34.

Peng, Z., Lu, Q. and Verma, D.P.S. (1996) Reciprocal regulation of Δ1-pyrroline-5-carboxylate synthetase and proline dehydrogenase genes controls proline levels during and after osmotic stress in plants. *Molecular and General Genetics* 253, 334–341.

Penyalver, R., Garcia, A., Ferrer, A., Bertolini, E., Quesada, J.M., Salcedo, C.I., Piquer, J., Perez-Panades, J., Carbinell, E.A., del Rio, C., Caballero, J.M. and López, M.M. (2006) Factors affecting *Pseudomonas savastanoi* pv. *savastanoi* plant inoculations and their use for evaluation of olive cultivar susceptibility. *Phytopathology* 96(3), 313–319.

Pereira, J.A., Casal, S., Bento, A. and Oliveira, M.B.P.P. (2002) Influence of olive storage period on oil quality of three Portuguese cultivars of *Olea europaea* 'Cobrancosa', 'Madural' and 'Verdeal Transmontana'. *Journal of Agriculture and Food Chemistry* 50, 6335–6340.

Perri, E., Parlati, M.V., Mullé, R. abd Fodale, A.S. (1994a) Attempts to generate haploid plants from *in vitro* cultures of *Olea europaea* L. anthers. *Acta Horticulturae* 356:, 47–50.

Perri, E., Parlati, M.V. and Rugini, E. (1994b) Isolation and culture of olive (*Olea europaea* L.) cultivar protoplasts. *Acta Horticulturae* 356, 51–53.

Petridou, M. and Voyiatzis, D.G. (1993) The beneficial effect of girdling, auxin, Tween-20 and paclobutrazol by an improved method of mount layering. *Acta Horticulturae* 356, 24–27.

Petridou, M. and Voyiatzis, D.G. (2002) Difficult to root cv. 'Kalamon' can easily be propagated by softwood layers with an improved method of mound-layering. *Acta Horticulturae* 586, 915–918.

Pilbeam, D.J. and Kirkby, A. (1992) Some aspects of the utilization of nitrate and ammonium by plants. In: Mengel, K. and Pilbeam, D.J. (eds) *Nitrogen Metabolism of Plants*. Clarendon Press, Oxford, UK, pp. 55–70.

Piney, K. and Polito, V.S. (1990) Flower initiation in 'Manzanillo' olive. *Acta Horticulturae* 286, 203–206.

Pontikis, C.A., Loukas, M. and Kousonis, G. (1980) The use of biochemical markers to distinguish olive cultivars. *Journal of Horticultural Science* 55, 333–343.

Porlingis, I.C. and Dogras, K. (1969) Time of flower differentiation of the olive as related to temperature and variety. *Annals of the Agricultural and Forestry School, Aristotelian University of Thessaloniki* 13, 321–341.

Porlingis, I.C. and Therios, I.N. (1976) Rooting response of juvenile and adult leafy olive cuttings to various factors. *Journal of Horticultural Science* 51, 31–39.

Porlingis, I.C. and Therios, I.N. (1979) The effect of the level of temperature on inflorescence induction, number of flowers per inflorescence and pistil development in the olive tree. *Scientific Annals of the School of Agriculture and Forestry – Section of Agriculture* 22(6), 176–193 [in Greek].

Porlingis, I.C. and Voyiatzis, D.G. (1999) Paclobutrazol decreases the harmful effect of high temperatures on fruit set in olive trees. *Acta Horticulturae* 474, 241–244.

Porras, A., Soriano, M.L. and Porras, A.P. (2003) Grafting olive cv. 'Cornicarba' on rootstocks tolerant to *Verticillium dahliae* reduces their susceptibility. *Crop Protection* 22(2), 369–374.

Prenzler, R.D., Lavee, S., Antolovich, M. and Robards, K. (2003) Quantitative changes in phenolic content during physiological development of the olive (*Olea europaea* L.) cultivar 'Hardy's Mammoth'. *Journal of Agriculture and Food Chemistry* 51(9), 2532–2538.

Preziosi, P., Proietti, P., Famiani, F. and Alfei, B. (1994) Comparison between monocone and vase training system on the olive cultivars 'Frantoio', 'Moraiolo' and 'Nostrale di Rigali'. *Acta Horticulturae* 356, 306–310.

Pritsa, T.S. and Voyiatzis, D.G. (2002) Effects of nutrient medium on the morphogenesis of cotyledonary explants of olive. *Acta Horticulturae* 586, 923–926.

Pritsa, T.S. and Voyiatzis, D.G. (2005) Correlation of ovary and leaf spermidine and spermine content with the alternate bearing habit of olive. *Journal of Plant Physiology* 162(11), 1284–1291.

Proietti, P. (2003) Changes in photosynthesis and fruit characteristics in olive in response to assimilate availability. *Photosynthetica* 41(4), 559–564.

Proietti, P. and Famiani, F. (2002) Diurnal and seasonal changes in photosynthetic characteristics in different olive (*Olea europaea* L.) cultivars. *Photosynthetica* 40(2), 171–176.

Proietti, P. and Tombesi, A. (1996a) Effects of gibberellic acid, asparagines and glutamine on flower bud induction in olive. *Journal of Horticultural Science and Biotechnology* 71(3), 383–388.

Proietti, P. and Tombesi, A. (1996b) Translocation of assimilates and source-sink influences on productive characteristics of the olive tree. *Advances in Horticultural Science* 10, 11–14.

Proietti, P., Boco, M., Famiani, F., Guelfi, P. and Tombesi, A. (2002a) Effetti dell'attacco *Spilocaea oleagina* Cast. sugli scambi gassosi delle foglie e sull'attività vegeto-produttiva dell'olivo. In: *Atti del Convegno Internazionale di Olivicoltura*, 22–23 April 2002, Spoleto, Italy, pp. 466–471.

Proietti, P., Famiani, F. and Tombesi, A. (1999) Gas exchange in olive fruit. *Photosynthetica* 36(3), 423–432.

Proietti, P., Famiani, F., Nasini, L. and Tombesi, A. (2002b) The influence of some agronomic parameters on the efficiency of mechanical harvest on young olive trees. *Acta Horticulturae* 586, 415–418.

Proietti, P., Nasini, L. and Famiani, F. (2006) Effect of different leaf-to-fruit ratios on photosynthesis and fruit growth in olive (*Olea europaea* L.). *Photosynthetica* 44(2), 275–285.

Proietti, P., Tombesi, A. and Boco, M. (1994) Influence of leaf shading and defoliation on oil synthesis and growth of olive fruit. *Acta Horticulturae* 356, 272–277.

Psilakis, N. (1996) *Mythology of Crete Island*. Karmanor, Heraclion, Greece, p. 428 [in Greek].

Psilakis, N. and Kastanas, H. (1999) *The Civilization of the Olive Tree. The Olive Oil*. Typocteta, Heraclion, Greece, p. 431 [in Greek].

Raese, J.T. (1987) Effect of calcium nutrition on freeze tolerance and fruit production of apple and d'Anjou pear trees. *HortScience* 22(4), 1043 (Abstr.).

Rains, D.W. (1972) Salt transport by plants in relation to salinity. *Annual Review of Plant Physiology* 23, 367–388.

Rallo, L. (1997) Fructificación y producción. In: Barranco, D., Fernández-Escobar, R. and Rallo, L. (eds) *El Cultivo del Olivo*. Junta de Andalucia, Andalucia, Spain, pp. 107–136.

Rallo, L. and Fernández-Escobar, R. (1985) Influence of cultivar and flower thinning within the inflorescence on competition among olive fruit. *Journal of the American Society of Horticultural Science* 110(2), 303–308.

Rallo, L. and Martin, G.C. (1991) The role of chilling in releasing olive floral buds from dormancy. *Journal of the American Society of Horticultural Science* 116(6), 1058–1062.

Rallo, L., Torreno, P., Vargas, A. and Alvarado, J. (1994) Dormancy and alternate bearing in olive. *Acta Horticulturae* 356, 127–134.

Ranalli, A., De Mattia, G., Patumi, M. and Projetti, P. (1999) Quality of virgin olive oil as influenced by origin area. *Grasas Y Aceites* 50(4), 249–259.

Rangel, B., Platt, K.A. and Thomson, W.W. (1997) Ultrastructural aspects of the cytoplasmic origin and accumulation of oil in olive fruit (*Olea europaea*). *Physiologia Plantarum* 101(1), 109–114.

Rapoport, H.F. and Martins, P.C. (2006) Flower quality in the olive: broadening the concept. *Proceedings of the 2nd International Seminar Olivebioteq*, 5–10 November 2006, Marsala-Mazara del Vallo, Italy, Special Seminars and invited lectures, 397–402.

Rapoport, H.F. and Rallo, L. (1991) Fruit-set and enlargement in fertilized and unfertilized olive ovaries. *HortScience* 26(7), 896–898.

Rapoport, H.F., Manrique, T. and Gucci, R. (2004) Cell division and expansion in the olive fruit. *Acta Horticulturae* 636, 461–465.

Rapoport, H.F., Rallo, L. and Polito, V.S. (1990) Pit hardening in the olive. In: *XXIIIth International Horticultural Congress*, Florence, Italy, Abstract 2364.
Raven, J.A., Handley, L.L. and Wollenweber, B. (2004) Plant nutrition and water use efficiency. In: Bacon, M.A. (ed.) *Water Use Efficiency in Plant Biology*. Blackwell Publishing/CRC Press, Boca Raton, Florida, 327 pp.
Ravetti, L.M. (2004) Evaluation of new olive mechanical harvesting technologies in Australia. In: *5th International Symposium on Olive Growing*, 29 September–2 October 2004, Izmir, Turkey.
Reale, S., Doveri, S., Diaz, A., Angiolillo, A., Lucentini, L., Pilla, F., Martin, A., Donini, P. and Lee, D. (2006) SNP-based markers for discriminatimg olive (*Olea europaea* L.) cultivars. *Genome* 49, 1193–1205.
Restrepo, M., Benlloch, M. and Fernández-Escobar, R. (2002) Influencia del esteès hidrico y del estado nutritivo del olivo en la absorcion foliar del potasio. Direcciòn General de Investigaciòn y Formaciòn Agraria y Resquera (ed.) *Jounadas de Investigaciòn y Transferencia de Technologia al Sector Oleicola*, Cordoba, Spain, pp. 307–310.
Ribeiro, H., Cunha, M. and Abreu, I. (2006) Comparison of classical models for evaluating the heat requirements of olive (*Olea europaea* L.) in Portugal. *Journal of Integrative Plant Biology* 48(6), 664–671.
Rinaldelli, E. and Mancuso, S. (1994) Cell transmembrane electropotentials in adventitious roots of *Olea europaea* L. cv. 'Frantoio' as related to temperature, respiration, external potassium, anoxia and 2,4-dinitrophenol treatments. *Advances in Horticultural Science* 8, 229–234.
Rinaldi, L.M.R. (2000) Germination of seeds of olive (*Olea europaea* L.) and ethylene production: effects of harvesting time and thidiazuron treatment. *Journal of Horticultural Science and Biotechnology* 75(6), 727–732.
Rio, C.D., Rallo, L. and Caballero, J.M. (1991) Effects of carbohydrate content on the seasonal rooting vegetative and reproductive cuttings of olive. *Journal of Horticultural Science* 66(3), 301–309.
Roberts, A.G. (ADAS) (2002) Magnetic water treatment trials. *Practical Hydroponics and Greenhouses*, pp. 15–20. (Publishing details unavailable)
Rodriguez Diaz, J.A., Weatherhead, E.K., Knox, J.W. and Camacho, E. (2007) Climate change impacts on irrigation water requirements in the Guadalquivir river basin in Spain. *Regional Environmental Change* 7, 149–159.
Romeo, F.V., De Luca, S., Piscopo, A. and Pojana, M. (2006) Table olive processing in South Italy. The safety problem: from tradition to innovation. In: *Proceedings of the 2nd International Seminar Olivebioteq*, 5–10 November 2006, Marsala-Mazara del Vallo, Italy, Seminars and invited lectures, 213–219.
Roselli, G. and Venora, G. (1990) Relationship between stomatal size and winter hardiness in the olive. *Acta Horticulturae* 286, 89–92.
Roselli, G., Benelli, G. and Morelli, D. (1989) Relationship between stomatal density and winter hardiness in olive (*Olea europaea* L.). *Journal of Horticultural Science* 64(2), 199–203.
Ross, J. (1981) *The Radiation Regime and Architecture of Plant Stands*. W. Junk Publishers, The Hague, the Netherlands.
Rugini, E. (1984) *In vitro* propagation of some olive (*Olea europaea* L.) cultivars with different root-ability, and medium development using analytical data from developing shoots and embryos. *Scientia Horticulturae* 24, 123–134.

Rugini, E. (1986) Olive (*Olea europaea* L.). In: Bajaj, Y.P.S. (ed.) *Biotechnology in Agriculture and Forestry*. Springer, Berlin, pp. 253–267.

Rugini, E. (1988) Somatic embryogenesis and plant regeneration in olive (*Olea europaea* L.). *Plant Cell, Tissue and Organ Culture* 14, 207–214.

Rugini, E. (1995) Somatic embryogenesis in olive (*Olea europaea* L.). In: Jain, S.M., Gupta, P.K. and Newton, R.J. (eds) *Somatic Embryogenesis in Woody Plants, Vol. II*. Kluwer Academic Publishers, Dordrecht, Netherlands, pp. 171–189.

Rugini, E. and Fedeli, E. (1990) Olive (*Olea europaea* L.) as an oilseed crop. In: Bajaj, Y.P.S. (ed.), *Biotechnology in Agriculture and Forestry Legume and Oilseed Crops, Vol. I*. Springer, Berlin, pp. 593–641.

Rugini, E. and Tarini, P. (1986) Somatic embryogenesis in olive (*Olea europaea* L.). In: Hennessy, M. (ed.) *Proceedings of the Conference on Fruit Tree Biotechnology*, Paris, p. 62.

Rugini, E., Bongi, G. and Fontanazza, G. (1982) Effects of ethephon on olive ripening. *Journal of the American Society of Horticultural Science* 107, 835–838.

Rugini, E., Di Francesco, G., Muganu, M., Astolfi, S. and Caricato, G. (1997) The effects of polyamines and hydrogen peroxide on root formation in olive and the role of polyamines as an early marker for rooting ability. In: Altman, A. and Waisel, Y. (eds) *Biology of Root Formation and Development*. Plenum Press, New York, pp. 65–74.

Rugini, E., Gutiérrez-Pesce, P. and Muleo, R. (2006) Overview in the olive biotechnologies. In: *Proceedings of the 2nd International Seminar Olivebioteq*, 5–10 November 2006, Marsala-Mazara del Vallo, Italy, Seminars and invited lectures, 317 pp.

Rugini, E., Lupino, M., De Agazio, M. and Grego, S. (1992) Endogenous polyamine and root morphogenesis variation under different treatment in cutting and *in vitro* explants of olive. *Acta Horticulturae* 300, 225–232.

Ruiz, N., Barranco, D. and Rapoport, H.F. (2006) Anatomical response of olive (*Olea europaea* L.) to freezing temperatures. *Journal of Horticultural Science and Biotechnology* 81(5), 783–790.

Ruiz-Barba, J.L., Garrido-Fernandez, A. and Jimenez-Diaz, R. (1991) Bactericidal action of oleuropein extracted from green olives against *Lactobacillus plantarum*. *Letters in Applied Microbiology* 12, 65–68.

Ryan, D., Robards, K. and Lavee, S. (1999) Changes in phenolic content of olive during maturation. *International Journal of Food Science and Technology* 34, 265–274.

Sano, H., Seo, S., Koizumi, N., Niki, T., Iwamura, H. and Ohashi, Y. (1996) Regulation by cytokinins of endogenous levels of jasmonic and salicylic acids in mechanically wounded tobacco plants. *Plant and Cell Physiology* 37, 762–769.

Sarmiento, R., Garcia, J.L. and Mazuelos, C. (1990) Free amino acids in easy- and difficult-to-root olive varieties. *Acta Horticulturae* 286, 105–108.

Savino, V. and Gallitelli, D. (1981a) Cherry leafroll virus in olive. *Phytopathologia Mediterranea* 20, 202–204.

Savino, V. and Gallitelli, D. (1981b) Isolation of cucumber mosaic virus from olive in Italy. *Phytopathologia Mediterranea* 22, 76–77.

Savino, V., Gallitelli, D. and Barba, M. (1983) Olive latent rigspot virus, a newly recognized virus infecting olive in Italy. *Annals of Applied Biology* 103, 243–249.

Savouré, A., Hua, X.J., Bertauche, N., van Montagu, M. and Verbuggen, N. (1997) Abscisic acid-independent and abscisic acid-dependent regulation of proline biosynthesis following cold and osmotic stress. *Molecular and General Genetics* 254, 104–109.

Schnathorst, W.C. (1981) Life cycle and epidemiology of *Verticillium*. In: Mace, M.E., Bell, A.A. and Beckman, C.H. (eds) *Fungal Wilt Diseases of Plants*. Academic Press, New York, pp. 81–111.
Sebastiani, L. and Tognetti, R. (2004) Growing season and hydrogen peroxide effects on root induction and development in *Olea europaea* L. (cvs 'Frantoio' and 'Gentile di Larino') cuttings. *Scientia Horticulturae* 100(1/4), 75–82.
Sebastiani, L., Minnocci, A., Scebba, F., Vitagliano, C., Panicucci, A. and Lorenzini, G. (2002a) Physiological and biochemical reactions of olive genotypes during site-relevant ozone exposure. *Acta Horticulturae* 586, 445–448.
Sebastiani, L., Minnocci, A. and Tognetti, R. (2002b) Genotypic differences in the response to elevated CO_2 concentration of one-year-old olive cuttings (*Olea europaea* L. cvs 'Frantoio' and 'Moraiolo'). *Plant Biosystems* 136, 199–208.
Sebastiani, L., Tognetti, R., Di Paolo, P. and Vitagliano, C. (2002c) Hydrogen peroxide and indole-3-butyric acid effects on root induction and development in cuttings of *Olea europaea* L. (cv. 'Frantoio' and 'Gentile di Larino'). *Advances in Horticultural Science* 16(1), 7–12.
Servili, M., Selvaggini, R., Esposto, S., Taticchi, A., Urbani, S. and Montedoro, G.F. (2006) The effect of oil mechanical extraction processes in the phenolic composition of virgin olive oil. *Proceedings of the 2nd International Seminar Olivebiotech*, 5–10 November 2006, Marsala-Mazara del Vallo, Italy, Special seminars and invited lectures, pp. 339–346.
Sessiz, A. and Özcan, M.T. (2006) Olive removal with pneumatic branch shaker and abscission chemical. *Journal of Food Engineering* 76, 148–153.
Sharpley, A.N., Meisinger, J.J., Power, J.F. and Suarez, D.L. (1992) Root extraction of nutrients associated with long-term soil management. *Advances in Soil Science* 19, 151–217.
Shehata, W.A., Abou-Elkhair, S.S., Stefanos, S.S., Youssef, A.A. and Nasr, F.N. (2003) Biological studies on the olive leaf moth, *Palpita unionalis* Hübner (Lepidoptera: Pyralidae), and the olive moth, *Prays oleae* Bernard (Lepidoptera: Yponomeutidae). *Journal of Pest Science* 76, 155–158.
Sibbett, G.S. (1994) Pruning mature bearing olive trees. In: Ferguson, L., Sibbett, G.S. and Martin, G.C. (eds) *Olive Production Manual*. University of California, Division of Agriculture and Natural Resources, Oakland, California, Publication 3353, pp. 57–60.
Sibbett, G.S. and Martin, G.C. (1981) *Olive Spray Thinning*. Division of Agricultural Sciences, University of California, Berkeley, California, Leaflet 2475, p. 4.
Sibbett, G.S. and Osgood, J. (1994) Site selection and preparation, tree spacing and design, planting and initial training. In: Ferguson, L., Sibbett, G.S. and Martin, G.C. (eds) *Olive Production Manual*. University of California, Division of Agriculture and Natural Resources, Berkeley, California, Publication 3353, pp. 31–37.
Simantirakis, B. (2003) The olive tree in Mediterranean history and culture. In: Stefanoudaki, E. (ed.) *Proceedings of the International Symposium on the Olive Tree and the Environment*, 1–3 October 2003, Chania, Greece, pp. 3–7.
Sisto, A., Cipriani, M.G. and Morea, M. (2004) Knot formation caused by *Pseudomonas savastanoi* subsp. *savastanoi* on olive plants is hrp-dependent. *Phytopathology* 94, 484–489.
Smirnoff, N. and Cumbes, Q.J. (1989) Hydroxyl radical scavenging activity of compatible solutes. *Phytochemistry* 28, 1057–1060.

Smyth, J. (2002) Perspectives of the Australian olive industry. *Advances in Horticultural Science* 16(3–4), 280–288.

Snobar, B. (1978) Maturity parameters of olives and the use of abscission chemicals. *Transactions of the American Society of Agricultural Engineers* 21(3), 465–468.

Soler-Rivas, C., Espin, J.C. and Wichers, H.J. (2000) Review. Oleuropein and related compounds. *Journal of the Science of Food Agriculture* 80, 1013–1023.

Sorrentino, G., Giorio, P. and d'Andria, R. (1999) Leaf water status of field-grown olive trees (*Olea europaea* L.) cv. 'Kalamata', under three water regimes. *Acta Horticulturae* 474, 441–444.

Spaniolas, S. (2007) Food forensics: the application of single nucleotide polymorphism technology for food authentication. PhD thesis, University of Nottingham, School of Biosciences, UK.

Spaniolas, S., Bazakos, C., Awad, M. and Kalaitzis, P. (2008a) Exploitation of the chloroplast *trn*L (UAA) intron polymorhisms for the authentication of plant oils by means of a Lab-on-a-Chip capillary electrophoresis system. *Journal of Agricultural and Food Chemistry* 56, 6886–6891.

Spaniolas, S., Bazakos, C., Ntourou, T., Bihmidine, S., Georgousakis, A. and Kalaitzis, P. (2008b) Use of γ-DNA as a marker to assess DNA stability in olive oil during storage. *European Food Research and Technology* 227 (1), 175–179.

Spaniolas, S., Bazakos, C., Tucker, G. and Kalaitzis, P. (2007). Development of single nucleotide polymorphism markers to authenticate extra virgin olive oil. In: *Proceedings of the 5th International Congress on Food Technology*, 9–11 March 2007, Thessaloniki, Greece, pp. 370–376.

Sparavano, L. and Graniti, A. (1978) Cutin degradation by two scab fungi, *Spilocaea oleagina* (Cast.) Hugh. and *Venturia inaequalis* (Cke) Wint. In: Kiraly, Z. (ed.) *Current Topics in Plant Pathology*. Akademiai Kidaò, Budapest, pp. 117–131.

Sparks, T.H. and Menzel, A. (2002) Observed changes in seasons: an overview. *International Journal of Climatology* 22, 1715–1725.

Standish, R. (1960) *The First Trees. The Story of the Olive*. Phoenix House, London.

Stefanoudaki, E. (2004) Factors affecting olive oil quality. PhD thesis, University of Cardiff, UK.

Stefanoudaki-Katzouraki, E. and Koustsaftakis, A. (1992) Studies on total polyphenols and chlorophyll content of olive oil during the ripening of olive fruits in the area of Crete. In: *Proceedings of the Olive Oil Quality Congress*, Florence, Italy, pp. 381–383.

Stoll, M., Loveys, B.R. and Dry, P.R. (2000) Hormonal changes induced by partial rootzone drying of irrigated grapevine. *Journal of Experimental Botany* 51, 1627–1634.

Stoneham, M., Goldacre, M., Seagroatt, V. and Gill, L. (2000) Olive oil, diet and colorectal cancer: an ecological study and a hypothesis. *Journal of Epidemiology and Community Health* 54, 756–760.

Stoop, J.M.H., Williamson, J.D. and Pharr, D.M. (1996) Mannitol metabolism in plants: a method for coping with stress. *Trends in Plant Science* 1, 139–144.

Suárez, M.P., Fernández-Escobar, R. and Rallo, L. (1984) Competition among fruits in olive. II. Influence of inflorescence or fruit thinning and cross-pollination on fruit set components and crop efficiency. *Acta Horticulturae* 149, 131–139.

Syvertsen, J.P., Lloyd, J., McConchie, C., Kriedemann, P.E. and Farquhar, G.D. (1995) On the site of biophysical constrains to CO_2 diffusion through the mesophyll of hypostomatous leaves. *Plant and Cell Environment* 18, 149–157.

Tabatabaei, S.J. (2006) Effects of salinity and N on the growth, photosynthesis and N status of olive (*Olea europaea* L.) trees. *Scientia Horticulturae* 108, 432–438.

Tapp, H.S., Defernez, M. and Kemsley, E.K. (2003) FTIR spectroscopy and multivariate analysis can distinguish the geographic origin of extra virgin olive oils. *Journal of Agriculture and Food Chemistry* 51(21), 6110–6115.

Tattini, M., Bertoni, P. and Caselli, S. (1992) Genotypic responses of olive plants to sodium chloride. *Journal of Plant Nutrition* 15, 1467–1485.

Tattini, M., Gucci, R., Coradeschi, M.A., Ponzio, C. and Everard, J.D. (1995) Growth, gas exchange and ion content in *Olea europaea* plants during salinity stress and subsequent relief. *Physiologia Plantarum* 95(2), 203–210.

Tattini, M., Gucci, R., Romani, A., Baldi, A. and Everand, J.D. (1996) Changes in non-structural carbohydrates in olive (*Olea europaea*) leaves during root zone salinity stress. *Physiologia Plantarum* 98(1), 117–124.

Tattini, M., Lombardini, L. and Gucci, R. (1997) The effect of NaCl stress and relief on gas exchange properties of two olive cultivars differing in tolerance to salinity. *Plant and Soil* 197, 87–93.

Tattini, M., Marzi, L., Tafani, R. and Traversi, M.L. (1999) A review on salinity-induced changes in leaf gas exchange parameters of olive plants. *Acta Horticulturae* 474, 415–418.

Tattini, M., Ponzio, C., Coradeschi, M.A., Tafani, R. and Traversi, M.L. (1994) Mechanisms of salt tolerance in olive plants. *Acta Horticulturae* 356, 181–184.

Tattini, M.V., Cimato, A., Bertoni, P. and Lombardo, M. (1990) Nitrogen nutrition of self-rooted olive in sand culture. Effect of NH_4^+-N/NO_3^--N on growth and nutritional status. *Acta Horticulturae* 286, 311–314.

Terral, J.F. (2000) Exploitation and management of the olive tree during prehistoric times in Mediterranean France and Spain. *Journal of Archaeological Science* 27(2), 127–133.

Testi, L., Villalobos, F.J., Orgaz, F. and Fereres, E. (2006) Water requirements of olive orchards. I. Stimulation of daily evapotranspiration for scenario analysis. *Irrigation Science* 24, 69–76.

Teviotdale, B.L., Ferguson, L. and Vossen, P.M. (2005) *UC IPM Pest Management Guidelines: Olive*. Diseases, University of California, Berkeley, California, Publication 3452, pp. 107–109.

Thanassoulopoulos, C.C., Biris, D.A. and Tjamos, E.C. (1979) Survey of *Verticillium* wilt of olive trees in Greece. *Plant Disease Reports* 63, 936–940.

Therios, I. (2005a) *Mineral Nutrition and Fertilizers*. Gartaganis Publications, Thessaloniki, Greece, 392 pp.

Therios, I. (2005b) *Olive Production*. Gartaganis Publications, Thessaloniki, Greece, 476 pp.

Therios, I. (2006) Mineral nutrition of olive trees. In: *Proceedings of the 2nd International Seminar Olivebioteq*, 5–10 November 2006, Marsala-Mazara del Vallo, Italy, Seminars and invited lectures, pp. 403–408.

Therios, I. and Karagiannidis, N. (1991) Effect of NaCl on growth and chemical composition of four olive cultivars. *Scientific Annals, School of Agriculture, Aristotle University of Thessaloniki* 28, 29–47.

Therios, I.N. (1981) Nitrate absorption by olive plants (*Olea europaea* L., cv. 'Chondrolia Chalkidikis'). IFIGESIA Thesis, pp. 85 [in Greek].

Therios, I.N. and Misopolinos, N.D. (1988) Genotypic response to sodium chloride salinity of four major olive cultivars (*Olea europaea* L.). *Plant and Soil* 106, 105–116.

Therios, I.N. and Sakellariadis, S.D. (1982) Some effects of varied magnesium nutrition on the growth and composition of olive plants (cultivar 'Chondrolia Chalkidikis'). *Scientia Horticulturae* 17, 33–41.

Therios, I.N. and Sakellariadis, S.D. (1988) Effects of nitrogen form on growth and mineral composition of olive plants (*Olea europaea* L.). *Scientia Horticulturae* 35, 167–177.

Therios, I.N., Weinbaum, S.A. and Carlson, R.M. (1979) Nitrate uptake effectiveness and utilization efficiency of two plum clones. *Physiologia Plantarum* 47, 73–76.

Thomashow, M.F. (1998) Role of cold-responsive genes in plant freezing tolerance. *Plant Physiology* 118, 1–8.

Thomashow, M.F. (1999) Plant cold acclimation: freezing tolerance genes and regulatory mechanisms. *Annual Review of Plant Physiology and Plant Molecular Biology* 50, 571–599.

Tisdale, S.L., Nelson, W.L., Beaton, J.D. and Havlin, J.L. (1993) *Soil Fertility and Fertilizers*. 5th edn, Macmillan Publishing Co., New York, 634.

Tjamos, E.C. (1993) Prospects and strategies in controlling *Verticillium* wilt in olive. *Bulletin OEPP/EPPO Bulletin* 23, 505–512.

Tognetti, R., d'Andria, R., Morelli, G. and Alvino, A. (2005) The effect of deficit irrigation on seasonal variations of plant water use in *Olea europaea* L. *Plant and Soil* 273, 139–155.

Tognetti, R., d'Andria, R., Morelli, G., Calandrelli, D. and Froignito, F. (2004) Irrigation effects on daily and seasonal variations of trunk sap flow and leaf water relations in olive trees. *Plant and Soil* 263, 249–264.

Tognetti, R., Sebastiani, L., Minnocci, A., Vitagliano, C. and Raschi, A. (2002) Foliar responses of olive trees (*Olea europaea* L.) under field exposure to elevated CO_2 concentration. *Acta Horticulturae* 586, 449–452.

Tognetti, R., Sebastiani, L., Vitagliano, C., Raschi, A. and Minnocci, A. (2001) Responses of two olive tree (*Olea europaea* L.) cultivars to elevated CO_2 concentration in the field. *Photosynthetica* 39, 403–410.

Tomati, V. and Galli, E. (1992) The fertilizing value of waste waters from the olive processing industry. In: Kubàt, I. (ed.) *Humus et Planta Proceedings*. Elsevier Science, Amsterdam, pp. 107–126.

Tombesi, A. (2006) Planting systems, canopy management and mechanical harvesting. In: *Proceedings of the 2nd International Seminar Olivebioteq*, 5–10 November 2006, Marsala-Mazara del Vallo, Italy, pp. 307–316.

Tombesi, A. and Cartechini, A. (1986) L'effetto dell'ombreggiamento della chioma sulla differenziazione delle gemme a fiore dell'olivo. *Rivista Ortoflorofrutticoltura Italiana* 70, 277–285.

Tombesi, A. and Ruffolo, M. (2006) Effects of olive cultivar and pruning intensity on attacks of *Spilocaea oleagina* (Cast.) Hughes. In: *Proceedings of the 2nd International Seminar Olivebioteq*, 5–10 November 2006, Marsala-Mazara del Vallo, Italy, 2, pp. 231–234.

Tombesi, A., Boco, M. and Pilli, M. (1999) Influence of light exposure on olive fruit growth and composition. *Acta Horticulturae* 474, 255–260.

Tombesi, A., Guelfi, P. and Nottiano, G. (1998) Ottimizzazione della raccolta delle olive e meccanizzazione. *Informatore Agrario* 46, 79–84.

Touraine, B., Clarkson, D.T. and Muller, B. (1994) Regulation of nitrate uptake at the whole plant level. In: Roy, J. and Garnier, E. (eds) *A Whole Plant Perspective on Carbon–Nitrogen Interactions*. SPB Academic Publishing, The Hague, Netherlands, pp. 11–30.

Tous, J., Romero, A. and Hermoso, J.F. (2006) High density planting systems, mechanization and crop management in olive. In: *Proceedings of the 2nd International Seminar Olivebioteq*, 5–10 November 2006, Marsala-Mazara del Vallo, Italy, Special seminars and invited lectures, pp. 423–430.

Tous, J., Romero, A., Planta, J. and Hermoso, J.F. (2004) Olive oil cultivars suitable for very high-density planting conditions. In: *5th International Symposium on Olive Growing*, Izmir, Turkey, Extracts, p. 31.

Tovar, M.J., Romero, M.P., Girona, J. and Motilva, M.J. (2002) L-phenylalanine ammonia-lyase activity and concentration of phenolics in developing olive (*Olea europaea* L. cv. 'Arbequina') fruit grown under different irrigation regimes. *Journal of Science and Food Agriculture* 82, 892–898.

Trichopoulou, A., Lagiou, P., Kuper, H. and Trichopoulos, D. (2000) Cancer and Mediterranean dietary traditions. *Cancer Epidemiology, Biomarkers and Prevention* 9(9), 869–873.

Trichopoulou, A., Toupadaki, N., Tzonou, A., Katsouyanni, K., Manousos, O., Kada, E. and Trichopoulos, D. (1993) The macronutrient composition of the Greek diet: estimates derived from six case-control studies. *European Journal of Clinical Nutrition* 47, 549–558.

Tripoli, E. Giammanco, M., Di Majo, D., Giammanco, S., La Guardia, M. and Grescimanno, M. (2006) The phenolic compounds of olive oil and human health. In: *Proceedings of the 2nd International Seminar Olivebioteq*, 5–10 November 2006, Marsala-Mazara del Vallo, Italy, Special Seminars and invited lectures, pp. 265–271.

Tripoli, E., Giammanco, M., Tabacchi, G., Di Majo, D., Giammanco, S. and La Guardia, M. (2005) The phenolic compounds of olive oil: structure, biological activity and beneficial effects on human health. *Nutritional Research Review* 18, 98–112.

Troncoso, A., Liñan, J., Cantos, M., Acebedo, M.M. and Rapoport, H.F. (1999) Feasibility and anatomical development of an *in vitro* olive cleft-graft. *Journal of Horticultural Science and Biotechnology* 74, 584–587.

Troncoso, A., Liñan, J., Prieto, J. and Cantos, M. (1990) Influence of different olive rootstocks on growth and production of 'Gordal Sevillana'. *Acta Horticulturae* 286, 133–136.

Troncoso, J., Liñan, J., Cantos, M., Garcia, J.L. and Troncoso, A. (2004) *In vitro* selection of salt-tolerant olive clones. In: *5th International Symposium on Olive Growing*, 27 September–2 October 2004, Izmir, Turkey, Abstract 222.

Tsambardoukas, V. (2006) Effects of N form on growth and mineral composition of the olive cv. 'Kalamon'. MSc thesis, Aristotelian University, School of Agriculture, Thessaloniki, Greece [in Greek with English summary].

Tsatsarelis, C.A. (1987) Vibratory olive harvesting: the response of fruit stem system to fruit-removing action. *Journal of Agricultural Engineering Research* 38(5), 77–90.

UC Cooperative Extension, Sonoma County (2006) *Olive Maturity Index*. University of California, Berkeley, California.

UCIPM (2001) *Pest Management Guidelines: Olive*. University of California, Department of Agriculture and Natural Resources, Berkeley, California, Publication 3552, p. 6.

UCIPM (2004) *Pest Management Guidelines 2004: Olive. Relative Toxicities of Insecticides and Miticides used in Olives to Natural Enemies and Honey Bees*. University of California, Berkeley, Calfornia, Publication 3452.

Ülger, S., Baktir, I. and Kaynak, L (1999) Determination of the effects of endogenous plant hormones on alternate bearing and flower bud formation in olives. *Turkish Journal of Agriculture and Forestry* 23(7), 619–623.

Ülger, S., Sonmez, S., Karkacier, M., Ertoy, N., Akdesir, O. and Aksu, M. (2004) Determination of endogenous hormones, sugars and mineral nutrition levels during the induction, initiation and differentiation stage and their effects on flower formation in olive. *Plant Growth Regulation* 42, 89–95.

Van Camp, W., Willenes, H., Bowler, C., Van Montagu, M., Inze, M., Reupold-Popp, P., Sandermann, H. and Langebartels, C. (1993) Elevated levels of superoxide dismutase protect transgenic plants against ozone damage. *Biotechnology* 12, 165–168.

Van Rensburg, L., Krüger, G.H.J. and Krüger, H. (1993) Proline accumulation as drought tolerance selection criterion: its relationship to membrane integrity and chloroplast ultrastructure in *Nicotiana tobacum* L. *Journal of Plant Physiology* 141, 188–194.

Van Steenwyk, R.A., Ferguson, L. and Zalom, F.G. (2002) *UC IPM Pest Management Guidelines. Olive (Insects and Mites)*. UCANR Publication 3452, University of Calfornia, Berkeley, California.

Vermeiren, I. and Jobling, J.A. (1980) *Localized Irrigation*. FAO, Rome, Paper 36, 202).

Vigh, L., Escriba, P.V., Sonnleitner, A., Sonnleitner, M., Piotto, S., Maresca, B., Horvath, I. and Harwood, J.L. (2005) The significance of lipid composition for membrane activity: new concepts and ways of assessing function. *Progress in Lipid Research* 44, 303–344.

Vigo, C. (1999) Effect of salinity (NaCl, Na_2SO_4, KCl) on growth and chemical composition of olive plants (*Olea europaea* L.) cultivar 'Chondrolia Chalkidikis' grown in a greenhouse. MSc thesis, Aristotelian University, School of Agriculture, Thessaloniki, Greece.

Vigo, C., Therios, I. and Bosabalidis, A. (2005) Plant growth, nutrient concentration and leaf anatomy in olive plants irrigated with diluted seawater. *Journal of Plant Nutrition* 28, 101–102.

Vigo, C., Therios, I., Patakas, A., Karatassou, A and Nastou, A. (2002) Changes in photosynthetic parameters and nutrient distribution of olive plants (*Olea europaea* L.) cultivar 'Chondrolia Chalkidikis' under NaCl, Na_2SO_4 and KCl salinities. *Agrochimica* 46, 33–46.

Visco, T., Molfese, M., Cipolletti, M., Corradetti, R. and Tombesi, A. (2004) The influence of vibration applied to the trunk and to the branches of different-sized olive trees on the efficiency of mechanical harvesting. In: *5th International Symposium on Olive Growing*, 29 September–2 October, Izmir, Turkey.

Visioli, F. and Galli, C. (1994) Oleuropein protects low density lipoprotein from oxidation. *Life Sciences* 55, 1965–1971.

Visioli, F. and Galli, C. (1997) Evaluating oxidation processes in relation to cardiovascular disease: a current review of oxidant/antioxidant methodology. *Nutrition, Metabolism and Cardiovascular Disease* 7, 459–466.

Visioli, F. and Galli, C. (1998a) Olive oils and their potential effects on human health. *Journal of Agriculture and Food Chemistry* 46, 4292–4296.

Visioli, F. and Galli, C. (1998b) The effect of minor constituents of olive oil on cardiovascular disease: new findings. *Nutritional Reviews* 56, 142–147.

Visioli, F., Bellomo, G. and Galli, C. (1998) Free radical-scavenging properties of olive oil polyphenols. *Biochemical and Biophysical Research Communications* 247, 60–64.

Visioli, F., Bellomo, G., Montedoro, G. and Galli, C. (1995a) Low density lipoprotein oxidation is inhibited *in vitro* by olive oil constituents. *Atherosclerosis* 117, 25–32.

Visioli, F., Caruso, D. Plasmati, E. Patelli, R., Mulinacci, N., Romani, A., Galli, G. and Galli, C. (2001) Hydroxytyrosol, as a component of the olive oil mill waste water, is dose-dependently absorbed and increases the antioxidant capacity of rat plasma. *Free Radical Research* 34(3), 301–305.

Visioli, F., Galli, C. and Bogani, P. (2006) Is olive oil good for you? In: *Proceedings of the 2nd International Seminar Olivebioteq*, 5–10 November 2006, Marsala-Mazara del Vallo, Italy, Special seminars and invited lectures, pp. 255–263.

Visioli, F., Romani, A., Mulinacci, N., Zarini, S., Conte, D., Vincieri, F.F. and Galli, C. (1999) Antioxidants and other biological activities of olive mill waste waters. *Journal of Agriculture and Food Chemistry* 47, 3397–3401.

Visioli, F., Vinceri, F.F. and Galli, C. (1995b) 'Waste waters' from olive oil production are rich in natural antioxidants. *Experiential* 51, 32–34.

Vitagliano, C., Minocci, A., Sebastiani, L., Panicucci, A. and Lorenzini, G. (1999) Physiological response of two olive genotypes to gaseous pollutants. *Acta Horticulturae* 474, 431–433.

Vlahov, G. (1992) Flavonoids in three olive (*Olea europaea*) fruit varieties during maturation. *Journal of Science and Food Agriculture* 58, 157–159.

Vlahov, G., Del Re, P. and Simone, N. (2003) Determination of geographical origin of olive oils using C^{-13} nuclear magnetic resonance spectroscopy. I. Classification of olive oils of the Puglia region with denomination of protected origin. *Journal of Agriculture and Food Chemistry* 51(19), 5612–5615.

Vossen, P. (2006) The potential for super-high-density olive oil orchards in California (from http://www.ucdavis.edu).

Voyiatzis, D.G. (1995) Dormancy and germination of olive embryos as affected by temperature. *Physiologia Plantarum* 95(3), 444–448.

Voyiatzis, D.G. and Porlingis, I.C. (1987) Temperature requirements for the germination of olive seeds (*Olea europaea* L.). *Journal of Horticultural Science* 62, 405.

Wahrburg, U., Kratz, M. and Cullen, P. (2002) Mediterranean diet, olive oil and health. *European Journal of Lipid Science and Technology* 104, 698–705.

Walker, M. (1996a) Antimicrobial attributes of olive leaf extract. *Townsend Letter for Doctors and Patients* 156, 80–85.

Walker, M. (1996b) Olive leaf extract. The new oral treatment to counteract most types of pathological organisms. *Explore* 7(4), 31–37.

Watad, A.E.A., Reuveni, M., Bressan, R.A. and Hasegawa, P.M. (1991) Enhanced net K^+ uptake capacity of NaCl-adapted cells. *Plant Physiology* 95, 1265–1269.

Weinbaum, S.A. (1984) Foliar application of urea to olive: Translocation of urea. Nitrogen as influenced by sink demand and nitrogen deficiency. *Journal of the American Society of Horticultural Science* 109(3), 356–360.

Weis, K.G., Goren, R., Martin, G.C. and Webster, B.D. (1988) Leaf and inflorescence abscission in olive. Regulation by ethylene and ethephon. *Botanical Gazette* 149(4), 391–397.

Weis, K.G., Webster, D.D., Goren, R. and Martin, G.C. (1991) Inflorescence abscission in olive. Anatomy and histochemistry in response to ethylene and ethephon. *Botanical Gazette* 152(1), 51–58.

Weiser, C.J. (1970) Cold resistance and acclimation in woody plants (a review). *HortScience* 5, 403–408.

Weiss, D. and Goldschmidt, E.E. (2003) Girdling affects carbohydrate-related gene expression in leaves, bark and roots of alternate-bearing citrus trees. *Annals of Botany* 92(1), 137–143.

Westwood, M.N. (1978) *Temperate Zone Pomology*. Freeman and Company, London, 428 pp.

Wilkinson, S. (2004) Water use efficiency and chemical signaling. In: Bacon, M.A. (ed.) *Water Use Efficiency in Plant Biology*. Blackwell Publishing/CRC Press, Boca Raton, Florida, 327 pp.

Wu, S.-J., Ding, L. and Zhu, J.-K. (1996) SOS1, a genetic locus essential for salt tolerance and potassium acquisition. *Plant and Cell* 8, 617–627.

Xiong, L., Schumaker, K.S. and Zhu, J.K. (2002) Cell signaling during cold, drought and salt stress. *Plant Cell* 14(1), 165–183.

Yamaguchi-Shinozaki, K. and Shinozaki, K. (1993) *Arabidopsis* DNA encoding two desiccation-responsive rd29 genes. *Plant Physiology* 101, 1119–1120.

Yoshida, Y., Kiyosue, T., Nakashima, K., Yamaguchi-Shinozaki, K. and Shinozaki, K. (1997) Regulation of levels of proline as an osmolyte in plants under water stress. *Plant Cell Physiology* 38, 1095–1102.

Yousfi, K., Cert, R.M. and Garcia, J.M. (2005) Changes in quality and phenolic compounds of virgin olive oils during objectively described fruit maturation. *European Food Research and Technology* 223(1), 117–124.

Zarrouk, M., Marzouk, B., Ben Miled Daoud, D. and Cherif, A. (1996) Oil accumulation in olives and effect of salt on their composition. *Olivae* 61, 41–45.

Zervakis, G. (2006) *Brief Report on Table Olive Cultivation and Industry*. http://www.prochile.cl/tarapaca/promotion_aceitunas_zervakis.pdf

Zhang, J. and Davies, W.J. (1989) Abscisic acid produced in dehydrating roots may enable the plant to measure the water status of the soil. *Plant Cell Environment* 12, 73.

Zhu, J.K., Liu, J. and Xiong, L. (1998) Genetic analysis of salt tolerance in *Arabidopsis*: evidence for a critical role of potassium nutrition. *Plant Cell* 10, 1181–1191.

Zuccherelli, G. and Zuccherelli, S. (2002) In vitro propagation of fifty olive cultivars. *Acta Horticulturae* 586, 931–934.

INDEX

Abscisic acid 16, 87, 88, 106, 130, 156,
    171–173, 214
Abscisic acid accumulation 173
Abscission
    chemicals 245
    zones of fruit 252
ACC (1-aminocyclopropane-1 carboxylic acid)
    172
ACC synthase 254
Acclimatization 144, 146
Acidity 284
Adramitini 219, 262, 267, 268, 290
Adulteration detection 286
    enzymatic methods 287, 288
AFLP markers 256
Agar 146
Agiou Orous 46, 260
Aglandau 267
Agrobacterium rhizogenes 146, 354
Alcohols and aldehydes 27, 28
Alkanes 27
Alsol 252
Alternate bearing 39, 105–107, 111, 211
Aminization 184
Aminoacids of olive fruit 237
Aminotriazole 330
Ammonification 184
Ammonium absorption 193, 194
Ammonium assimilation, enzymes 196
Ammonium incorporation into aminoacids
    186
Amplified Fragment Length Polymorphism
    (AFLP) 256, 288
Amygdalolia 268
Anatomy of olive tree 15
    buds 19
    flowers 19, 20

fruit 20
inflorescences 19, 20
leaves 17, 18
main branches 17
root system 15–17
trunk 17
Anchorage 15, 16
Anthocyanins 25, 26, 29, 238, 240, 241
Antioxidant activity 30, 31, 317
Antioxidants 145, 238, 295, 298, 299, 303,
    310, 312, 313
Apical dominance 22, 214, 222, 223
Appropriate soil 52
Arbequina 47, 122, 126, 150, 155, 205,
    219, 249, 265, 266
Arbequina IRTA-1-18 122
Arbosana 122, 147, 249
Armillaria mellea 114, 259, 335, 349, 350
Aromatic compounds 238, 239
Arpa 54
Ascolana tenera 269–271
Ascolano 17, 44, 51, 269–271, 349
Ascorbic acid 145
Aspidiotus nerii 344
Assessment of frost tolerance 58
    chlorophyll fluorescence 59
    Ec measurement 58–61
    electrolyte leakage 59
    stomata size and density 59
    visual observations 58, 59, 62

Authenticity 286, 288
Authenticity standards 283

Bacillus thuringiensis 320–322, 355
Bacterial diseases 351, 352

Bactrocera oleae 259, 265, 267, 269, 270, 322, 335–338, 355
Bare soil and frost 62
Bark grafting 141
Bark splitting 55, 56, 139
Barnea ('K-18') 106, 265
Basammon 190
Biennial bearing 105, 182, 183, 245
Biodiversity 77, 295, 323, 324
Biological olive culture 319, 320, 323
Biological produce 50
Biotechnological aspects 353
Bitter index 165
Bitterness index 241
Black olives in brine 271
Black ripe olives 272, 274, 276
Blood pressure lowering 304
Blue baby 187
Boron 166, 185, 189, 200, 321, 353
Boron accumulation in olive rootstock 47, 48
Boron adsorption 200
Boron concentration seasonal variation 202
Boron toxicity 201
Botanical species of *Olea* 11
Bridge grafting 142
Brine-cured olives 275
Bronze age 5
Browning
    fruits 28, 240
    wood and cambium 54–56
Bruised olives 274
b-sitosterol 238
Budding 43, 46, 138

Ca/Mg ratio 199
Ca/N ratio 146, 353
Caffeic acid 29, 30, 107, 236, 242
Caffeoylquinic acid 106, 107
Calcium
    freezing 57
    salt tolerance 214
Calcium channels 214
Calcium roles 57, 199, 214
Californian-type black ripe olives 276
Callus 128, 129, 139
Cambial zone 139
Campesterol 28, 238
Canvases 247
Capillary pore space 64
Carbon dioxide 70, 74
Cardiovascular system 6
Carotenoids 25, 165
    olive oil 237, 238, 240, 241, 285, 317

Catechole oxidase 240
CEC 64
Cell division 109
Cell wall softening 240
Chemical composition variation 183
Chemical scarification 127
Chemical signaling 172
Chemical weed management 325
Chill-heating model 94
Chilling 19, 81
Chilling requirements 53, 54, 81
Chip budding 138
Chip grafting 138
Chlorogenic acid 106
Chloroplast pigments changes 240
Chondrolia Chalkidikis 46, 81, 87, 127, 129, 157, 190, 202, 205, 217, 219, 267, 268, 272, 276, 290
Chromatography 287
Cinnamic acid 107
Classification criteria of olive cultivars 255
Clay minerals 63
Clay soil 52, 64
Cleft grafting 140, 141
Climate 51
Climate change 73, 75, 93
Climate change adaptability 80
Climate change side effects 76
Climatic impact 79
    irrigation 79
    phenology 79
Cold acclimation 57, 58
Cold frame 128
Cold Regulated Protein (COR) 354
Compatible solutes 175
Compost from moist pomace 297
Conductivity 61, 152
Consumer countries of table olives 37, 38
Continuous cycle 279
Conventional cultivation 319, 323
Conventional methods 286–288
Coratina 58, 219, 266
Countries of cultivation 10, 11
Cover crops 321, 322
Cretan civilization 2
Crop coefficient 154
Cross-pollination 53, 99
Crude pomace oil 281
Cryopreservation 143, 149, 150
Cryoprotective protein (CPR) 354
Cryoscopy 69
Cryprotective substances 57
Cultivar selection 113

Cultivars taxonomy 256
Culture areas 78
Culture initiation 144
Curante virgin olive oil 280
Cuticle 18
Cutting types 132
Cuttings 5, 125–133
Cycle length 77
Cycloartenol 28
Cytokinins 16, 130, 174, 214, 215
  ethylene 215
  plant stress 214
  role 214
Cytoplasmic male sterility 94

Dark respiration (R_D) 72
Darkness 146
Deficit irrigation 163
Denitrification 185
Dense planting varieties 122
Density of planting 121
Desertification 78, 295
De-stoning 127, 271
Differential thermal analysis (DTA) 59, 61, 62
Dimethyloleuropein 28, 29
Discontinuous processing system 279
Disease control 319, 322
Disease growth 78
Disinfection 144
Distal end 129
Diuron 329, 330
DNA polymorphism 288
Dormancy 22, 57, 126, 128
Double sigmoid 22, 103
Double-bladed knife 143
Drip irrigation 161, 322
Driselase 149
Drupe 20, 22, 25, 229, 230
Dual purpose cultivars 23
Dwarfing rootstocks 17, 43, 44, 114

Ecodormancy 22
Efficiency of transpiration 167
Egg cell 100
Electrical conductance 68, 211
Electrical resistance blocks 68
Electrolyte release 61
Electronic leaf 133
Embryo dormancy 127
Embryo sac 22, 100
Endocarp (stone) 20, 22, 25, 26, 28, 229, 230, 235

Endodormancy 22
Endogenous hormones 106, 222, 224
Environmental problems 295
Epan 154
Epicarp 25, 27, 28, 230
Ethephon and ripening 230, 252
Ethrel 252
Etiolation 131, 354
European Union table olive consumption 41
European Union table olive production 40
Evapotranspiration 154
  estimation methods 154, 162
Exocarp 20, 103
Explant rooting 146
Extra virgin olive oil 30, 38, 280, 281, 313

Farming systems 49
Fenton's reaction 298, 299
Fermentation 164, 271–273, 275, 276
  aerobic 276
  anaerobic 276
Fertilization 22, 52, 180, 186, 319, 321, 324, 331, 332, 345
Fertilization optimal rate 208
Field capacity 66
Flashpoint 284
Floranid 190
Florets 100
Flower bud induction 20, 22, 82
Flower distribution 97
Flower induction 81
Flower fertilization 93, 102
Flower initiation 81
Flower phenology 93
Flowering bud induction 16, 53, 54, 82
  date of harvesting 90
  defoliation 91
  differentiation of flowers 82
  fruit thinning 90
  growth regulators 87, 88
  juvenility 85
  LAS index 84
  light quality 84
  light shading 84
  morphological changes 82
  sugars – natural elements 89
  temperature – chilling 86
  variation of temperature 87
Flowering date prediction 79, 94
Flowering period 97
Fluorescence 73

Fog system 132
Foliar application of nutrients 209
Foliar uptake 209
Frankliniella occidentalis 335, 341, 342
Frantoio 58, 126, 212, 213, 262, 263, 266, 353
Freezing point depression 69
Freezing symptoms 54
Fresh olives storage 278
Frost-hardy cultivars 63
Frost sensitivity 51
Frost-susceptible cultivars 63
Frost tolerance 57, 58, 62
    calcium 62
    olive varieties 62
Fruit acceleration 251
Fruit chemical composition 27
Fruit color 26, 229, 241
Fruit detachment force (FDF) 229, 245, 246, 252–254
Fruit growth 22, 52, 103, 229
    assimilate supply 229
Fruit load 20, 22, 105
    ripening 229
Fruit ripening
    olive oil quality 241
Fruit set 22, 47, 83, 100, 102, 223, 350
Fruit shrinkage 54, 55
Fruit size and ripening precocity
    orchard orientation 234
    tree vigor 234
Fruit structure 25
Fruit thinning 109
FS-17 122, 185, 263
Fungal diseases 346

Gaidurolia 267
Galatsaniki 260
Gas spectroscopy 287
Gathering tools 247
Gene regulated by salt stress 219, 354
Genic male sterility 94
Genotypic difference 185
Gibberellic acid 16, 22, 87, 88, 130, 146
Girdling 73, 106, 221
Glaciers 9
Global table olive consumption 40
Global table olive production 36
Global warming 76
Globe system 120
Glutamine synthetase 186

Glutamine-2 oxoglutarate transminase (GS-GOGAT) 186
Gordal 47, 268
Gordal Sevillano 269
Graft union 138, 139
Grafting 43–45, 125, 138, 139
    recommended for 138
Grafting methods 140–143
Grafting time 140
Grass-cutting machines 331
Greek olive cultivars 28, 258
Green maturation 229
Green ripe olives 230, 236, 272, 274
Greenhouse effect 74
Grossa di Spagna 270
Ground water 152
Growth flushes 105
Growth phases 230

Habit 13, 223
Hand-harvesting 245
Hardwood cutting 130, 132
Harvesting 227, 241, 245–254, 272, 285, 322, 324, 352
Harvesting criteria 229
Harvesting nets 247
Harvesting tools 247
    cylinder harvester 248
    scissors-type harvester 247
    tweezers-type harvester 248
Heat stress 52, 53
Heat waves 76
Hectareage of olives 33
Hemiberlesia lataniae 335, 343, 344
Herbicides 295, 324, 325–333
    absorption 325, 326
    biological control 333
    contact 325, 329, 330
    hormonal action 325, 330
    orchard age 327–329
    post-emergence 325, 328–331
    pre-emergence 325, 328–330
High-density system 47, 50, 121
    appropriate varieties 122
History of olive 1
Hurricanes 76
Hydraulic conductivity 67
Hydraulic resistance 158
Hydrogen production 301
Hydroxytyrosol 29–31, 239, 241, 298, 299, 303, 305, 311–313, 315, 317
Hydroxytyrosol levels 234

Index

IAA production 130, 222–224
IBA concentration 137
IBA solution 132, 133, 146
Imperfect flower 84, 183
In vitro culture 143–150
   micropropagation aims 143, 144
Inactivation of cytochrome P_{450} 305
Incompatibility 43, 138
Indirect organogenesis 144
Infiltration 153
Infrared thermography 161
Inorganic N sources 146
Insects 335–346
Integrated management 323, 324
Integrated olive culture 319, 323, 324
Integrated weed management 331, 333
Internal CO_2 (Cj) 72
Inter-simple Sequence repeats ((ISSR) 289
Iron 199
   phosphorous solubility 199
Iron imbalance 199
Irrigation 15, 151–166, 183, 186, 211, 241, 319, 323, 324, 345, 327, 328, 330, 350
Irrigation methods 28, 156
Irrigation need 156
   root system 156
Irrigation programming methods 160
Irrigation quality 164

Juvenile cuttings 131, 133
Juvenility 129

K^+-Na^+ antagonism 214
K_{232} 164, 240
K_{270} 164, 240, 241
Kalamon 46, 81, 138, 185, 204, 205, 213, 219, 261, 262, 267, 271–273, 275, 290, 353
Karydolia 267, 268
King David 5
K_m 191, 192
Kolovi 87, 262
Konservolia 267
Koroneiki 46, 47, 54, 58, 81, 87, 106, 122, 127, 157, 185, 186, 204, 219, 230, 246, 249, 257–259, 268, 290
Kothreiki 219, 261, 290

Ladolia 256
Lag phase 190

Laminar air flow 146, 147
Lampante virgin olive oil 280
Large-fruited olive cultivars 267–270
Latent buds 19, 137
Leaf abscission 16, 58, 71, 346, 352
Leaf characteristics 18, 255
Leafy cuttings 43, 45, 129, 132
Leaf fossils 1
Leaf analysis 181, 324
Leaf orientation 53
Leaf rolling 69
Leaf sampling season 181
Leaf surface characteristics 53
Leaf water potential (Ψ_{leaf}) 160, 172
Leaf:fruit ratio 110, 229
Lianolia Kerkiras 219, 259
Lianolia Patron 260
Light interception 71
Lineweaver – Burk plot 191
Linolenic acid 236
Liquid nitrogen 150
Local cultivars worldwide 257
Loosening agents 252
Low chilling requirements 51, 356
Low temperature stress 54
Low vigor cultivars 114, 122
Low volume irrigation 161
Low-frequency deficit irrigation 164
Luteoxanthin 238

Magnesium 198, 199
Main olive cultivars 258
Malondialdehyde (MDA) 216
Manganese 199
Manganese deficiency 199
Manganese form 200
Mannitol 25, 54, 146, 175, 176, 216
   families containing 216
Mannitol accumulation 54, 175
Mannitol content 177, 216
Mannitol degradation 176
Manzanillo 17, 44–46, 106, 157, 205, 262–264, 268, 271–273, 276
   genotypes 265
Mastoidis 81, 157, 204, 260, 290
Matric potential (Ψm) 54, 65
Maturation stage evaluation 242
Maturity index (MI) 243
Mechanical harvesters 248, 249, 253
   over-row harvester 124, 249
   tree characteristics 251

Mechanical harvesting efficiency 253
  orchard characteristics 251
Mechanical pruning 226
Mechanical scarification 127
Mediterranean diet 316, 317
  ageing 316
  cancer prevention 317
  minor components 317
Medium-fruited olive cultivars 260–267
Mefluidine 62
Megaritiki 46, 58, 81, 157, 204, 219, 260, 268
Meski 219, 264
Mesocarp (flesh) 20, 25, 45, 103, 235
Methaemoglobinaemia 187
Methane 75
Methods to reduce growth vigor 47
Methods to solve nutritional problems 181
Micro-capsules 148
Micronutrients 199–203
Micropropagation stages 144
Mineral nutrition 44, 105, 130, 179
Mineral requirements 180
Minoan civilization 2, 4
Mission 17, 44–46, 264, 271, 273
Mist system 132, 133
Mixed shoots 17
Modified Penman equation 154
Moisture
  effect on callus 139
  equivalent 66
Molecular markers 255, 288, 353
Monocone system 118, 253
Morphology and physiology of olive 22
Morphology of olive 13
Mount layering 130, 131
MS 145
Mycenaean civilization 4
Mycocentrospora cladiosporioides 335, 350, 351

^{15}N-labeled urea 209
N, P, K and Ca requirements 179, 180
N:P ratio 185
NAA as thinning agent 112
NaOH treatment 30, 128, 276
Neolithic period 5
Net assimilation (Pn) 71
Net assimilation rate 168
Neutron probe 68
Nitrate absorption 179, 188, 190–194
  accompanying cations 193
  calcium 192

  kinetics 190–194
  light intensity 194
  nitrogen deficiency 193
  pH 192
  temperature 192
Nitrate accumulation 187
Nitrate assimilation 186
Nitrate concentration and absorption rate 190
Nitrate leaching 187–189
Nitrate reductase (NR) 173, 174, 186, 190
Nitrate reduction 196
Nitrate transport systems 194
Nitrification 184
  inhibition 188
Nitrogen 16, 20, 22, 52, 102, 179–296, 215
  excess and olive oil 186
  translocation in soils 184
Nitrogen assimilation 179, 95, 195, 196
Nitrogen fertilization time 183
Nitrogen fertilizers 183, 184
Nitrogen form and growth 194
Nitrogen mobilization 185
Nitrogen and photosynthesis 185
Nitrogen and polyphenols 186
Nitrogen quantity 183
Nitrogen soil forms 184
Nitrous oxide 74
Non conventional methods 288, 289
Non-saponifiable components of virgin olive oil 236
N-serve 188
Nuclear Magnetic Resonance (NMR) 287
Number of olives 33, 34
Nutrient
  concentration in olive fruits 204
  concentration with the various plant components 181
  waste water 166
  water use efficiency 169
Nutrient deficiency 181, 182, 185
Nutrient excess 182, 185
Nutrient sufficiency 181, 182

Ocean acidification 76
o-diphenol 240
Off year 87
Oil bottling 281
Oil categories 280, 281
Oil stability 165
Oil storage 281
Olea 9, 12, 14, 43
Olea chrysophylla 10, 17, 43, 114, 127

Index

Olea europaea 5, 9, 14, 29, 30, 43, 57, 138, 255, 312
Olea excelsa 10
Olea noti 1
Olea oblonga 17, 43–46, 114, 349
Olea oleaster 10, 14
Olea sativa 10, 14, 113, 139, 141
Olea sylvestris 127
Olea verrucosa 43, 114, 127
Oleic acid 122, 212, 236, 241, 285, 307, 309–311, 356
    antiongogenic action 307, 310
Oleuropein 28–31, 89, 90, 235–237, 241, 242, 272, 298, 299, 303–305, 312, 315, 317
    antimicrobial activity 303, 304
    antimycoplasmal activity 315
    antioxidant 303
    atherogenetic disease 304
Oleuropein biosynthesis 29
Oleuropein degradation 29
Olive leaf extract 303–306
    anti-HIV activity 306
Olive Medium (OM) 145, 146
Olive mill by-products utilization 296
Olive mill products 289, 295
Olive mill waste water 296–301
    characteristics 300
Olive oil
    Alzheimer's disease 310
    breast cancer 307
    cardiovascular diseases 309
    chlorophyll a and b 237, 233
    colonic cancer 308
    diabetes 310
    endometrial cancer 308
    human cancer 307
    minor constituents 242
    obesity 309
    ovarian cancer 308
    stone removal 242
Olive oil certification 282
Olive oil colour 285
Olive oil exports 38
Olive oil extraction systems 279, 295
Olive oil flavor 242, 282, 285, 311
Olive oil pollution 285
Olive oil standards 282
Olive oil synthesis 229, 236
Olive oil varieties 255
Olive oil world production 34
Olive oils characteristics 280
Olive pomace 296, 297, 301

Olive products 271, 274
Olive stone 125, 126
Olive stone germination 53
Olives darkened by oxidation 274
Olivir 305, 306
Olympic games 2
On year 87
Optimum temperature 52, 53, 347, 351
Organic acid
    composition of virgin olive oil 235
    olive oil 235
Organic cultivation 319
Organic farming 78
Organic matter 63, 188, 189, 295, 296, 297, 300, 322, 326
Organic olive orchards 189
Organogenesis 143, 144
Origin of olive 9
Osmoprotective function 175
Osmoregulatory capacity 156, 159
Osmotic potential (Ψ_s) 65, 212
Osmotin 57, 58, 354
Ovary size 101
Ovule 20, 22, 101, 125
Oxidative stability (K_{225}) 164
Oxyenus Maxwelli 345
Ozone 70, 354

Packing and labeling 277
Palmette system 120
Palpita unionalis 341
Panicle 83
Parenchymal cells 129
Parlatoria oleae 260, 264, 270, 342, 343
Partial root-zone drying 163, 170–172
Patch budding 142, 143
    season 142
Peat moss 134
Peltate trichome 18
Perfect flower 20, 22, 81–84, 100, 183
Perlite 134, 135
Perlite properties 135
Permanent wilting percentage (PWP) 66
Permitted phytoprotective materials 320
Peroxide index 240, 241, 284
Pest control 50, 319, 320, 322, 335, 339
Pharmaceutical capsules 149
Phenolic acids 106
Phenolic compound 28, 30, 241, 242, 311, 312, 317
    anticancer potential 311
    olive mill waste water 299, 315

Phenols 28, 235, 239, 241, 242, 279, 297, 312, 313, 315
Phloeotribus scarabeoides 344, 345
Phoenician civilization 4
Phosphate fertilization 198
Phosphorous 166, 185, 197, 198
Phosphorous deficiency 185
Phosphorous inorganic soil 198
Phosphorous ionic form 198
Phosphorous organic soil 198
Phosphorous roles 197
Photo-oxidation 298
Photosynthetic active radiation (PAR) 71
Phytofluene 238
Phytoprotective materials 320
Pistil receptivity 94
Piston flow 188
Plant material choice 130
Plantation system 113, 114, 116
  intensive 116
  organic 117
  semi-intensive 116
  traditional 116
Planting 44, 113–124, 319, 325, 327, 328
Planting date 137
Planting distances 114, 116, 137
Planting season 131, 133
Planting systems 44, 113, 324
Plastic nets 248, 322
Pleistocene period 9
Pn 156, 157, 204, 212, 213
Pneumatic beating poles 246–248
  linear poles 246
  T-shaped poles 246, 247
Polarity 129
  graft union 140
Pollen 1, 13, 53, 94–100
  antioxidative enzymes 94
Pollen dissemination 98
Pollen germination 97, 100
Pollen incompatibility 102
Pollen viability 96
Pollination 22, 51, 54, 83, 93–102, 111, 230
Pollinators 96, 353
Polyconic system 118
Polygamous 84
Polyphenol content 164, 356
Polyphenol levels 233
Pomace olive oil 30, 281, 284, 286, 295
  solvent extraction 297
Pore space 63

Poseidon 2
Potassium 52, 166, 196, 197, 321
  drought tolerance 208
Potassium-exchangeable 197
Potassium needs 197
Potassium-readily available 197
Potassium roles 196
Potassium-soil forms 197
Prays oleae 225, 270, 320, 322, 335, 339, 340, 355
Preformed roots 128
Pressure potential (Ψ_p) 65
Processed olive types 273
Processing equipments 273
Proliferation media 147
Proline accumulation 175, 219
Proline biosynthesis, enzymes 176
Proline synthesis 175, 176
Propagation 125–150
Protected Designation of Origin (PDO) 282, 289
Protected Geographical Indication (PGI) 283, 289
Protection of human erythrocytes from oxidative damage 313, 314
Protein denaturation 29
Protoplasts 149
Proximal end 129, 140
Pruning cost 222, 224, 225
Pruning effects
  flower bud induction 223
  hormones 224
  photosynthesis 223
  shading 223
  shoot length 224
Pruning frequency 227
Pruning intensity 222, 227
Pruning late 227
Pruning machinery 226
Pruning mature trees 225
Pruning objectives 221, 225
Pruning rules 222
Pruning season 227
Pruning shapes 226
Pruning techniques 221
Pruning tools 225
Pruning young trees 225
Pseudomonas syringae pv. *savastanoi* 122, 224, 259, 264, 265, 267, 269, 322, 335, 351, 352
Pungency 241
Purity standards 283

Index

Q_{10} quotient 192

Radio waves-advantages 216
Random Amplified Polymorphic DNA (RAPD) 256, 264, 288, 290
RAPD markers 45, 256, 288
Raw olives 28, 272
  processing 273
Raw olives storage 272
Reactive nitrogen species 314, 315
Reactive oxygen species (ROS) 203, 216, 312–314
Reducing sugars 233
Refined olive oil 281, 284, 312
Refined pomace oil 281
Regulated deficit irrigation 164, 170–172
Rejuvenation 144, 226
Resistance to pests and diseases 255, 353, 355
Restriction Fragment Length Polymorphysm (RFLP) markers 256
Ring budding 138
Ripening changes 240
Ripening index (RI) 243
Ripening period 23
Root grafts 141
Root length 52, 156
Root as storage tissue 16
Rooted cutting 16, 43, 133
Rooting media 147
Rooting percentage 137
Rooting substrate 132
Rootstock 15–17, 43–45, 139, 140, 349, 350, 355
Rootstock description 45
  Allegra 45
  cultivars 45, 46
  Olea oblonga 45
  Wild olive 43, 45, 46
Rootstock roles 44
  fruit size 44
  salt tolerance 44
  tree size 44
Rootstock selection 113
Rootstock vigour 46
Rubisco 72
Rubra 54

Saddle grafting 138
Saissetia oleae 225, 260, 264, 266, 267, 269, 270, 322, 335, 338, 339

Salinity 43, 44, 211–220, 345
  cytokinin metabolism 215
  cytokinin transport 215
  mannitol content 216
  oxidative stress 216
  phenols 212
  plant growth 211
  water absorption 212
Salinization and pollen viability 211
Saloneque 266
Salt accumulation 16, 152
Salt semi-tolerant cultivars 219
Salt sensitive cultivars 219, 213
Salt tolerance 44, 211, 214, 217, 354, 355
  cultivars 219
  genes 219, 220, 354
  mechanism 219
  potassium levels 216, 219
  radio waves 216
  SOS1, SOS2, SOS3 214, 219, 354
Sandy loam 52, 64
Santa Caterina 270
Santo Agostino 270
Sap flow 157, 161
Sap pH 173
Saturated flow 67
Sea level rising 76
Secondary processing 271, 277
Seed coat dormancy 127
Seed collection 126
Seed dormancy 127
Seed propagation 125
Seed storage 126
Seedbeds 128
Seedlings 16, 17, 43–45, 125, 128, 141
Self-breaking beds 149
Self-pollination 99
Self-sterility 94
Semi-hardwood cutting 132
Sensing low temperature 57
Sensory evaluation 281, 282
Sequence Characterized Regions (SCAR) 288
Sevillano 17, 44–46, 51, 268, 271–273, 276
Shield budding 142
Shikimic acid 107
Shoot bearing shoots 17
Shoot elongation 144, 146
Shoot proliferation 144, 145
Shriveled black olives 274
Side veneer grafting 138
Sigmasterol 238
Simazine 330
Simple Sequence Repeats (SSR) 288, 290

Single Nucleotide Polymorphism (SNP) 256, 289, 290
Slow release-fertilizers 189, 190
Small-fruited olive cultivars 256–260
Sodium alginate 148
Sodium hypochlorite 144
Sodium translocation 213
Softwood cutting 132
Soil aggregates 64, 326
Soil condition 15, 16, 51, 63, 255, 285, 324
Soil cultivation 321, 322, 324, 331, 333
Soil fertility spacial variability 208
Soil horizons 65
Soil particles 64
Soil pH 52
Soil porosity 64
Soil properties 16, 63, 326
Soil structure 64
Soil texture 64, 326
Soil water measurement 67
    direct method 67
    indirect method 68
Soil water readily available 66
Somatic embryogenesis 144, 147, 148
Sowing 128
Spanish-type green olives 275
Specific Absorption Rate (SAR) 152, 153
Specific ion toxicity 153
Sperm cells 100
Sphaeroblast 15, 130, 137
Spilocaea oleaginea 335, 346–348, 350, 355
Splice grafting 142
Split olives 275
Split root system 156, 163, 167
Spring vegetative wave 22
Squalene 236, 242, 311, 317
    cancer protection 311
Squalene concentration 286
SSR markers 256
Staminate flower 20, 81, 96
Stem water potential 160
Sterols 28, 235, 238, 287
Stock plants 144
Stomata 18, 20, 167, 223, 326, 351
Stomatal conductance (Gs) 18, 71, 156–158, 186,
Stomatal mechanism 152
Streptomycin sulfate 144
Subculturing 145
Suberin 128
Suckers 125, 137, 138
Sugars of olive fruits 235, 236
Sulfur dioxide 70

Sulfuric acid treatment 127
Supercooling 54, 61
Super-high-density orchards 123, 222, 249
Superoxide dismutase (SOD) 95, 354
Sustained deficit irrigation 164
Synthetic seeds 144, 148

Table olives
    characteristics 243, 244
    most important varieties 271
    ripening criteria 244
Table olives quality 278
Tanche 266
Taxonomy of olive 11, 13
T-budding 142
Temperature
    effect on callus 139
    fruit set 102
    maximum 51
    minimum 51
Temperature requirements 51
Temperature stress 52, 54
Tensiometry 68
Thasitiki 261
Thermal conductivity 68
Thermal requirements 53
Thermocouple psychometry 69
Thidiazuron 146, 148, 215
Thinning methods 110, 111, 112
    chemical thinning 111
    hand thinning 110
    mechanical thinning 111
Tissue oxidation 144
Tocopherol concentrations of certain oils 239
Tocopherols 186, 235, 238, 239
    α-tocopherol 30, 165, 242, 317
Tolerance
    abiotic stress 353, 354
    heat stress 52
    salt 204, 353
    sea water 205
    water stress 52, 168, 353
Tongue grafting 138
Toxic ions in waste water 166
Training 113, 117, 221
Transpiration 133, 151, 168
Transpiration quotient 151, 168
Transpiration rate 72
Tree crown 137, 349, 350
Triglycerids 236
Triterpenoids 28
Trunk diameter variation 160

Trunk girth 156
Trunk shakers efficiency 253
Trunk shrinkage 160
Tsunati 259
Tyrosol 28–30, 239, 241, 312, 313, 315

Ultraviolet (UV) absorption 286
Ultraviolet (UV) light absorbance 284
Ultraviolet (UV) spectroscopy 287

Vase system 117
Vasebush system 119
Vegetative buds 17, 22, 58
Vegetative shoots 17
Verdale 266
Vermiculite 134, 135
Vernalization 51
Verticillium dahliae 45, 122, 225, 260, 263, 264, 266, 335, 348, 349, 355
Vibration characteristics 251
Viral diseases 352
Virgin olive oil 30, 31, 38, 241, 280, 281, 283, 289, 295, 312, 314, 316, 317
V_{max} 191, 192

Wagga Verdale 266
Waste water 30, 31, 165, 166, 297
  extraction and their hypocholesterolaemic effects 315
  use as fertilizer 299
Water application scheduling 160
Water content 151
Water criteria 153
Water cured olives 275
Water deficit avoidance 167
Water infiltration 67
Water logging 67, 69
Water potential (Ψ_{soil}) 65, 156, 157, 212

Water quality 152, 153
Water requirements 152, 153
Water sprouts 17, 225, 227
Water stress 43, 52, 58, 100, 101, 130, 164, 175, 180, 215, 227, 259, 325, 354
  critical period 162
Water use efficiency (WUE) 73, 156, 157, 163, 167–174, 223
  definition 168
Water vapor 74
Waxing 141
Wedge grafting 138, 142
Weeds 134, 322, 323, 325–333
  common names 327
  scientific names 327
  sensitivity to current herbicides 332
Wetting zone 67
What is pruning 221
Whip grafting 141
Why to prune 221, 222
Wilting 66
Wind-affected areas 52, 324
Wound roots 128
Wounding 131
WPM 145
Wrapping 143
Wreath 2
WUE
  methods to improve 170–172
  regulated deficit irrigation 170, 171
WUE measurement 169
WUE_e 168
WUE_t (transpiration efficiency) 168

Xylopode 15

Zeatin 146, 172, 174
Zeuzera pyrina 345, 346
Zygote 100